Manufacturing Processes
Reference Guide

Manufacturing Processes Reference Guide

by Robert H. Todd and Dell K. Allen,
Brigham Young University,
and Leo Alting,
Technical University of Denmark

Industrial Press Inc.
New York

Library of Congress Cataloging in Publication Data

Todd, Robert H., 1942–
     Manufacturing processes reference guide / by Robert H. Todd, Dell
K. Allen, and Leo Alting.—1st ed.
   512 p. 10.7 x 25.4 cm.
     Includes bibliographical references and index.
     ISBN 0-8311-3049-0 :
     1. Manufacturing processes—Handbooks, manuals, etc.   I. Allen,
Dell K., 1931–   . II. Alting, Leo, 1939–   . III. Title.
TS183.T63   1993
671—dc20
                                                             93-31767
                                                             CIP

INDUSTRIAL PRESS INC.
200 Madison Avenue
New York, New York 10016-4078

MANUFACTURING PROCESSES REFERENCE GUIDE
First Edition, 1994

10 9 8 7 6 5 4 3 2 1

Contents

Conversion Factors

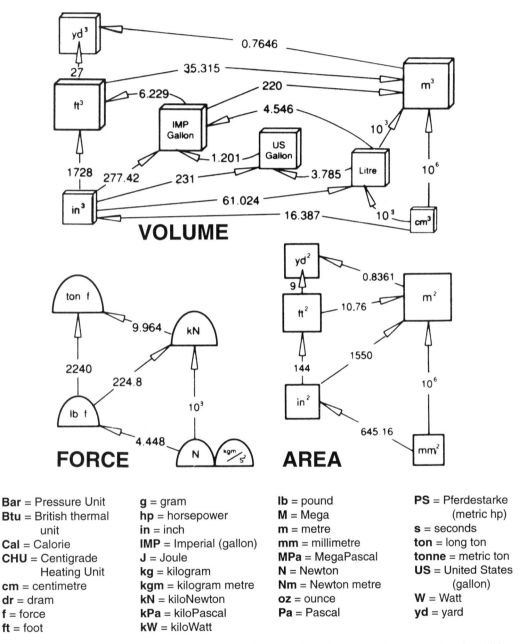

VOLUME

yd³ — 0.7646 — m³
27
ft³ — 35.315 — m³
ft³ — 6.229 — IMP Gallon
IMP Gallon — 220 — m³
IMP Gallon — 4.546 — Litre
1728
277.42
IMP Gallon — 1.201 — US Gallon
US Gallon — 231 — in³
US Gallon — 3.785 — Litre
in³ — 61.024 — Litre
in³ — 16.387 — cm³
10^3
10^6
10^3

FORCE

ton f
2240
9.964 — kN
lb f — 224.8 — kN
N $\frac{kgm}{s^2}$ — 10^3 — kN
lb f — 4.448 — N

AREA

yd² — 0.8361 — m²
9
ft² — 10.76 — m²
144
in² — 1550 — m²
in² — 645.16 — mm²
mm² — 10^6 — m²

Bar = Pressure Unit
Btu = British thermal unit
Cal = Calorie
CHU = Centigrade Heating Unit
cm = centimetre
dr = dram
f = force
ft = foot

g = gram
hp = horsepower
in = inch
IMP = Imperial (gallon)
J = Joule
kg = kilogram
kgm = kilogram metre
kN = kiloNewton
kPa = kiloPascal
kW = kiloWatt

lb = pound
M = Mega
m = metre
mm = millimetre
MPa = MegaPascal
N = Newton
Nm = Newton metre
oz = ounce
Pa = Pascal

PS = Pferdestarke (metric hp)
s = seconds
ton = long ton
tonne = metric ton
US = United States (gallon)
W = Watt
yd = yard

NOTE: Unit sizes generally increase with vertical upward movement on charts. Units vertically above each other are in the same series. Arrows generally point to larger units, always in direction of conversion.

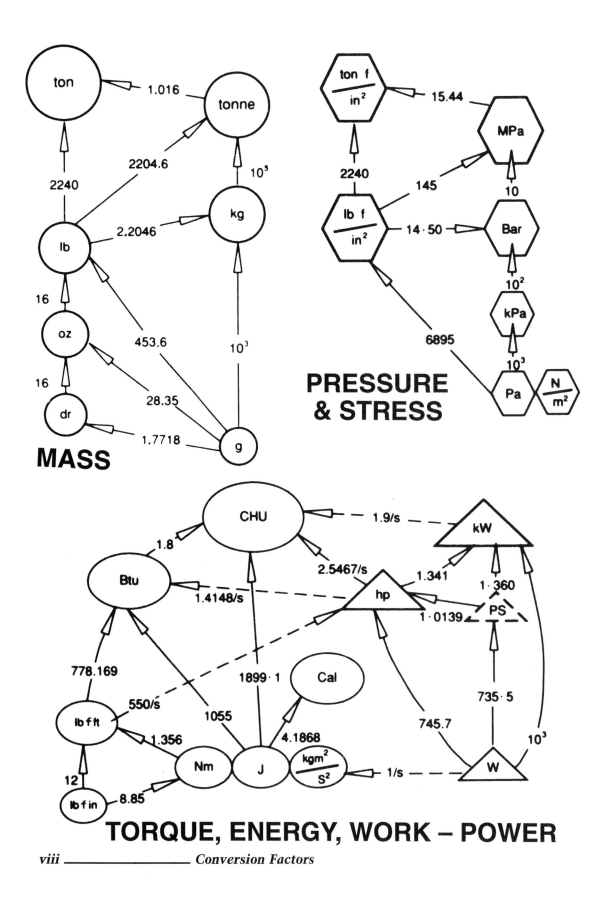

MASS

PRESSURE & STRESS

TORQUE, ENERGY, WORK – POWER

Preface

The *Manufacturing Processes Reference Guide* has been developed as a tool to assist students of engineering and others interested in manufacturing to become familiar with over 130 manufacturing processes used in industry today. It is not an exhaustive treatment in that many additional processes listed in the manufacturing processes taxonomy are not treated in this guide.

Each of the processes that are treated in this guide are divided into three categories: (1) knowledge, (2) application, and (3) development. Some processes are treated in all three categories; others are only treated in one or two.

One of the challenges facing industry and engineering education today is the *integration* of design and manufacturing engineering activities. This reference guide has been developed to provide those in industry and education with a ready reference of information regarding often used manufacturing processes. A significant first step for someone who is charged with the responsibility of designing a device or improving a device's design is to be familiar with the processes that will be used to manufacture it. This familiarity almost always leads to more practical, less expensive, and better designed products for the consumer.

Acknowledgments ———————————————

The material contained in the *Manufacturing Processes Reference Guide* is an abridgement of a seventeen-volume set of instructional materials developed through the sponsorship of the Manufacturing Consortium, which was organized in 1980 under the direction of Professors Dell K. Allen of Brigham Young University and Leo Alting of Technical University of Denmark.

The Manufacturing Consortium, which is made up of industrial representatives from Black & Decker, Boeing, Caterpillar, General Motors, Grumman, Tektronix, Texas Instruments, Westinghouse, and Xerox, worked approximately five years to provide information, review, and advise in the development of the seventeen volumes. Their work along with that of many other faculty and students from Brigham Young University is gratefully acknowledged as the basis of this effort.

Special recognition goes to Lori Calico, Ron Goodson, Kim Greenburg, Louise Lindorf, Jason Zoolakis, and the secretarial staff of the Department of Manufacturing Engineering and Engineering Technology of Brigham Young University (especially Lori Calico, who typed the entire manuscript) for their monumental effort in helping this reference guide to be developed.

Robert H. Todd
January, 1994
Provo, Utah

Taxonomy of Manufacturing Processes _____

The Manufacturing Processes Taxonomy, based on the process classification system initially developed at Brigham Young University[1] and later adapted by members of the Manufacturing Consortium, provides a concise roadmap of some 300 processes used for modifying geometry or properties of engineering materials.

It has been said that students can learn twice as much in half the time when the material to be studied has been classified and the critical attributes have been clearly identified. In this text, we attempt to do both.

Processes used for modifying workpiece *geometry* are called "shaping" processes. Processes used for modifying *properties* of materials are called "nonshaping" processes. Shaping processes have been grouped into "mass-conserving" processes, "mass-reducing" processes, and mass-increasing, or "joining," processes. Nonshaping processes have been grouped into processes dealing with heat treatment and with surface finishing. Each of these processes has been further subdivided into fourteen major "families" of processes, as shown on the first sheet of the Manufacturing Processes Taxonomy chart. In turn, each of the process families has been subdivided into unique individual processes.

---

[1] Allen, Dell K. and Paul R. Smith, *Process Classification*, Monograph No. 5, Computer Aided Manufacturing Laboratory Brigham, Brigham Young University, Provo, Utah, January 1980.

Manufacturing Processes

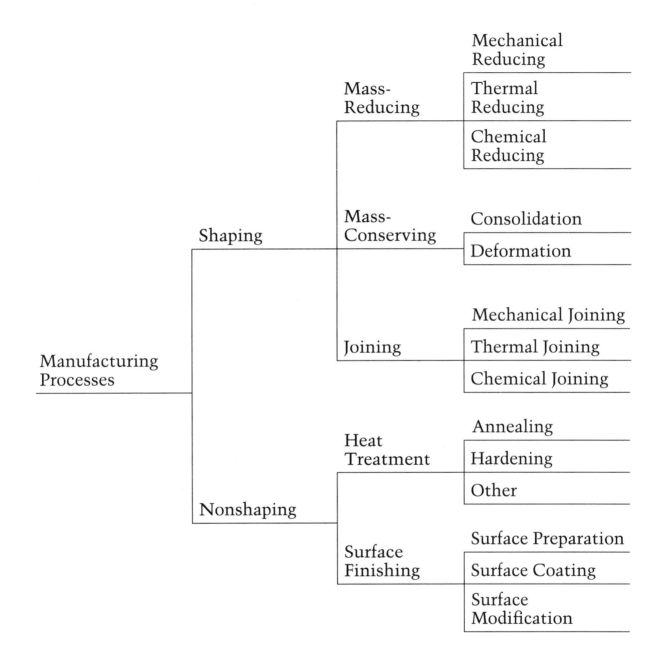

Mass-Reducing

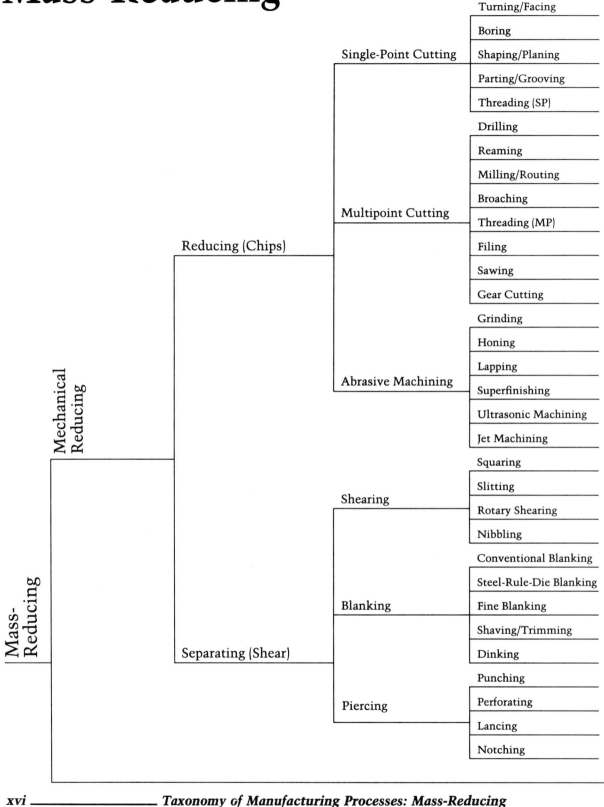

		Single-Point Cutting	Turning/Facing
			Boring
			Shaping/Planing
			Parting/Grooving
			Threading (SP)
		Multipoint Cutting	Drilling
			Reaming
			Milling/Routing
	Reducing (Chips)		Broaching
			Threading (MP)
			Filing
			Sawing
			Gear Cutting
		Abrasive Machining	Grinding
			Honing
Mechanical Reducing			Lapping
			Superfinishing
			Ultrasonic Machining
			Jet Machining
		Shearing	Squaring
			Slitting
			Rotary Shearing
			Nibbling
		Blanking	Conventional Blanking
			Steel-Rule-Die Blanking
	Separating (Shear)		Fine Blanking
			Shaving/Trimming
			Dinking
		Piercing	Punching
			Perforating
			Lancing
			Notching

Mass-Reducing

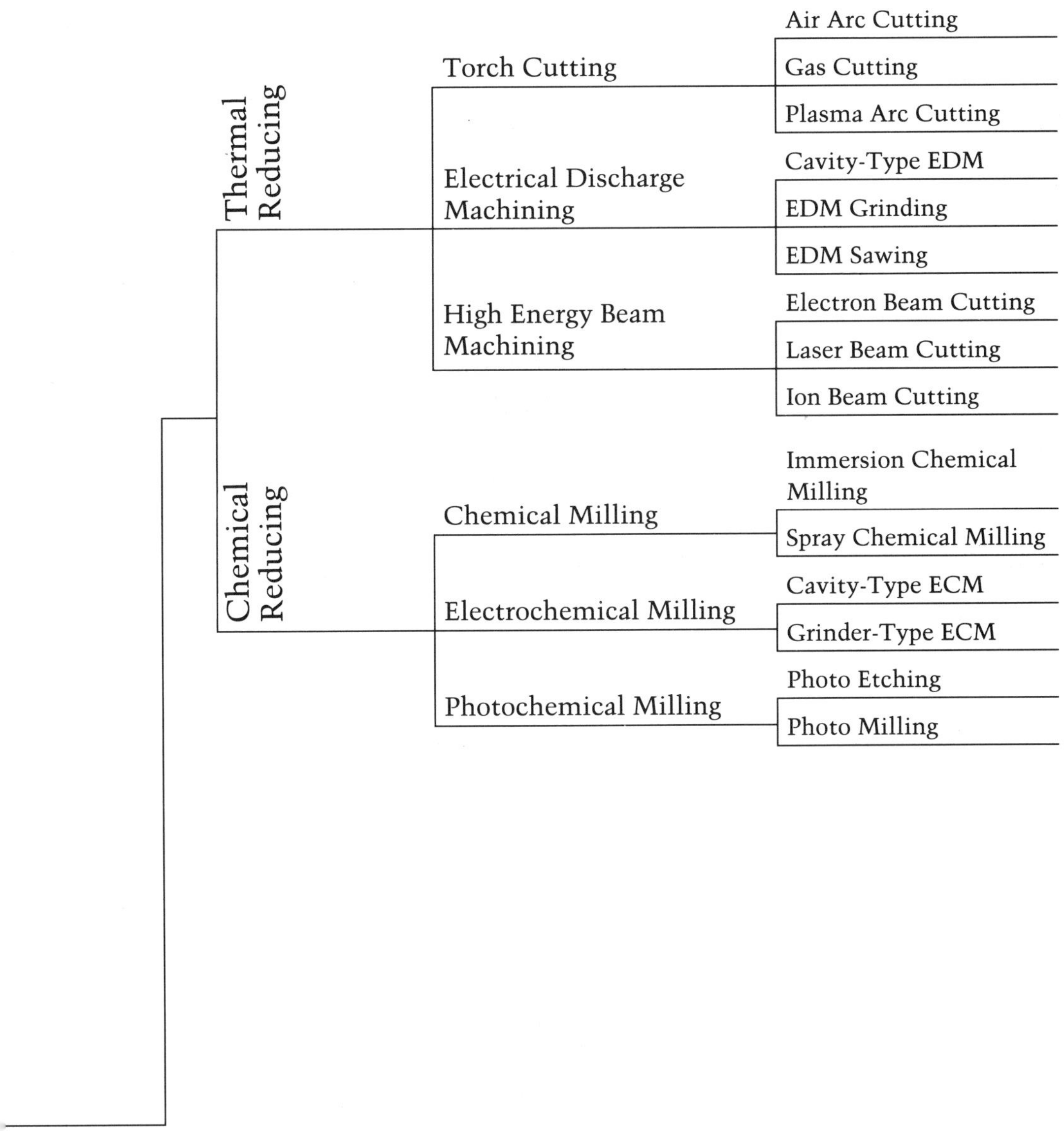

Thermal Reducing

Torch Cutting
- Air Arc Cutting
- Gas Cutting
- Plasma Arc Cutting

Electrical Discharge Machining
- Cavity-Type EDM
- EDM Grinding
- EDM Sawing

High Energy Beam Machining
- Electron Beam Cutting
- Laser Beam Cutting
- Ion Beam Cutting

Chemical Reducing

Chemical Milling
- Immersion Chemical Milling
- Spray Chemical Milling

Electrochemical Milling
- Cavity-Type ECM
- Grinder-Type ECM

Photochemical Milling
- Photo Etching
- Photo Milling

Mass-Conserving

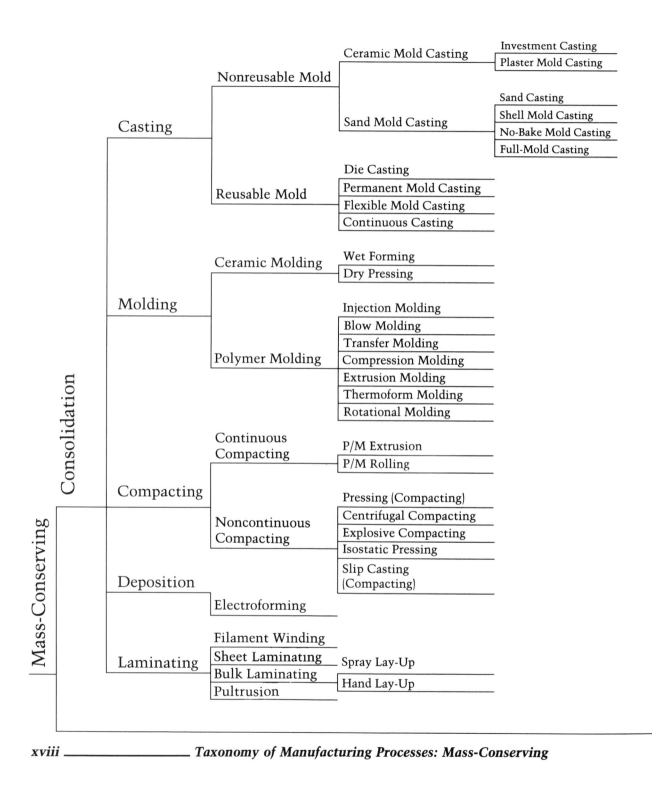

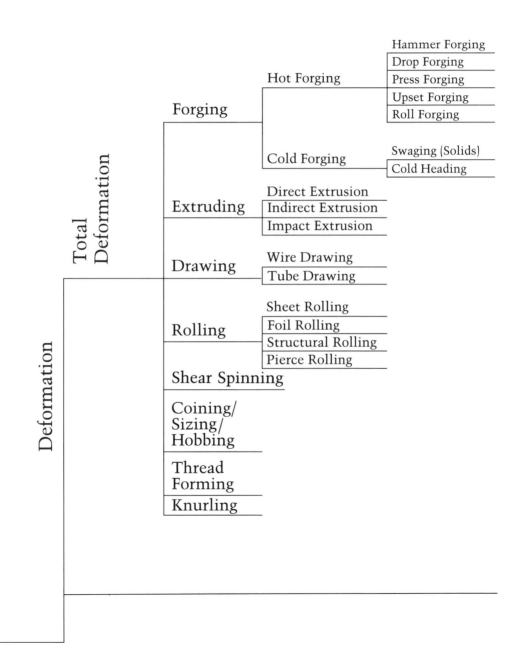

Deformation

Total Deformation

Forging
- Hot Forging
  - Hammer Forging
  - Drop Forging
  - Press Forging
  - Upset Forging
  - Roll Forging
- Cold Forging
  - Swaging (Solids)
  - Cold Heading

Extruding
- Direct Extrusion
- Indirect Extrusion
- Impact Extrusion

Drawing
- Wire Drawing
- Tube Drawing

Rolling
- Sheet Rolling
- Foil Rolling
- Structural Rolling
- Pierce Rolling

Shear Spinning

Coining/ Sizing/ Hobbing

Thread Forming

Knurling

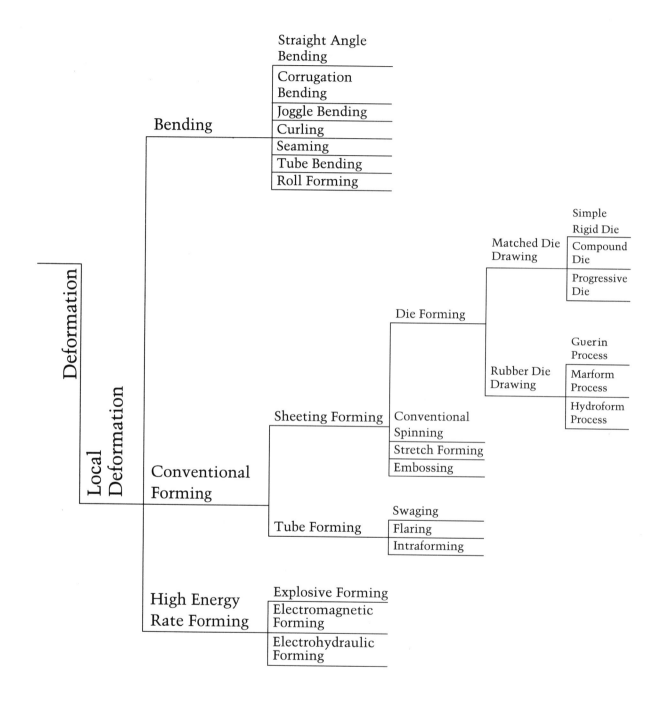

Joining

Mechanical Joining

- Pressure (Cold) Welding
- Friction (Inertial) Welding
- Ultrasonic Welding
- Explosive Welding

Thermal Joining

Thermal Welding

Electric Arc Welding
- Shielded Metal Arc Welding
- Gas Metal Arc (MIG) Welding
- Gas Tungsten Arc (TIG) Welding
- Submerged Arc Welding
- Carbon Arc Welding
- Stud Welding

Electrical Resistance Welding
- Spot Welding
- Seam Welding
- Projection Welding
- Butt Welding
- Percussion Welding
- Electroslag Welding

Gas/Chemical Welding
- Combustible Gas Welding
- Atomic Hydrogen Welding

Braze Welding
- Gas Brazing
- Carbon Arc Brazing

Diffusion Bonding

High Energy Beam Welding
- Electron Beam Welding
- Laser Beam Welding
- Plasma Arc Welding

Brazing
- Infrared Brazing
- Resistance Brazing
- Torch Brazing
- Dip Brazing
- Furnace Brazing
- Induction Brazing

Chemical Joining

Soldering
- Friction/Ultrasonic Soldering
- Induction Soldering
- Infrared Soldering
- Dip Soldering
- Iron Soldering
- Resistance Soldering
- Torch Soldering
- Wave Soldering

Adhesive Bonding

Heat Treatment

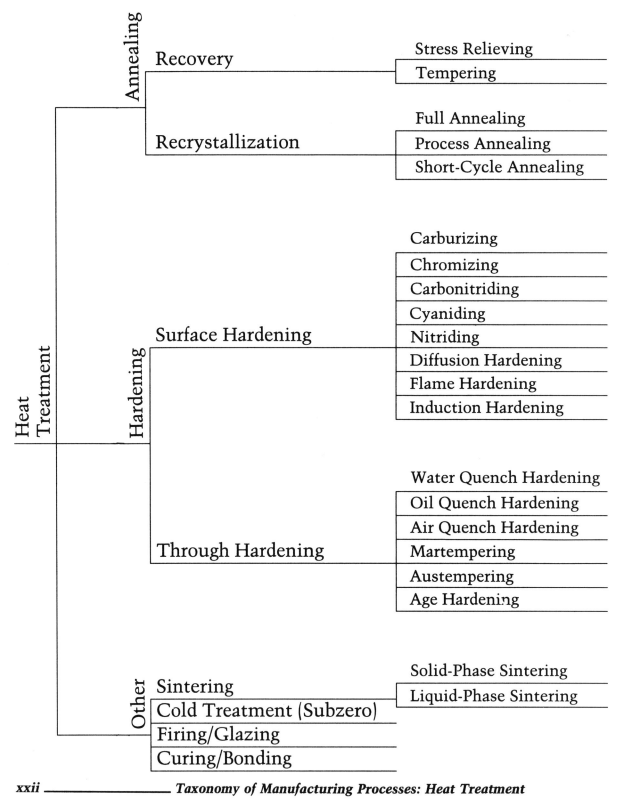

Annealing
- **Recovery**
  - Stress Relieving
  - Tempering
- **Recrystallization**
  - Full Annealing
  - Process Annealing
  - Short-Cycle Annealing

Hardening
- **Surface Hardening**
  - Carburizing
  - Chromizing
  - Carbonitriding
  - Cyaniding
  - Nitriding
  - Diffusion Hardening
  - Flame Hardening
  - Induction Hardening
- **Through Hardening**
  - Water Quench Hardening
  - Oil Quench Hardening
  - Air Quench Hardening
  - Martempering
  - Austempering
  - Age Hardening

Other
- **Sintering**
  - Solid-Phase Sintering
  - Liquid-Phase Sintering
- Cold Treatment (Subzero)
- Firing/Glazing
- Curing/Bonding

Surface Finishing

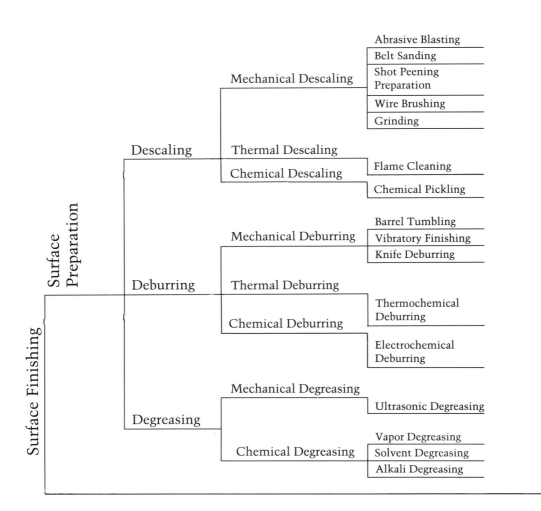

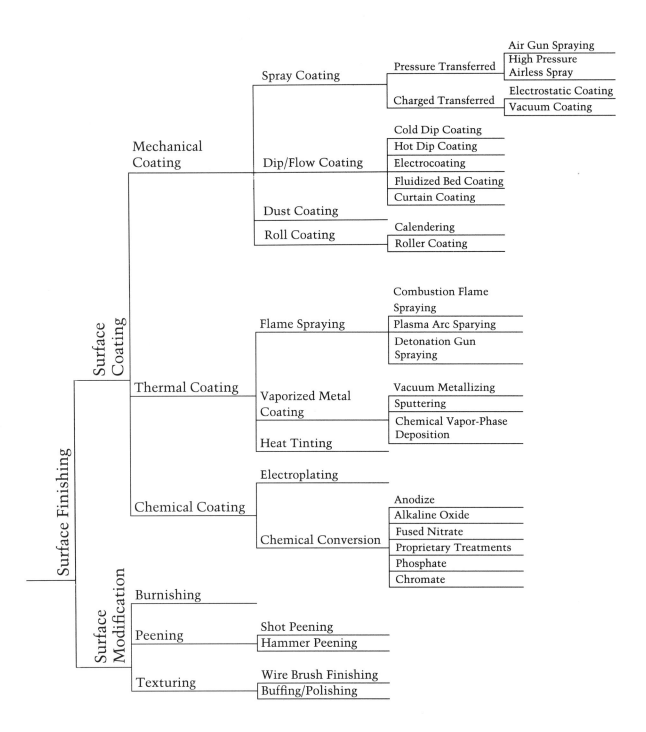

1.
Mechanical Reducing

Abrasive Jet Machining

Abrasive jet machining is a finishing process that removes material from workpieces by focusing a high speed stream of abrasive particles carried in an air jet.

Process Characteristics

* Uses a high velocity stream of abrasive particles carried in an air or gas jet
* Is used for machining delicate or very hard materials
* Produces no heat damage to workpiece surface
* Nozzles are usually made of tungsten or sapphire to resist abrasion
* Produces a taper in deep cuts
* Distance of nozzle from workpiece affects the size of the machined area and the removal rate

Process Schematic

Material is removed by fine abrasive particles, usually about 0.001 in. in diameter, contained in a high velocity air stream. The process is used mainly to cut materials that are sensitive to heat damage and thin sections of hard materials that chip easily. A hood exhaust is used to collect the chips and abrasive grit.

Abrasive material	Grit size	Orifice diameter (in.)
Aluminum oxide	10 to 50 μ	0.005 to 0.018
Silicon carbide	25 to 50 μ	0.008 to 0.018
Dolomite	2500 μ	0.026 to 0.05
Glass beads	0.025 to 0.05 in.	0.026 to 0.05

μ = micron; in. = inch.

Workpiece Geometry

Hand-held polishing, deburring, etching, and radiusing are readily accomplished. However, constant motion of the nozzle is necessary to prevent excessive erosion or grooves in the workpiece. Intricate shapes and holes can be cut in heat-sensitive, brittle, thin, or hard material. Cleaning is usually needed on crevices and internal passageways.

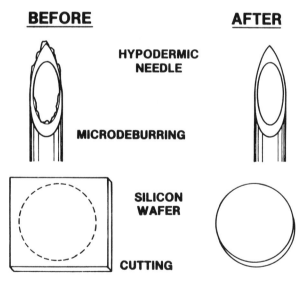

BEFORE **AFTER**

HYPODERMIC NEEDLE

MICRODEBURRING

SILICON WAFER

CUTTING

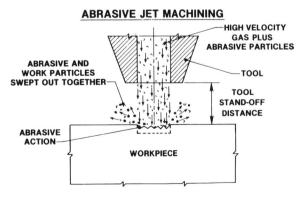

ABRASIVE JET MACHINING

HIGH VELOCITY GAS PLUS ABRASIVE PARTICLES

TOOL

ABRASIVE AND WORK PARTICLES SWEPT OUT TOGETHER

TOOL STAND-OFF DISTANCE

ABRASIVE ACTION

WORKPIECE

Setup and Equipment

Commercial bench-mount units usually are used. The setup usually consists of a power supply and mixer, exhaust hood, nozzle, and gas supply. The nozzle can be hand-held or mounted in a fixture for automatic operation. The workpiece or the nozzle can be moved by some suitable mechanism in automated production.

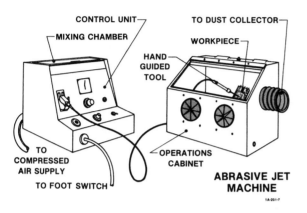

ABRASIVE JET MACHINE

1A-201-7

Typical Tools and Geometry Produced

Nozzles must be highly resistant to abrasion and are made of tungsten carbide or synthetic sapphire. For average material removal, nozzles of tungsten carbide have a useful life of 12 hr to 30 hr (hours), and nozzles of sapphire last about 300 hr. The distance of the nozzle from the workpiece affects the size of the machined area and the rate of material removal.

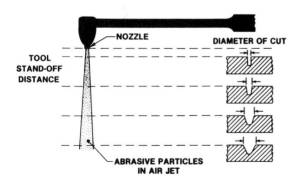

Geometrical Possibilities

This process is adaptable to most geometries, including the microdeburring of gears or other complex surfaces and the cutting of silicon wafers. Workpiece size is limited only by the cabinet size, but workpieces usually range from 5 in.^{2} to 25 in.^{2} (square inches). The removal rate is about 0.001 in./min (inches per minute).

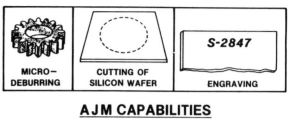

| MICRO-DEBURRING | CUTTING OF SILICON WAFER | ENGRAVING |

AJM CAPABILITIES

	.0001	.001	.1	1	10	100
WORKPIECE SIZE (IN²)				▨	■	▨
MATERIAL REMOVAL (IN)	■	▨				

■ TYPICAL RANGE ▨ FEASIBLE RANGE

Tolerances and Surface Finish

Typical tolerances are ±0.005 in. Feasible surface finish ranges from 4 to 63 microinches, depending on workpiece material and abrasive media.

TOLERANCES		
ABRASIVE JET MACHINING	**TYPICAL**	**FEASIBLE**
POSITION/SIZE	±0.005	±0.001

SURFACE FINISH								
PROCESS	**MICROINCHES (A.A.)***							
ABRASIVE JET MACHINING	2	4	8	16	32	63	125	250
		▨	■	■	■	▨		

■ TYPICAL RANGE ▨ FEASIBLE RANGE

* A.A. = arithmetic average.

Tool Style

There are several types of guns and orifices. A right-angle head is the most commonly used. A straight head is used for more detailed work. A rectangular orifice can carry heavier abrasives, which results in increased removal rates. A round orifice is used for fine abrasives, detailed work, and close tolerances.

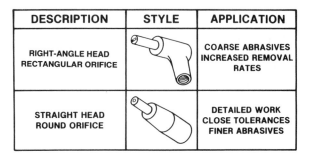

DESCRIPTION	STYLE	APPLICATION
RIGHT-ANGLE HEAD RECTANGULAR ORIFICE		COARSE ABRASIVES INCREASED REMOVAL RATES
STRAIGHT HEAD ROUND ORIFICE		DETAILED WORK CLOSE TOLERANCES FINER ABRASIVES

Workholding Methods

Small vises are used to hold awkward work-pieces. For smaller workpieces, clips or clamps can be used. These clips can be designed to meet the needs of the specific workpiece.

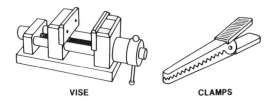

VISE   CLAMPS

Effects on Work Material Properties

The effects of this process are microfine surface finishes on hard workpieces and surface imbedding of abrasive particles for softer materials.

Work material properties	Effects of abrasive jet machining
Mechanical	* Microfine surface finish * Surface imbedding for soft metals
Physical	* Little effect
Chemical	* Little effect

Typical Workpiece Materials

Machinability ratings on hard materials, such as ceramics, germanium, glass, mica, and silicon, are good to excellent. For softer materials, such as steel and aluminum, the ratings are poor to good.

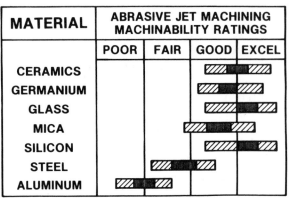

MATERIAL	ABRASIVE JET MACHINING MACHINABILITY RATINGS			
	POOR	FAIR	GOOD	EXCEL
CERAMICS			▨▨ ▥ ▨▨	
GERMANIUM			▨▨ ▥	▨▨
GLASS			▨▨ ▥	▨▨
MICA			▨▨ ▥ ▨▨	
SILICON			▨▨ ▥	▨▨
STEEL		▨▨ ▥	▨▨	
ALUMINUM	▨▨ ▥	▨▨		

▥ TYPICAL RANGE   ▨▨ FEASIBLE RANGE

Abrasive Material

Listed below are several types of abrasives and their applications. Aluminum oxide is the most widely used for cutting hard materials. Silicon carbide is used for cutting ceramics and other extra hard materials. Glass beads are used for fine deburring, light cleaning, and polishing. Sodium bicarbonate is used for cleaning precision parts. Calcium or magnesium compounds are used for light cleaning and for engraving.

Abrasive material	Application
Aluminum oxide	* Most widely used * Cutting most hard materials
Silicon carbide	* Fast cutting on ceramics and other extra hard materials
Glass beads	* Fine deburring * Extremely light cleaning and polishing
Sodium bicarbonate	* Extra fine cleaning of precision parts
Calcium/magnesium	* Light cleaning and etching

Factors Affecting Process Results

Tolerance and surface finish depend upon the following:

* Workpiece material
* Abrasive material and size
* Nozzle to workpiece distance
* Air pressure
* Orifice size and shape
* Working angles

Abrasive Material Information

Listed are the grit sizes for different types of abrasive materials and orifice diameters. Grit diameter sizes range from 10 microns to as large as 0.05 in. Orifice diameters range from 0.005 in. to 0.05 in.

Abrasive material	Grit size	Orifice diameter (in.)
Aluminum oxide	10 to 50 μ	0.005 to 0.018
Silicon carbide	25 to 50 μ	0.008 to 0.018
Dolomite	2500 μ	0.026 to 0.05
Glass beads	0.025 to 0.05 in.	0.026 to 0.05

Process Conditions

The distance from the nozzle to the workpiece for engraving is from 0.4 in. to 0.6 in. and for finishing 0.03 in. to 0.6 in. The removal rates are also given for the different types of materials. They range from 16 mgpm (milligrams per minute) for ceramics to as much as 35 mgpm for silicon.

Conditions	Engraving	Finish
Workpiece to nozzle distance (in.)	0.4 to 0.6	0.03 to 0.6

Workpiece material	Abrasive jet machining removal rate (mgpm)
Ceramics	16 to 24
Germanium	20 to 30
Glass	16 to 24
Mica	20 to 32
Silicon	22 to 35

Dust Collection

Abrasive jet machining requires an enclosed workspace and a dust collection system to filter out fine dust particles prior to returning air to the working environment.

Power Requirements

Power requirements are given in the form of air pressure ranges and compressor horsepower requirements. The abrasive flow rate ranges from 1 to 25 grams per minute with gas pressures ranging from 25 psi to 130 psi (pounds per square inch). Horsepower requirements range from 0.5 hp to 4.0 hp (horsepower) depending on air pressure and orifice nozzle size.

Abrasive flow rate (grams/min)	Air pressure (psi)	Horsepower requirements*
1	25	0.5
5	45	1.2
10	65	1.9
15	85	2.6
20	105	3.3
25	130	4.0

* Horsepower requirements are to maintain a steady pressure and flow.

Cost Elements

* Setup time
* Machining time
* Part cleaning time
* Abrasive change time
* Direct labor rate
* Overhead rate
* Amortization of equipment and tooling costs

Time Calculations

To calculate machining time, the weight of the material to be removed is divided by the removal rate. To calculate the total time, the setup time is added to the machining time and the cleaning time.

Removal rate (mgpm) = R
Removal amount (grams) = W

$$\text{Machining time} = \frac{W}{R}$$

$$\text{Total time} = T_s + T_m + T_c$$

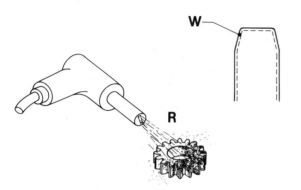

Safety Factors

The following risks should be taken into consideration:

* Personal
  - Injury from propelled abrasive particles
  - Pulmonary diseases
  - Eye damage
* Environmental
  - Abrasive dust

Arbor Milling

Arbor milling is a cutting process in which material is removed by a rotating multiple-tooth cutter. Cutting usually takes place on a surface *parallel* to the axis of tool rotation. The chips formed are swept away by the rotation of the cutter.

Process Characteristics

* Utilizes a rotating, multipoint cutter, which produces discontinuous chips
* Cutters are supported on a mandrel
* Produces flat, contoured, or shaped surfaces (grooves, gears, etc.)
* Is a versatile process with a high metal-removal rate

Process Schematic

An arbor milling tool progressively generates a surface by removing chips from a workpiece as it is fed into a rotating tool. In some cases, the workpiece remains stationary while the tool is fed into the workpiece. In conventional milling, the cutter rotates opposite the direction of the workpiece feed, while in climb milling, the tool rotates in the direction of the workpiece feed.

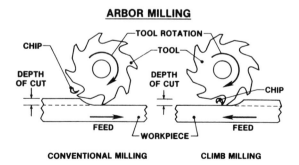

Workpiece Geometry

Shown are typical workpieces before and after milling. Flat or formed surfaces may be ma-chined with excellent finish and accuracy. In some cases, the work is completed in one pass of the workpiece. Arbor milling is most efficient when the workpiece is no harder than Rockwell C25. However, harder steels may be successfully milled.

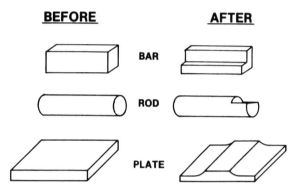

Setup and Equipment

Arbor milling is commonly performed on a horizontal milling machine like the one shown. The tool is mounted on an arbor (like an axle) that is suspended between the spindle and arbor support. This type of machine allows the tool to be placed in numerous positions in relation to the workpiece.

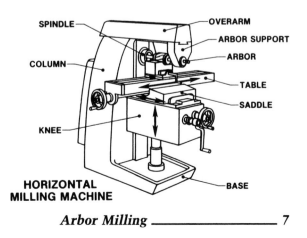

Typical Tools and Geometry Produced

A wide variety of arbor milling cutters are used, including double angle, form relieved, plane, and staggered tooth. Each of these tools is designed to produce forms on the workpiece. Arbor milling is a versatile process because a large variety of cutters are available.

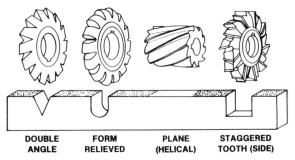

MILLING CUTTERS

| DOUBLE ANGLE | FORM RELIEVED | PLANE (HELICAL) | STAGGERED TOOTH (SIDE) |

Geometrical Possibilities

Shown are some of the shapes that can be produced by arbor milling cutters. Mostly concave or convex shapes are produced. Cutters used in arbor milling are typically capable of making cuts from 0.25 in. to 6 in. wide and from 0.02 in. to 0.5 in. deep. Smaller (or larger) depths and widths are feasible with the proper equipment.

END VIEW	END VIEW	SIDE VIEW	END VIEW

ARBOR MILLING CAPABILITIES

	.001	.01	.1	1	10	100
WIDTH OF CUT (IN)						
DEPTH OF CUT (IN)						

■ TYPICAL RANGE ▨ FEASIBLE RANGE

Tolerances and Surface Finish

For most arbor milling applications, tolerances can be held within ±0.005 in. For precision applications, tolerances can be held within ±0.001 in. Feasible surface finishes may range from 32 to 500 microinches, with a typical range between 63 and 200 microinches. Finish cuts will generate surfaces near 32 to 63 microinches, roughing cuts near 200 microinches.

TOLERANCES		
ARBOR MILLING	TYPICAL	FEASIBLE
DEPTH	±0.005	±0.001

SURFACE FINISH									
PROCESS	MICROINCHES (A.A.)								
	2	4	8	16	32	63	125	250	500
ARBOR MILLING									

■ TYPICAL RANGE ▨ FEASIBLE RANGE

Tool Style

Arbor mounted milling tools are available in a wide range of sizes and shapes. Shown are four common types of arbor milling cutters that can generate a variety of shapes on a workpiece. In addition to the wide assortment of cutters that may be purchased, cutters can be interlocked or "ganged" together to machine a large area or varied shapes.

DESCRIPTION	STYLE	APPLICATION
DOUBLE ANGLE		VARIOUS ANGLES ON SLOTS AND EDGES
FORM RELIEVED		CONCAVE, CONVEX SURFACES AND GEAR TEETH
PLANE (HELICAL)		FLAT SURFACES OR SLAB MILLING
STAGGERED TOOTH (SIDE)		DEEP SLOTS AND CONCAVE SURFACES

Workholding Methods

Shown are two frequently used workholding devices, the vise and strap clamp. The pneumatic clamp is used in applications where short load/unload time is important. The strap clamp exerts a holding force principally in the vertical direction, to hold the workpiece securely on the machine table.

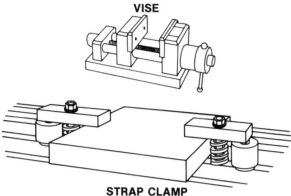

VISE

STRAP CLAMP

Effects on Work Material Properties

During arbor milling, the mechanical properties of the workpiece may be affected by a built-up edge on the cutter or a dull tool. Arbor milling may cause an untempered martensitic layer on the workpiece surface when milling heat-treated alloy steels. The physical properties of the workpiece are affected very little by arbor milling.

Work material properties	Effects of arbor milling
Mechanical	* A built-up edge on cutter causes a rough workpiece surface * Dull tools cause severe surface damage and high residual stresses
Physical	* Little effect
Chemical	* An untempered martensitic layer (0.001 in.) may be produced when milling heat-treated alloy steels

Typical Workpiece Materials

Aluminum, brass, cast iron, mild steel, and thermoset plastics have good machinability ratings. Stainless steel has a fair rating because it tends to work harden when machined, even though it is initially ductile.

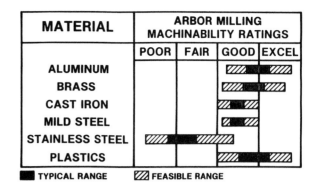

MATERIAL	ARBOR MILLING MACHINABILITY RATINGS			
	POOR	FAIR	GOOD	EXCEL
ALUMINUM			▨■	■▨
BRASS			▨■	■▨
CAST IRON			▨■	
MILD STEEL			▨■	
STAINLESS STEEL	▨■	■▨		
PLASTICS			▨■	■▨

■ TYPICAL RANGE   ▨ FEASIBLE RANGE

Tool Materials

High speed tool steel has been used for some time, but is being replaced in many applications by carbide, ceramic, and diamond tooling. Carbide inserts are easily replaceable and are long lasting. Ceramic tools can withstand high temperatures, which makes high speed machining possible, but are rather brittle. Diamond tools produce a superior surface finish and are used for nonferrous and nonmetallic materials.

Tool materials	Applications
High speed steel	* Special tool shapes * Low production
Carbides (inserts)	* High production
Ceramics (inserts)	* High speed machining * High production * Uninterrupted cuts
Diamonds (inserts)	* High surface qualities, fine tolerances * Nonferrous or nonmetallic materials

Factors Affecting Process Results

Tolerance and surface finish depend upon the following:

* Tool geometry and sharpness
* Cutting speed and feed rate
* Rigidity of tool, workpiece, and machine
* Alignment of machine components and fixtures
* Cutting fluid

Tool Geometry

Geometry is determined by angles on the tool. Shown are the four angles that are most often used. Generally, the axial and radial rake angles are high for soft materials and low for hard materials. Negative rake angles are often recommended on carbide cutters. The axial and radial relief angles prevent the tool from creating too much friction and resistance upon contact with the workpiece. Shown are some recommended ranges for these angles when using a high speed steel cutter.

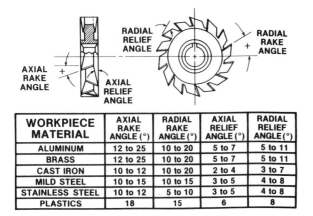

WORKPIECE MATERIAL	AXIAL RAKE ANGLE (°)	RADIAL RAKE ANGLE (°)	AXIAL RELIEF ANGLE (°)	RADIAL RELIEF ANGLE (°)
ALUMINUM	12 to 25	10 to 20	5 to 7	5 to 11
BRASS	12 to 25	10 to 20	5 to 7	5 to 11
CAST IRON	10 to 12	10 to 20	2 to 4	3 to 7
MILD STEEL	10 to 15	10 to 15	3 to 5	4 to 8
STAINLESS STEEL	10 to 12	5 to 10	3 to 5	4 to 8
PLASTICS	18	15	6	8

Process Conditions

Shown are the suggested ranges for cutting speeds and feed rates using high speed tool steel under dry cutting conditions at a 0.015 in. depth of cut. Generally speaking, the cutting speeds and feed rates are high for soft materials and low for hard materials. Both cutting speeds and feed rates can be substantially increased when coolants are used and carbide tooling is substituted for steel tooling.

Typical speeds and feeds			
Workpiece material	Hardness (Hardness Brinell)	Cutting speed (sfpm)	Feed rate (ipt)
Aluminum	70 to 125	300 to 500	0.006 to 0.010
Brass	60 to 100	110 to 275	0.007 to 0.009
Cast iron	250 to 320	30 to 55	0.005 to 0.006
Mild steel	275 to 325	60 to 80	0.006
Stainless steel	275 to 325	40 to 55	0.006
Plastics	· · ·	150 to 350	0.006

Lubrication and Cooling

Typical cutting fluids include mineral, synthetic, and water-soluble oils. Application of these cutting fluids is usually done by spraying, misting, or flooding the workpiece. The main function of cutting fluids is to cool the tool, which increases tool life and makes high cutting speeds and feed rates possible. A secondary function of cutting fluids is to lubricate the tool/workpiece interface, which contributes to a better workpiece surface finish.

Work material	Cutting fluid	Application
Aluminum	None, mineral oil, fatty oil	Spray, flood
Brass	Mineral oil, specialty fluid	Spray, flood
Cast iron	Soluble oil, chemical and synthetic oil, none	Spray, flood
Mild steel	Chemical and synthetic oil, soluble oil	Spray, flood
Stainless steel	Sulfurized mineral oil, fatty soluble oil, chemical and synthetic oil	Spray, flood
Plastics	Mineral oil, soluble oil, cold air, none	Spray, flood, air jet

Power Requirements

Shown is the unit power needed to mill certain materials with different hardnesses. The formula for unit power is horsepower divided by cubic inches of material removed per minute (in.^{3}/min) at 80% efficiency. For example, when milling mild steel with an HB hardness of 330 to 370, the horsepower needed to mill 10 cubic inches of material per minute is 15 hp. Fifteen horsepower divided by 10 in.^{3}/min = 1.5 unit power.

Machine hp = unit-power × removal rate (in.^{3}/min)

Material	Hardness (HB)	Unit power*
Aluminum	30 to 150	0.3
Brass	50 to 145	0.6
	145 to 240	1.0
Cast iron	110 to 190	0.6
	190 to 320	1.1
Mild steel	85 to 200	1.1
	330 to 370	1.5
	485 to 560	2.1
Stainless steel	135 to 275	1.4
	275 to 430	1.5
Plastics	N/A	0.05 est.

* Unit power based on: ● high speed steel (HSS) and carbide tools ● feed of 0.005 to 0.012 ipt ● 80% efficiency.

Time Calculations

It is important to know the milling setup time so that actual milling time and positioning time can be calculated. The elements involved in calculating the milling time are diameter of cutter, overtravel, approach distance, length of workpiece, depth of cut, feed rate, rapid traverse rate, and rapid traverse distance.

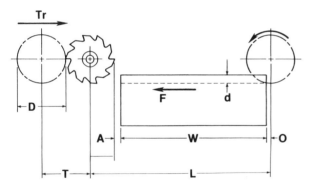

Diameter of cutter (in.) = D
Depth of cut (in.) = d
Length of workpiece (in.) = W
Length of cut (in.) = L
Rapid traverse distance (in.) = T
Rapid traverse rate (ipm) = Tr
Number of teeth in cutter = N
Cutter feed rate (ipm) = F
Cutting speed (sfpm) = V
Feed per tooth = f
Approach distance (in.) = A
Overtravel = O
rpm = revolutions per minute

$$\text{Milling time} = \frac{L}{F}$$

$$\text{Traverse time} = \frac{T}{Tr}$$

$$\text{rpm} = \frac{4 \times V}{D} \text{ (approx.)}$$

$$\text{Feed rate} = f \times N \times \text{rpm}$$

Cost Elements

The following is a list of cost elements:

* Setup time
* Load/unload time
* Idle time
* Cutting time
* Tool costs
* Direct labor rate
* Overhead rate
* Amortization of equipment and tooling

Safety Factors

The following risks should be taken into consideration:

* Personal
  – Rotating tool
  – Hot and sharp chips
  – Eye and skin irritation from cutting fluids

Band Filing

Band filing is a multipoint cutting process in which a workpiece is fed into a continuously moving file, and chips are removed by cutting teeth that are arranged in succession along the file surface. Fine, accurate work can be produced on the band filing machine.

Process Characteristics

* Workpieces are fed into a continuously moving band
* Uses a multipoint tool
* Produces fine, accurate work
* Is a finishing operation in which small amounts of material are removed
* Produces fine feed marks

Process Schematic

The band file consists of interconnecting file segments that are attached to a spring steel band. As the band file rotates, the individual teeth on each file segment cut very small chips from the workpiece surface. A band machine will produce a uniformly smooth internal or external surface on virtually any kind of metal workpiece.

Workpiece Geometry

Shown are typical workpieces used in band filing. Deburring, file reliefs, and squaring sides are some of the operations performed.

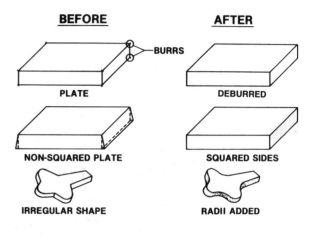

BEFORE   AFTER

PLATE   DEBURRED

NON-SQUARED PLATE   SQUARED SIDES

IRREGULAR SHAPE   RADII ADDED

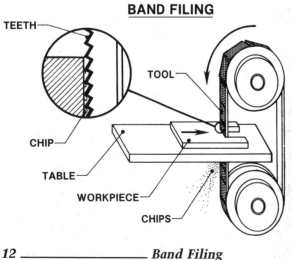

BAND FILING

TEETH
TOOL
CHIP
TABLE
WORKPIECE
CHIPS

Setup and Equipment

Band filing machines provide continuous cutting action. Most band filing is done on contour band sawing machines by means of a special band file that is substituted for the usual band saw blade. The continuous band file runs on wheels located in the base and in the head.

Typical Tools and Geometry Produced

Some files that are used in band filing are shown in the illustration. The shape of the file used depends upon the shape of the workpiece, its material, size, hardness, and the desired finish. The coarseness of cut is determined by the number of teeth on the file, which ranges from 10 to 24 teeth per inch.

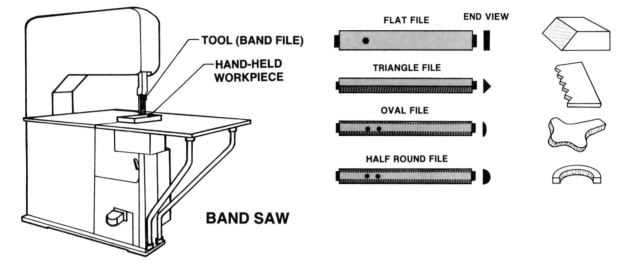

TOOL (BAND FILE)

HAND-HELD WORKPIECE

BAND SAW

FLAT FILE   END VIEW

TRIANGLE FILE

OVAL FILE

HALF ROUND FILE

Band Sawing

Band sawing is a multipoint cutting process in which a workpiece is advanced into a moving continuous band that has teeth along one edge. Parts may be severed from a workpiece. Band sawing can likewise produce irregular or curved cuts.

Process Characteristics

* Uses a flexible steel band with a toothed edge
* Workpieces are fed into the cutting edge on vertical machines
* Produces fine chips, a narrow cut/kerf, and fine feed marks
* Can produce straight, irregular, or curved cuts
* Produces uniform cutting action as a result of an evenly distributed tooth load

Process Schematic

In band sawing, the workpiece is pushed into the blade and the direction of the cut is guided manually or mechanically. There is no restriction on the possible shape of the finished workpiece. A continuous band runs vertically at the point of the cut, where it passes through a work table on which the workpiece rests.

Workpiece Geometry

Shown are typical workpieces before and after bandsawing. Contour sawing of dies, jigs, cams, templates, and numerous other parts can be done with band saws. Formerly, these had to be made either by hand or by other machine tools. A further advantage is that the band saw cuts a very narrow kerf, which minimizes the amount of material wasted.

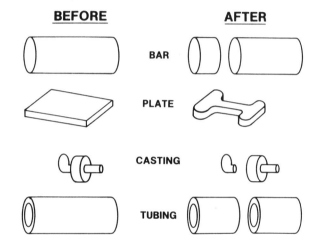

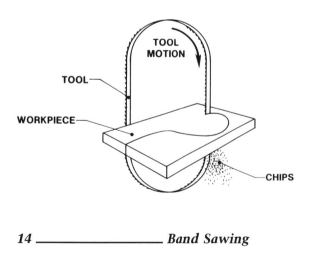

Setup and Equipment

Metal band sawing machines resemble those used for wood but differ in saw cutting speeds and types of blades. Most machines are designed with the saw operating in a vertical position. The work is supported on a horizontal table, which has a tilting adjustment for cutting angles. Horizontal machines may also be used.

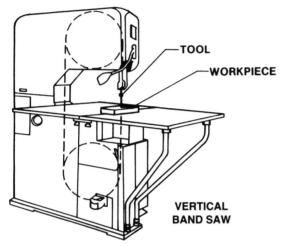

Typical Tools and Geometry Produced

The three most common tooth forms are shown: the precision tooth form, which gives accurate cuts with a smooth finish; the buttress tooth form, which provides faster cutting and large chip loads; and the claw tooth form, which gives additional chip clearance for fast cuts and soft materials. At least two teeth must be in contact with the workpiece at all times to avoid stripping off the teeth.

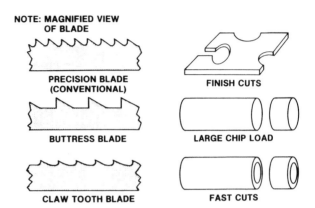

Broaching

Broaching is a cutting operation where accurate sizing and finishing of surface or shape is achieved by a single pass of a multipoint cutter (the broach). The stock removal of the broach is built into the tool by having each successive tooth cut deeper into the material. Thus, both roughing and finishing cuts may be built into the same tool.

Process Characteristics

* Uses a single pass for finished shapes or sizes
* Produces close tolerances and good surface finish
* Uses a multipoint cutting tool (broach)
* Has the roughing and finishing teeth on the same tool
* Machine surfaces are parallel to axis of tool motion

Process Schematic

A broach is composed of a series of single-point cutting edges projecting from a rigid bar. Each tooth protrudes a greater distance from the axis of the bar than the preceding one, up to and including the first finished tooth. All finishing teeth are the same size.

BROACHING

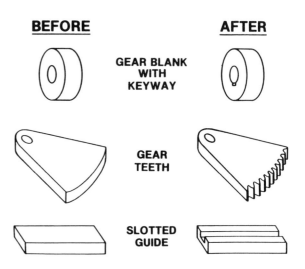

Workpiece Geometry

Shown are the typical workpieces before and after broaching. Broaching originally was developed for cutting keyways. It quickly led to mass production machining of various surfaces, such as flats, internal and external cylindrical surfaces, and many irregular surfaces. Almost any shape hole can be broached. Broaching usually produces better accuracy and finish than milling or reaming.

BEFORE   **AFTER**

GEAR BLANK WITH KEYWAY

GEAR TEETH

SLOTTED GUIDE

Setup and Equipment

A broaching machine typically consists of a workholding device, a drive mechanism, and a supporting frame. Shown is a vertical push down machine. The broaching tool is pushed through or across the workpiece, which is held stationary. Other types of broaching machines include pull broach, surface broach, and continuous broach. These broaching machines are either horizontal or vertical. Most machines are hydraulically powered, but some are driven electromechanically.

BROACHING

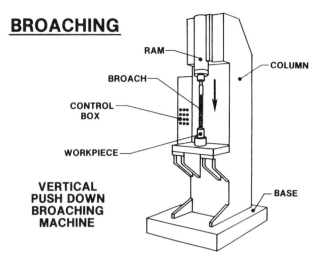

RAM
COLUMN
BROACH
CONTROL
BOX
WORKPIECE
BASE

**VERTICAL
PUSH DOWN
BROACHING
MACHINE**

Typical Tools and Geometry Produced

Broaching tools are different from other production tools because they are used for only one particular operation. There are two types of broaches used: pull and push broaches. Most internal broaching is done with pull broaches because they remove more material and take longer cuts than push broaches. Holes are sized mainly by push broaches in heat-treated parts and for short-run production.

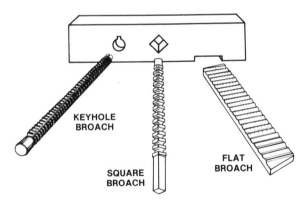

KEYHOLE
BROACH

SQUARE
BROACH

FLAT
BROACH

Geometrical Possibilities

Almost any shape can be broached both internally and externally. Just a few of the possibilities are shown in this figure. Internal broaches usually make cuts between 0.125 in. and 6 in.; however, cuts can range from 0.05 in. to 12 in. in diameter. Surface broaches usually make cuts between 0.075 in. and 10 in.; however, cuts can range from 0.02 in. to 20 in. in width.

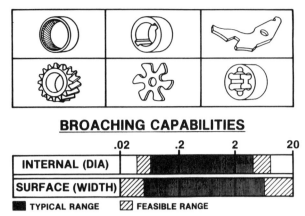

BROACHING CAPABILITIES

	.02	.2	2	20
INTERNAL (DIA)				
SURFACE (WIDTH)				

■ TYPICAL RANGE ▨ FEASIBLE RANGE

Tolerances and Surface Finish

For most broaching applications, tolerances are held within ±0.002 in. For precision applications, tolerances can be held within ±0.0005 in. Surface finishes may range from 8 to 125 microinches. The typical range is between 16 and 63 microinches.

TOLERANCES		
BROACHING	**TYPICAL**	**FEASIBLE**
	±0.002	±0.0005

SURFACE FINISH		
PROCESS	**MICROINCHES (A.A)**	
BROACHING	2 4 8 16 32 63 125 250 500	

■ TYPICAL RANGE ▨ FEASIBLE RANGE

Tool Style

Shown are four different types of tools used for broaching. Solid internal broaches are used for small diameter cuts, whereas the shell type is used for larger diameter cuts. Surface broaches are usually made to a specific shape for external broaching. A keyway broach is used for both internal and external keyways, gears, and bushings.

DESCRIPTION	STYLE	APPLICATION
INTERNAL SOLID		MANY SHAPES: ROUNDS, SQUARES, KEYWAYS, GEARS, BUSHINGS, HELIX, SPLINES, ROTARY SHAPES, SERRATIONS
INTERNAL SHELL		
SURFACE		MANY SHAPES: FLATS, CONCAVE, CONVEX
KEYWAY		KEYWAYS, SLOTS, GEAR TEETH

Typical Workpiece Materials

Aluminum, brass, and plastics have good to excellent ratings, while cast iron and mild steel have good ratings. Stainless steel has a fair rating because of its hardened structure and its tendency to work harden when machined.

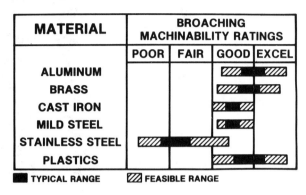

MATERIAL	BROACHING MACHINABILITY RATINGS			
	POOR	FAIR	GOOD	EXCEL
ALUMINUM			▨	▨
BRASS			▨	▨
CAST IRON			▨ ▨	
MILD STEEL			▨ ▨	
STAINLESS STEEL	▨	▨		
PLASTICS			▨	▨

■ TYPICAL RANGE   ▨ FEASIBLE RANGE

Workholding Methods

The internal indexing fixture is used for broaching internal shapes, such as gears or slots. Once one gear or slot is broached, the workpiece is automatically or manually indexed for the next gear or slot. Also shown is a surface broaching fixture that must be on an adjustable table so that the broach can pass by without hitting the table.

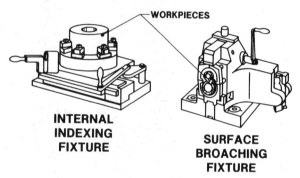

WORKPIECES

INTERNAL INDEXING FIXTURE

SURFACE BROACHING FIXTURE

Tool Materials

Hardened high speed steel is the most common broaching material and is used on all types of workpiece materials. However, where carbide inserts can be used, such as on surface broaches, higher production rates are possible and closer tolerances are produced.

Tool materials	Applications
Hardened high speed tool steels	* Most common broach material * All types of workpiece material * External and internal surfaces * Special tool shapes—solid and shell
Carbide inserts	* Usually used on cast iron * Surface broaches * High production rates * Extremely close tolerances

Effects on Work Material Properties

During broaching, the mechanical properties of the workpiece that are affected include hardening on unhardened materials and softening on hardened materials. The physical properties are localized welding of tool and work material. There is little effect on the chemical properties.

Work material properties	Effects of broaching
Mechanical	* May cause work hardening on ferrous materials
Physical	* May cause localized welding of tool and work material
Chemical	* Little effect

Factors Affecting Process Results

Tolerance and surface finish depend upon the following:

* Workpiece material
* Tool geometry
* Cutting speed and chip load per tooth
* Rigidity of tool, workpiece, and machine
* Alignment of machine components and fixtures
* Cutting fluid

Tool Geometry

Broaching tool geometry is designated by angles on the tool. The hook angle is between −5° and 20°, depending upon the material, and the clearance angle is rarely less than 1° or greater than 3°. Also shown are the pitch and land lengths, the tooth depth, and the root radius.

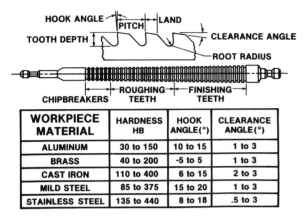

WORKPIECE MATERIAL	HARDNESS HB	HOOK ANGLE(°)	CLEARANCE ANGLE(°)
ALUMINUM	30 to 150	10 to 15	1 to 3
BRASS	40 to 200	-5 to 5	1 to 3
CAST IRON	110 to 400	6 to 15	2 to 3
MILD STEEL	85 to 375	15 to 20	1 to 3
STAINLESS STEEL	135 to 440	8 to 18	.5 to 3

Process Conditions

Shown are the suggested ranges for cutting speeds and chip load per tooth for broaching with a high speed steel tool and cutting fluids. Generally speaking, the cutting speeds and chip load per tooth are low for hard materials, such as cast iron and stainless steel. These ranges are a little higher for softer materials, such as aluminum, brass, and mild steel.

Workpiece material	Hardness (HB)	Cutting speed (fpm)	Rough chip load (ipt)*
Aluminum	30 to 150	35 to 45	0.005 to 0.007
Brass	40 to 200	20 to 30	0.004 to 0.005
Cast iron	120 to 320	10 to 30	0.002 to 0.005
Mild steel	100 to 275	25 to 40	0.003 to 0.004
Stainless steel	135 to 375	10 to 20	0.002 to 0.003

* Tooth load decreases as fineness of finish increases.

Lubrication and Cooling

Typical cutting fluids for broaching include one or more of the following: kerosene, mineral oil, water, soluble oil, mineral lard oil, and sulfurized mineral oil. The main function of cutting fluids is to cool the tool, which increases tool life and makes high cutting speeds possible. A secondary function is to lubricate the tool/workpiece interface, which contributes to a better surface finish.

Work material	Cutting fluid	Application
Aluminum	Kerosene, mineral oil, sulfurized mineral oil	Flood
Brass	Water, soluble oil, mineral lard oil	Flood
Cast iron	Water, soluble oil, sulfurized mineral oil	Flood, submerge
Steel (all types)	Water, soluble oil, sulfurized mineral oil, mineral lard oil	Flood, submerge, jet spray

Power Requirements

Shown is the formula for calculating the machine horsepower needed to broach different materials. The chart gives the unit power chip load per tooth for the different types of materials. The information the user needs to provide is the metal removal rate and the efficiency of the broach drive.

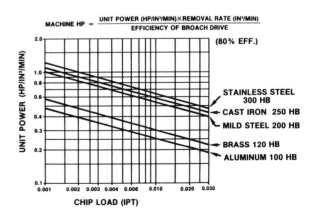

Time Calculations

It is important to understand the broaching setup so that actual broaching time, retract time, and positioning time can be calculated. The elements involved in calculating broaching time are length of broach teeth and cutting speed. For calculating retract time, the elements are retract distance and retract rate. For positioning time, the elements involved are approach distance and positioning rate.

Broaching Setup

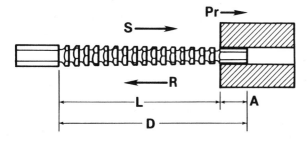

Length of broach teeth (in.) = L
Cutting speed (sfpm) = S
Retract distance (in.) = D
Retract rate (ipm) = R
Approach distance (in.) = A
Positioning rate (ipm) = Pr

$$\text{Broaching time} = \frac{L}{S}$$

$$\text{Retract time} = \frac{D}{R}$$

$$\text{Positioning time} = \frac{A}{Pr}$$

Cost Elements

The following are cost elements:

* Setup time
* Load/unload time
* Idle time
* Cutting time
* Tool change time
* Tool costs
* Direct labor rate
* Overhead rate
* Amortization of equipment and tooling

Safety Factors

The following risks should be taken into consideration:

* Personal
  - Reciprocating tool
  - Hot and sharp chips
  - Eye and skin irritation from cutting fluids
  - Tool breakage

Centerless Grinding

Centerless grinding is an abrasive machining process by which small chips of material are removed from the external surface of a cylindrical metallic or nonmetallic workpiece. This process relies on the relative rotations of the grinding wheel and regulating wheel to rotate the workpiece. The process does not require chucking or locating the workpiece between centers for rotation.

Process Characteristics

* Requires no chucking or mounting of the workpiece
* Produces close tolerances and smooth surfaces
* Is applicable for cylindrical, stepped, formed, and conical workpieces
* Is most efficient for through-feed production grinding operations
* Requires coolant
* Is primarily a finishing process

Process Schematic

The workpiece is located between the regulating and grinding wheels and is supported by a rest blade. The grinding wheel drives the workpiece, and the regulating wheel controls workpiece rotation. The difference in rotating speed between the grinding wheel and the workpiece determines the material removal.

Workpiece Geometry

Workpiece geometries that can be machined include plain cylindrical surfaces, surfaces with radii and fillets, stepped parts, and tapered surfaces. Close dimensional tolerances can be achieved as well as smooth surface finishes.

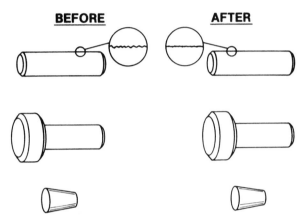

Setup and Equipment

Centerless grinding machines basically consist of a grinding wheel, a regulating wheel, a work rest blade, and coolant capabilities. Machines are available in a range of sizes and models, many of which are capable of performing several different types of centerless grinding operations such as taper grinding, step grinding, and through-feed grinding of solid or tubular stock.

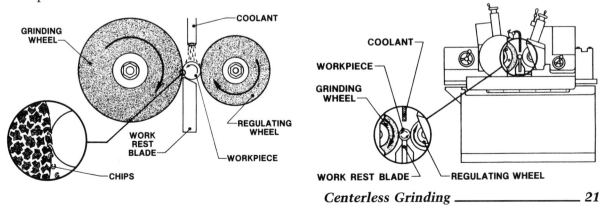

Typical Tools and Geometry Produced

Most applications use plain wheels that impart no special contours to the workpiece surface. Grinding a formed surface requires the use of specially formed grinding and regulating wheels. The geometry produced on the workpiece is a mirror image of the grinding wheel shape. A grinding wheel with a flat, cylindrical periphery will produce a flat, cylindrical surface on the workpiece. Typical workpiece sizes range from 0.6 in. to 20 in. in diameter and 2 in. to 40 in. in length. However, with through-feed grinding, it is possible to grind workpieces that are between 0.003 in. and 40 in. in diameter and from 0.5 in. to nearly any length.

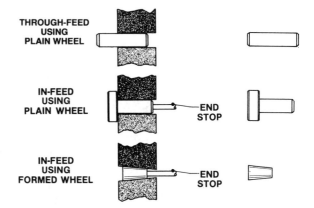

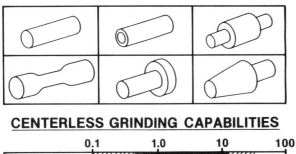

CENTERLESS GRINDING CAPABILITIES

	0.1	1.0	10	100
DIAMETER (IN)				
LENGTH (IN)				

▮ TYPICAL RANGE   ▨ FEASIBLE RANGE

Geometrical Possibilities

Shown are several types of centerless grinding. Through-feed grinding is best for long, straight workpieces. In-feed grinding is best for plunge grinding formed workpieces. In-feed/through-feed grinding is best for plunge grinding multiple diameter workpieces. End-feed grinding is best for tapered workpieces.

Workholding Methods

When a workpiece is in the loaded position, it rests on two blades without making contact with the grinding or regulating wheels. When a workpiece is in the grinding position, it rests on only one blade and makes contact with both wheels, creating a clearance between the workpiece and the other blade. Often just a single blade is used for a workholding device.

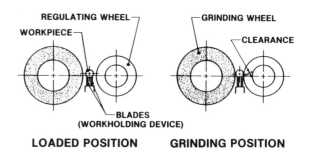

LOADED POSITION   **GRINDING POSITION**

Tool Style

Tools are available in a wide range of shapes, depending on their application. Shown are straight, tapered, and formed grinding wheels. Formed wheels are used for "in-feed" or "plunge grinding," whereas plain wheels are most often used for "through-feed" or "continuous grinding." Specialized shapes may be formed on the face of the grinding wheel by a dressing unit that uses hard, abrasive material to grind a desired shape on the wheel. The wheel may also be formed by using a crush roll device, which molds the wheel by exerting great pressure with a shaped roll.

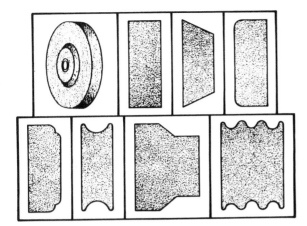

TOLERANCES		
CENTERLESS GRINDING	TYPICAL	FEASIBLE
DIAMETRIC	±.001	±.00005
ROUNDNESS	±.002	±.00001

SURFACE FINISH	
PROCESS	MICROINCHES (A.A.)
CENTERLESS GRINDING	2 4 8 16 32 63 125 250 500

▓ TYPICAL RANGE   ▨ FEASIBLE RANGE

Centerless Grinding Variations

Through-feed grinding is best for long, straight workpieces. In-feed grinding is best for plunge grinding formed workpieces. In-feed/through-feed grinding is best for plunge grinding multiple diameter workpieces, and end-feed grinding is best for grinding tapered workpieces.

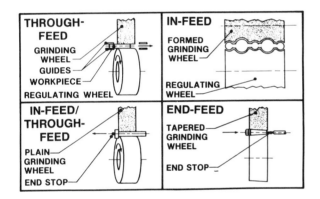

Tolerances and Surface Finish

For most applications, tolerances are held within ±0.001 in. for diameter and ±0.002 in. for roundness. For precision applications, tolerances for the diameter are as high as ±0.00005 in. and ±0.00001 in. for roundness. Surface finish may range from 2 to 125 microinches, with the typical range between 8 and 32 microinches.

Effects on Work Material Properties

High grinding temperatures may cause residual stresses and a thin martensite layer. This thin layer of martensite may cause a reduction in material fatigue strength. If the grinding temperature is elevated above the curie temperature, a loss of magnetic properties may occur in ferromagnetic materials. Cylindrical grinding may reduce corrosion resistance due to the resulting highly stressed surface.

Work material properties	Effects of centerless grinding
Mechanical	* Residual surface stresses * A thin martensitic layer may form on the part surface * Fatigue strength may be reduced
Physical	* Magnetic properties may be lost on ferromagnetic materials
Chemical	* May increase susceptibility to corrosion

Tool Geometry

Shown below is the standard marking system used for grinding wheels. The marking identifies five characteristics of a grinding wheel.

1. The abrasive material is identified by a letter (A—aluminum oxide, C—silicon carbide, etc.)
2. Grain size is indicated by a number from 8 (coarse) to 600 (very fine)
3. The grade is denoted by a letter from A (hard wheel) to Z (soft wheel)

Centerless Grinding ——————— **23**

4. The spacing between abrasive grains is labeled by a number from 1 (dense spacing) to 16 (open spacing)

5. Six common grinding wheel bonds are vitrified (V), resinoid (B), silicate (S), shellac (E), rubber (R), and oxychloride (O)

The manufacturer may add additional non-standardized identification letters or numbers.

STANDARD MARKING SYSTEM

A 36 L 6 V

ABRASIVE MATERIAL — GRAIN SIZE — GRADE — STRUCTURE — BOND

ABRASIVE MATERIAL	ALUMINUM OXIDE(A), SILICON CARBIDE(C)
GRAIN SIZE	COARSE ⟶ FINE 8 to 600 46 to 60 MOST COMMON
GRADE (BOND STRENGTH)	J–O SOFT WORKPIECES H–K HARD WORKPIECES J–N MOST COMMON
STUCTURE (GRAIN SPACING)	DENSE ⟶ OPEN 1 to 16
BOND MATERIAL	VITRIFIED(V), RESINOID(B), SILICATE(S), SHELLAC(E), RUBBER(R), OXYCHLORIDE(O)

Process Conditions

This figure indicates typical in-feed, wheel speed, work speed, identification number (ID), and work speed ranges. The type of grinding wheel used depends on the workpiece material and desired surface finish. Regulation wheels often use a rubber bonded abrasive to provide sufficient friction to keep the workpiece from "reaching-away" from the grinding wheel.

Workpiece material	Hardness (HB)	In-feed (in./pass)
Aluminum	30 to 150	0.005 to 0.0015
Brass	40 to 200	0.005 to 0.0015
Cast iron, ductile	520 max	0.005 to 0.0015
Mild steel	500 max	0.005 to 0.0015
Stainless steel	150 to 200	0.005 to 0.0015
Thermoplastic	—	0.020 to 0.003

Wheel ID No. ANSI	
Rough	Finish
C 54 J 9 V	C 54 J 9 V
C 54 K 9 V	C 54 K 9 V
C 54 K 7 V	C 54 K 7 V
A 60 M 7 V	A 60 M 7V
A 60 L 7 V	A 60 L 7 V
C 60 K 9 V	C 60 K 9 V

Wheel speed: 5500 sfpm to 6500 sfpm (5000 to 6000 sfpm for plastics); work speed: 50 ipm to 150 ipm (inches per minute) (5 to 30 ipm for plastics); max = maximum.

Typical Workpiece Materials

Aluminum, brass, and plastics have poor to fair ratings, whereas cast iron and mild steel have good machinability characteristics. On the other hand, stainless steel has a poor to fair rating because of its toughness and tendency to work harden.

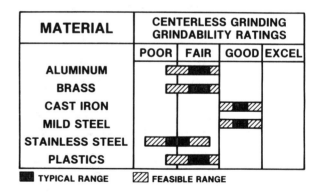

MATERIAL	CENTERLESS GRINDING GRINDABILITY RATINGS			
	POOR	FAIR	GOOD	EXCEL
ALUMINUM				
BRASS				
CAST IRON				
MILD STEEL				
STAINLESS STEEL				
PLASTICS				

■ TYPICAL RANGE ▨ FEASIBLE RANGE

Tool Materials

Aluminum oxide, silicon carbide, diamond, and cubic boron nitride (CBN) are four commonly used abrasive materials for centerless grinding wheels. Of these materials, aluminum oxide is the most common. Because of cost, diamond and CBN grinding wheels are generally made with a core of less expensive material surrounded by a layer of diamond or CBN. Diamond and CBN wheels are very hard and wear-resistant and are capable of grinding materials, such as carbides and ceramics, that cannot be economically ground by aluminum oxide or silicon carbide wheels.

Abrasive materials	Applications
Aluminum oxide (A)	* Best for most steels and steel alloys * Low cost * High volume applications
Silicon carbide (C)	* Best for cast iron, nonferrous metals, and nonmetallic metals * Low cost
Diamond (D, MD, SD)	* Can be natural or man-made * Best for most carbides and some nonmetallic materials * Very high cost
Cubic boron nitride (B)	* Superior for high speed steel * Long life * Cool cutting * Very high cost

Factors Affecting Process Results

Tolerance and surface finish depend upon the following:

* Workpiece material
* Grinding wheel material
* Cutting speed and feed rate
* Proper grit, grade, and bond of wheel
* Cutting fluid
* Rigidity of tool, workpiece, and machine

Lubrication and Cooling

The use of grinding fluids is essential to cool the wheel and workpiece, lubricate the wheel/ workpiece interface, and remove chips. Among the commonly used grinding fluids are water-soluble chemical fluids, water-soluble oils, synthetic oils, and petroleum-based oils. It is important that the grinding fluid be applied directly to the cutting area to insure that the fluid is not carried away by the fanlike action of the rapidly rotating grinding wheel.

Work material	Cutting fluid	Application
Aluminum	Light duty oil	Flood
Brass	Light duty oil	Flood
Cast iron	Heavy duty emulsifiable oil, light duty chemical and synthetic oil	Flood
Mild steel	Heavy duty water-soluble oil	Flood
Stainless steel	Heavy duty emulsifiable oil, heavy duty chemical and synthetic oil	Flood
Plastics	Water-soluble oil, dry, heavy duty emulsifiable oil, light duty chemical and synthetic oil	Flood

Time Calculations

General rules for traverse grinding feed rates are as follows: for roughing, 1/2 to 2/3 of the grinding wheel width per revolution of the workpiece; for semifinishing, 1/3 to 1/2 of the grinding wheel width per revolution of the workpiece; and for finishing, 1/4 to 1/2 of the grinding wheel width per revolution of the workpiece are usually satisfactory.

TRAVERSE GRINDING

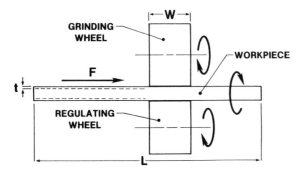

Length of cut (in.) = L
Feed rate (ipm) = F
Approach time (min) = A
Spark-out time (min) = s
Width of grinding wheel (in.) = W
Depth of cut per pass (in.) = d
Thickness of material
to be removed (in.) = t
Number of passes required = N

$$\text{Grinding time} = N \times \frac{L}{F}$$

$$\text{Number of passes} = \frac{t}{d}$$

Time Calculations

In plunge grinding, the cutting time is the length of the cut divided by the feed rate. Total process time is the sum of the approach time, cutting time, and spark-out time.

PLUNGE GRINDING

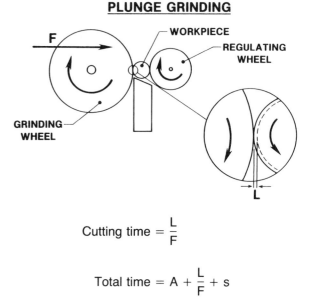

$$\text{Cutting time} = \frac{L}{F}$$

$$\text{Total time} = A + \frac{L}{F} + s$$

Power Requirements

The metal removal rate in cubic inches per minute and the efficiency of the grinding wheel drive mechanism determine power requirements. Horsepower may be computed using the chart shown. In this example a 2 in. diameter 8620 alloy steel is to be ground at a feed rate of 50 in./min. By connecting the points indicated, metal removal rate and horsepower requirements may be quickly ascertained. In the example shown the power requirement is 3.5 hp.

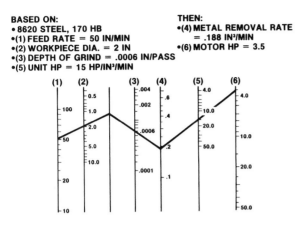

BASED ON:
- 8620 STEEL, 170 HB
- (1) FEED RATE = 50 IN/MIN
- (2) WORKPIECE DIA. = 2 IN
- (3) DEPTH OF GRIND = .0006 IN/PASS
- (5) UNIT HP = 15 HP/IN³/MIN

THEN:
- (4) METAL REMOVAL RATE = .188 IN³/MIN
- (6) MOTOR HP = 3.5

Cost Elements

Cost elements include the following:

* Setup time
* Load/unload time
* Idle time
* Grinding time
* Tool change time
* Tool costs
* Direct labor rate
* Overhead rate
* Amortization of equipment and tooling

Safety Factors

The following risks should be taken into consideration:

* Personal
  - Rotating workpiece and grinding wheel
  - Eye irritation from fine chips
  - Eye and skin irritation from cutting fluids
  - Grinding wheel disintegration
* Environmental
  - Cutting fluid disposal
  - Grinding sludge disposal

Circular Sawing —————————

Circular sawing is a multipoint cutting process in which a circular tool is advanced against a stationary workpiece to sever parts or produce narrow slots.

Process Characteristics

* Uses thin circular blades with teeth on periphery
* Rotating blade is fed into a stationary workpiece
* Produces a narrow cut and a good surface finish
* Has good dimensional accuracy and repeatability
* Burrs are usually produced, which must be subsequently removed

Process Schematic

In circular sawing, the workpiece usually is held securely in a vise, while the circular blade slowly advances into the work. Chips are produced by a succession of cutting teeth arranged around the periphery of the saw blade. Each tooth forms a chip as it passes through the workpiece. The chip is contained between two successive teeth until these teeth leave the workpiece.

Workpiece Geometry

Shown are typical workpieces before and after sawing. Circular saws are very useful in precision cut-offs where length must be held as close to tolerance as possible. The blade cuts a narrow slot. A minimal amount of material is wasted in the cut. Circular saws can make straight and angular cuts in materials with different cross-sections.

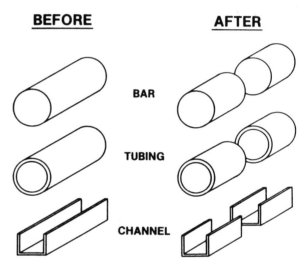

BEFORE   **AFTER**

BAR

TUBING

CHANNEL

CIRCULAR SAWING

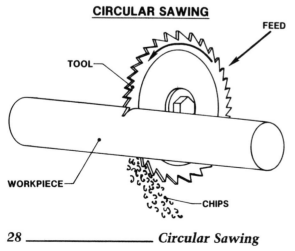

FEED

TOOL

WORKPIECE

CHIPS

Setup and Equipment

Machines for circular saws are commonly known as "cold sawing" machines. The saw blades are fairly large in diameter and operate at low rotational speeds. The cutting action is basically the same as that of a milling cutter. The saw is fed horizontally into the workpiece. Cutting fluid is recommended for cold sawing operations to cool the blade and wash away the chips.

Typical Tools and Geometry Produced

The three most common types of circular saw blades are shown. The diameter of a circular saw blade must be proportional to the cross-sectional area of the workpieces. Solid-tooth blades are used primarily for small materials. Larger saws use inserted-tooth blades. Only the teeth on these blades are made of high speed steel or tungsten carbide; the rest of the disk is made of less expensive steel.

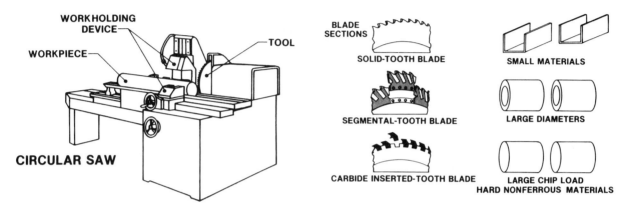

WORKHOLDING DEVICE

WORKPIECE

TOOL

CIRCULAR SAW

BLADE SECTIONS

SOLID-TOOTH BLADE

SEGMENTAL-TOOTH BLADE

CARBIDE INSERTED-TOOTH BLADE

SMALL MATERIALS

LARGE DIAMETERS

LARGE CHIP LOAD
HARD NONFERROUS MATERIALS

Conventional Blanking

Conventional blanking is a shearing process in which a workpiece is separated from the parent material when the punch enters the die.

Process Characteristics

* Shears the workpiece from the parent stock as the punch enters the die
* Produces burnished and sheared section on the cut edge
* Produces burred edges
* Quality is controlled by the punch and die clearance

Process Schematic

A schematic of this process shows a punch, which has sheared the part from the workpiece. Shear forces are determined to a large extent by the clearance between the punch and die. A typical die-cut edge on sheet metal has four distinctive attributes: "roll-over," "burnish," "fracture," and "burr" (illustrated in the enlarged area).

CONVENTIONAL BLANKING

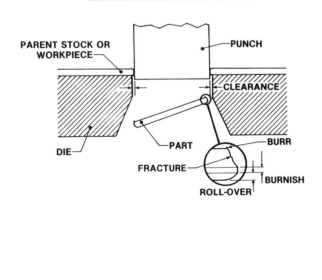

Workpiece Geometry

Both coil and sheet stock are commonly used workpieces. Even though blanking is fast and has a relatively low cost, it does result in considerable waste material. Care should be taken by the tool designer to "nest" the parts, if possible, to reduce unnecessary waste.

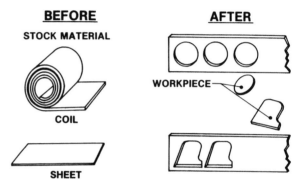

Setup and Equipment

This process is commonly done on mechanical and hydraulic presses. The press basically consists of a frame, a ram, a bolster, die posts, and a motor. A punch is mounted on the ram, and the die is mounted on a bolster plate. Usually, sheet or coil steel is used in high production.

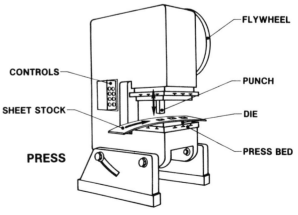

Typical Tools and Geometry Produced

Blanking dies consist of one or more mating pairs of rigid punches and dies. Dies are expensive to build and thus are usually used for high production runs. Blanking dies are carefully constructed to retain their accuracy, even after long production runs. Shown are some of the typical geometries produced with conventional blanking. In general, most conventional blanking parts are less than 1/4 in. thick. Part peripheries range from several inches to as much as 32 feet.

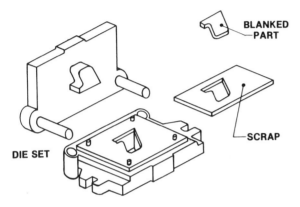

Tolerances and Surface Finish

For most conventional blanking operations, tolerances are typically held within ±0.010 in. However, tolerances of approximately ±0.002 in. can be obtained by proper control of process variables. Surface finishes may range from 125 to as much as 2000 microinches on the fractured edge and are typically between 250 and 1000 microinches.

TOLERANCES		
CONVENTIONAL BLANKING	TYPICAL	FEASIBLE
	±0.010	±0.002

SURFACE FINISH	
PROCESS	MICROINCHES (A.A.)
CONVENTIONAL BLANKING	32 63 125 250 500 1000 2000

▥ TYPICAL RANGE   ▨ FEASIBLE RANGE

Tool Style

There are basically two types of tool systems. A "drop-through" die allows the workpiece to fall through for easy removal. A return-type die returns the blanked workpiece to the level at which it was cut, whereupon it is removed by an air blast or gravity feed.

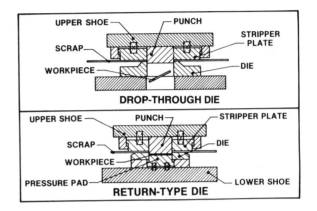

Workholding Methods

The sheet metal workpiece is held between a stripper plate and a die plate. The punch blanks out the workpiece, and what is left becomes the scrap. The stripper plate also strips the scrap material from around the punch.

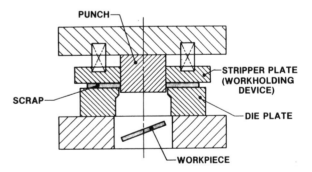

Effects on Work Material Properties

Effects of this process may include workhardening on the edge of the blanked part, along with some residual stresses and cracks. Too much clearance may cause excessive roll-over and the formation of large burrs. Conventional blanking has little effect on either the physical or chemical properties of the workpiece.

Work material properties	Effects of conventional blanking
Mechanical	* Workhardening on edge of blanked part * Blanked edges may have residual cracks * Excess clearance may cause roll-over and burr
Physical	* Little effect
Chemical	* Little effect

Typical Workpiece Materials

Shown are some typical workpiece materials and their respective blanking ratings. Aluminum has an excellent rating, whereas brass, bronze, and mild steel have lower ratings because of their high shear strength. Stainless steel is rated from fair to good. Plastics are rated from poor to excellent, depending on the type of plastic, as some shear cleanly and others either shatter or cause excessive burring.

MATERIAL	CONVENTIONAL BLANKING RATINGS			
	POOR	FAIR	GOOD	EXCEL
ALUMINUM				▨▧▨
BRASS			▨▧▨	
BRONZE			▨▧▨	
MILD STEEL			▨▧▨	
STAINLESS STEEL		▨▧▨		
PLASTICS	▨▧▨▨▧▧▨▧▨▨			

▧ TYPICAL RANGE   **▨ FEASIBLE RANGE**

Tool Materials

Tool steels in grades A2, D2, and D4 are best for low to high production runs of up to one million parts. Carbides are best for higher production runs (over one million parts) and intricate shapes.

Tool materials	Applications
Tool steels (A2, D2, D4)	* Low to high production (up to one million parts)
Carbides	* Higher production (over one million parts) * Intricate shapes

Process Conditions

It is sometimes helpful to reduce punching force requirements by cutting a slope in the end of the tool. As this slope (shear angle) increases, the required punching force decreases. As this angle is increased, less of the tool comes in contact with the workpiece. This causes a concentration of the force and reduces the total force needed to pierce the workpiece. In the figure, the height of the shear angle is related to the thickness of the workpiece. By increasing the shear angle, the original punching force has been reduced by as much as two-thirds. Shown is the amount of tool contact when the tool is halfway through the workpiece. Punching forces for the given conditions are also listed. It should be noted that blanking with shear angles produces curved slugs, which can only be used for scrap metal.

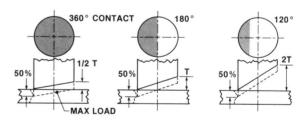

Material (1 in. dia × 0.125 in. thick)	1/2 T (Shear)	T (Shear)	2T (Shear)
	Force in tons		
Aluminum	2.2	1.1	0.72
Brass	6.5	3.2	2.2
Bronze	7.1	3.5	2.4
Mild steel	8.6	4.3	2.9
Stainless steel	11.2	5.6	3.7

Tool Geometry

Typical die dimensions and clearance values are shown. The clearance should be 8% to 10% of the stock thickness. Clearance varies somewhat for different materials and their hardness or cold-worked condition.

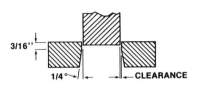

Workpiece materials	Clearance* % of stock thickness
Mild steel	8 to 10
Aluminum, copper	5 to 10
Hard steel	14 to 16
Plastic	3 to 15

* For 6% to 8% roll-over and a fracture angle of 7 to 11°.

Force Requirements

Approximate blanking forces required to shear different types of materials are given in the table. In each case, the perimeter of the workpiece was 10 in., and the thickness of the workpiece was 0.050 in.

Workpiece material	Shear strength (psi)	Required blanking force (lbs)
Low-carbon steel	45,000	22,500
Aluminum	40,000	20,000
Copper	26,000	13,000
Nickle steel	100,000	50,000
Nickel	60,000	30,000

Required force = shear strength × workpiece thickness × blank perimeter

Time Calculation

Processing time includes feed time, closing time, blanking time, and opening time. Feed time is a function of feed rate and distance between the outer lines of blanked components.

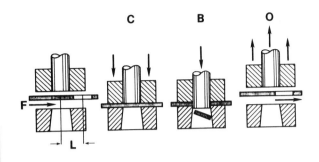

Feed time (sec) = F
Closing time (sec) = C
Blanking time (sec) = B
Opening time (sec) = O
Feed rate (in./sec) = R
Distance between center of blanked holes (in.) = L

$$\text{Feed time} = \frac{L}{R}$$

$$\text{Blanking cycle time} = F + C + B + O$$

Productivity Tip

The tool designer should look for ways to design dies to improve quality and productivity. If the production quantity is sufficient to afford two sets of dies, significant material savings can be affected, as shown in the illustrations in the right-hand column.

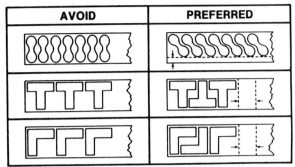

USE STOCK LAYOUT TECHNIQUES THAT MAXIMIZE MATERIAL USAGE

AVOID	PREFERRED

Safety Factors

The following risks should be taken into consideration:

* Personal
  - Contact with moving parts
  - Sharp edges
  - Noise
* Environmental
  - Vibration

Cylindrical Grinding

Cylindrical grinding is an abrasive machining process in which material is removed from the external surface of a metallic or nonmetallic cylindrical workpiece by rotating the grinding wheel and workpiece in opposite directions while they are in contact with one another. The workpiece is mounted between centers and is rotated by means of a workpiece holder (grinding dog or center driver).

Process Characteristics

* Produces straight, tapered, and formed workpieces
* Is used only for cylindrical workpieces
* Produces highly accurate surfaces and smooth finishes
* Is primarily a final machining process

Process Schematic

The workpiece is held between centers. The workpiece and grinding wheel rotate in opposite directions. Small chips of material are removed from the workpiece as it passes longitudinally across the abrasive grinding wheel.

CYLINDRICAL GRINDING

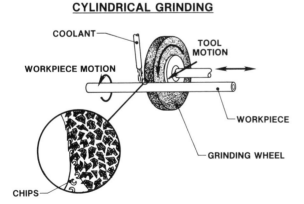

Workpiece Geometry

Shown are some typical workpiece geometries. Close dimensional tolerances can be achieved as well as smooth surface finishes.

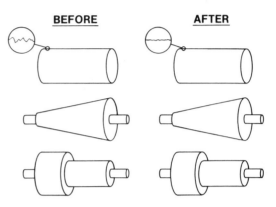

Setup and Equipment

The main components of a cylindrical grinding machine include a grinding wheel, two centers between which the workpiece is suspended, and a grinding dog, chuck, or other driving mechanism. The wheel and workpiece move relative to one another in both the longitudinal and radial directions to allow the workpiece to be ground to the desired depth along its entire length. Most cylindrical grinding machines allow the table or grinding wheel to swivel for grinding tapers.

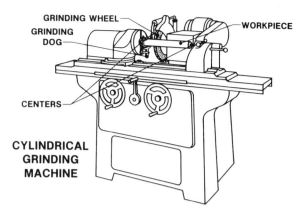

CYLINDRICAL
GRINDING
MACHINE

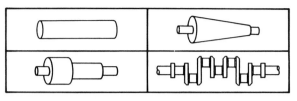

CYLINDRICAL GRINDING CAPABILITIES

	0.1	1.0	10	100
DIAMETER (IN)				
LENGTH (IN)				

▓ TYPICAL RANGE   ▨ FEASIBLE RANGE

Typical Tools and Geometry Produced

Many shapes of abrasive wheels may be used. Plain wheels are used for grinding conventional straight or tapered workpieces. Formed wheels are used for grinding cylindrically shaped workpieces. This process has less vibration than other processes.

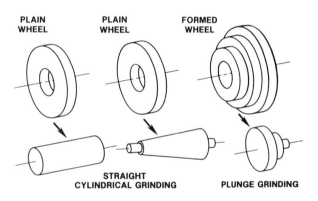

Geometrical Possibilities

Shown are typical shapes produced with straight and formed wheels. The geometry produced on the workpiece is a mirror image of the grinding wheel shape. A grinding wheel with a straight cylindrical periphery will produce a straight cylindrical surface on a workpiece. Typical workpiece sizes range from 0.75 in. to 20 in. in diameter and 0.80 in. to 75 in. in length. However, it is possible to grind workpieces that are between 0.25 in. and 60 in. in diameter and between 0.30 in. and 100 in. in length.

Tolerances and Surface Finish

Tolerances are held within ±0.0005 in. for diameter and ±0.0001 in. for roundness. For precision applications, tolerances for the diameter are as high as ±0.00005 in. and ±0.00001 in. for roundness. Surface finish may range from 2 to 125 microinches, with the typical range between 8 and 32 microinches.

TOLERANCES		
CYLINDRICAL GRINDING	TYPICAL	FEASIBLE
DIAMETER	±.0005	±.00005
ROUNDNESS	±.0001	±.00001

SURFACE FINISH										
PROCESS	MICROINCHES (A.A.)									
CYLINDRICAL GRINDING	2	4	8	16	32	63	125	250	500	

▓ TYPICAL RANGE   ▨ FEASIBLE RANGE

Tool Style

Shown are several grinding wheel face contours that have been standardized by the Grinding Wheel Manufacturers' Association. Specialized shapes may be formed on the face of the grinding wheel by a dressing unit that uses hard, abrasive material to wear a desired shape on the wheel. The wheel may also be formed by using a crush roll device, which molds the wheel by exerting great pressure with a shaped roll.

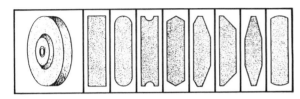

Workholding Methods

The workpiece is supported between centers and is rotated by a lathe dog, which is driven by a faceplate. The lathe dog is manually clamped to the end of the workpiece. In some cases, special drive centers may be used to engage the end of a part, thus permitting grinding of the entire workpiece surface in one setup.

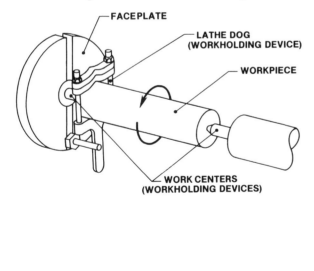

Effects on Work Material Properties

Some changes in mechanical properties result from residual stresses. A thin martensitic layer on a part surface may be caused by high grinding temperatures. This thin layer of martensite may cause a reduction in material fatigue strength due to microcracks. If the grinding temperature is elevated above the curie temperature, a loss of magnetic properties may result in ferromagnetic materials. Cylindrical grinding may reduce corrosion resistance due to high surface stresses.

Work material properties	Effects of cylindrical grinding
Mechanical	* Residual surface stresses * A thin martensitic layer may form on the part surface * Fatigue strength may be reduced
Physical	* Magnetic properties may be lost on ferromagnetic materials
Chemical	* May increase susceptibility to corrosion

Typical Workpiece Materials

Aluminum, brass, and plastics have poor to fair ratings, whereas cast iron and mild steel have good machinability characteristics. Conversely, stainless steel has a poor to fair rating because of its toughness and tendency to work harden.

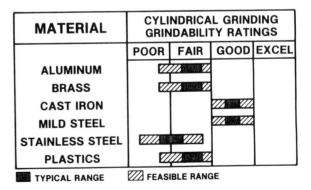

Abrasive Materials

Aluminum oxide, silicon carbide, diamond, and cubic boron nitride (CBN) are four commonly used abrasive materials for cylindrical grinding wheels. Of these materials, aluminum oxide is the most common. Because of cost, diamond and CBN grinding wheels are generally made with a core of less expensive material surrounded by a layer of diamond or CBN. Diamond and CBN wheels are very hard and capable of economically grinding materials, such as ceramics and carbides, that cannot be ground by aluminum oxide or silicon carbide wheels. This figure illustrates the standard marking system used for grinding wheels. Five characteristics of a grinding wheel are identified:

1. The abrasive material is identified by a letter (A—aluminum oxide, C—silicon carbide, etc.)
2. Grain size is indicated by a number from 8 (coarse) to 600 (very fine)
3. The grade is denoted by a letter from A (hard wheel) to Z (soft wheel)
4. The spacing between abrasive grains is labeled by a number from 1 (dense spacing) to 16 (open spacing)
5. Six common grinding wheel bonds are vitrified (V), resinoid (B), silicate (S), shellac (E), rubber (R), and oxychloride (O)

The manufacturer may add additional non-standardized identification letters or numbers.

STANDARD MARKING SYSTEM

A 36 L 6 V
ABRASIVE MATERIAL — GRAIN SIZE — GRADE — BOND — STRUCTURE

ABRASIVE MATERIAL	ALUMINUM OXIDE(A), SILICON CARBIDE(C)
GRAIN SIZE	COARSE → FINE 8 to 600 46 to 60 MOST COMMON
GRADE (BOND STRENGTH)	J–O SOFT WORKPIECES H–K HARD WORKPIECES J–N MOST COMMON
STRUCTURE (GRAIN SPACING)	DENSE → OPEN 1 to 16
BOND MATERIAL	VITRIFIED (V), RESINOID(B), SILICATE(S), SHELLAC(E), RUBBER(R), OXYCHLORIDE(O)

Factors Affecting Process Results

Tolerance and surface finish depend upon the following:

* Workpiece material
* Grinding wheel material
* Cutting speed and feed rate
* Proper grit, grade, and bond of wheel
* Cutting fluid
* Rigidity of tool, workpiece, and machine

Process Conditions

Wheel speed for cylindrical grinding may range from 5500 to 6500 surface feet per minute. The work speed varies from 70 to 100 inches per minute (50–100 ipm for aluminum). The type of wheel used depends on the workpiece material type, hardness, and on the desired surface finish.

Abrasive materials	Applications
Aluminum oxide (A)	* Best for most steels and steel alloys * Low cost * High volume applications
Silicon carbide (C)	* Best for cast iron, nonferrous metals, and nonmetallic materials * Low cost
Diamond (D, MD, SD)	* Can be natural or man-made * Best for most carbides and some nonmetallic materials * Very high cost
Cubic boron nitride (B)	* Superior for high speed steel * Long life * Cool cutting * Very high cost

Workpiece material	Hardness (HB)	In-feed (in./pass)	Wheel ID No. ANSI	
			Rough	Finish
Aluminum	30 to 150	0.002 to 0.0005	C 54 J 9 V	C 54 J 9 V
Brass	40 to 200	0.002 to 0.0005	C 46 J 9 V	C 46 J 9 V
Cast iron. ductile	520 max	0.002 to 0.0005	C 54 J 7 V	C 54 J 7 V
Mild steel	500 max	0.002 to 0.0005	A 60 L 7 V	A 60 L 7 V
Stainless steel	150 to 200	0.002 to 0.0005	A 60 J 7 V	A 60 J 7 V
Thermoplastic	—	0.005 to 0.0005	C 60 K 9 V	C 60 K 9 V

in./pass = inches per pass.

Lubrication and Cooling

The use of grinding fluids is essential to cool the wheel and workpiece, lubricate the wheel/workpiece interface, and aid in removing chips. Among the commonly used grinding fluids are water-soluble chemical fluids, water-soluble oils, synthetic oils, and petroleum-based oils. It is important that grinding fluid be applied directly to the cutting area to insure that the fluid is not carried away by the fanlike action of the rapidly rotating grinding wheel.

Work material	Cutting fluid	Application
Aluminum	Light duty oil	Flood
Brass	Light duty oil	Flood
Cast iron	Heavy duty emulsifiable oil, light duty chemical and synthetic oil	Flood
Mild steel	Heavy duty water-soluble oil	Flood
Stainless steel	Heavy duty emulsifiable oil, heavy duty chemical and synthetic oil	Flood
Plastics	Water-soluble oil, dry, heavy duty emulsifiable oil, light duty chemical and synthetic oil	Flood

Power Requirements

The metal removal rate in cubic inches per minute and the efficiency of the grinding wheel drive mechanism determine power requirements. The graph shown may be used to compute required horsepower quickly.

BASED ON:
- 8620 STEEL, 170 HB
- (1) FEED RATE = 50 IN/MIN
- (2) WORKPIECE DIA. = 2 IN
- (3) DEPTH OF GRIND = .0006 IN/PASS
- (5) UNIT HP = 15 HP/IN³/MIN

THEN:
- (4) METAL REMOVAL RATE = .188 IN³/MIN
- (6) MOTOR HP = 3.5

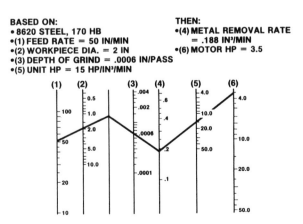

Time Calculations

Understanding the grinding setup is important in calculating grinding time and feed rates. Shown are the relative motions and positions of the workpiece, grinding wheel, and feed directions. Given is the setup for plunge grinding and traverse grinding with corresponding parameters. Plunge grinding is different from traverse grinding in that the feed is directed toward (perpendicular to) the workpiece instead of running parallel to it.

Length of cut (in.) = L
Feed rate (ipm) = F
Approach time (min) = A
Spark-out time (min) = s
Width of grinding wheel (in.) = W
Depth of cut per pass (in.) = d
Thickness of material
to be removed (in.) = t
Number of passes required = N

PLUNGE GRINDING

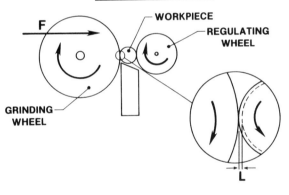

Plunge Grinding

$$\text{Cutting time} = \frac{L}{F}$$

$$\text{Total time} = \frac{A + L}{F + s}$$

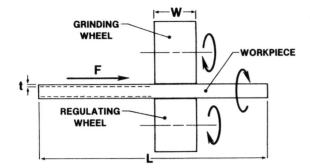

TRAVERSE GRINDING

Traverse Grinding

$$\text{Grinding time} = \frac{N \times L}{F}$$

$$\text{Number of passes} = \frac{t}{d}$$

Cost Elements

Cost elements are as follows:

* Setup time
* Load/unload time
* Idle time
* Grinding time
* Tool change time
* Tool costs
* Direct labor rate
* Overhead rate
* Amortization of equipment and tooling

Safety Factors

The following risks should be taken into consideration:

* Personal
  - Rotating workpiece and grinding wheel
  - Eye irritation from fine chips
  - Eye and skin irritation from cutting fluids
  - Grinding wheel disintegration
* Environmental
  - Cutting fluid disposal
  - Grinding sludge disposal

Die Threading

Die threading is a machining process that uses a cluster of multipoint cutting tools (chasers) to produce uniform helical threads on the external surface of a cylinder. The chasers move parallel to the axis of the workpiece. When the cutting is finished, the chasers automatically retract, permitting return of the head without reversing the spindle.

Process Characteristics

* Uses a cluster of four multipoint chasers
* Produces helical threads on the external surface of cylindrical workpieces
* Has automatic retraction of chasers upon thread completion
* Produces chips that fall out upon chaser retraction
* Requires axial chaser feed to match thread lead
* Is used for production threading

Process Schematic

The self-opening die heads are designed for stationary or rotary applications. Shown is a stationary die head being fed in axially as the workpiece rotates. True thread forms are usually cut into the chasers for reproduction on the workpiece. Either a right-hand or left-hand thread is cut, depending on the rotation of the workpiece and the position of the chasers.

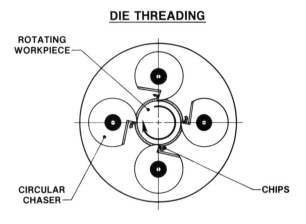

DIE THREADING

Workpiece Geometry

Shown are typical workpieces. Coarse or fine threads can be easily cut by changing the chasers in the die head. The practice for cutting taper pipe threads does not differ greatly from standard screw threads. Die threading can also be done on drill presses and special threading machines.

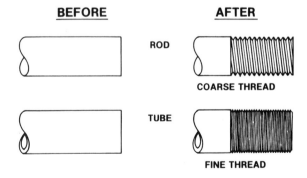

Setup and Equipment

Shown is an engine lathe, which consists of a work spindle, bed, and tailstock. The work-piece is gripped firmly in the chuck and the stationary die head is mounted in the tailstock.

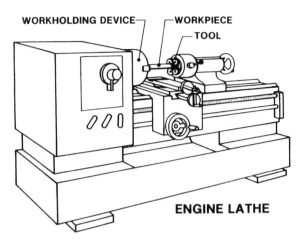

WORKHOLDING DEVICE — WORKPIECE
— TOOL

ENGINE LATHE

Typical Tools and Geometry Produced

Two types of chasers are shown. They can be used in either stationary or rotary die heads. Each of these chasers has four adjustable, multipoint cutters that can be removed for sharpening or interchanged for different thread sizes. Circular chasers are used in high production for all metals. Radial chasers are usually restricted to soft, easy-to-machine materials. Tangential chasers (not shown) are suitable for threading steel and other fairly hard materials.

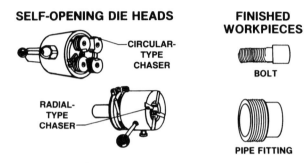

SELF-OPENING DIE HEADS

CIRCULAR-TYPE CHASER

RADIAL-TYPE CHASER

FINISHED WORKPIECES

BOLT

PIPE FITTING

Drilling

Drilling is a cutting process in which a hole is originated or enlarged by means of a multipoint, fluted, end cutting tool. As the drill is rotated and advanced into the workpiece, material is removed in the form of chips that move along the fluted shank of the drill. One study showed that drilling accounts for nearly 90% of all chips produced.

Process Characteristics

* Uses a multipoint, fluted, end cutting tool (drill)
* Cutting tools or workpieces are rotated and advanced relative to each other
* Creates or enlarges nonprecision holes
* May produce coarse, helical feed marks, depending on machining parameters (feed, speed, tool geometry, coolant, etc.)
* Creates small burrs on entry and coarse burrs on exit

Process Schematic

Drilling involves relative axial and rotational motions between the drill and the workpiece. Usually, the drill rotates and advances into the workpiece, but sometimes the opposite is true. Chips are removed by flowing along grooves or flutes in the drill. Although long spiral chips usually result from drilling, adjustment of the feed rate can result in chips with a range of shapes and sizes.

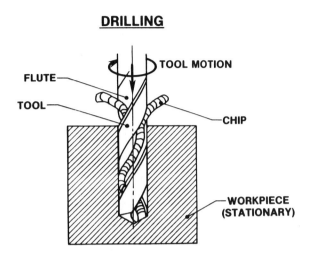

DRILLING

Workpiece Geometry

Shown are typical workpieces before and after drilling. A drilled hole may be distinguished from one produced by piercing, casting, molding, torch cutting, etc., by the presence of helical feed marks inside the hole and small burrs on the workpiece as the drill enters and exits. Drilled holes are usually quite sharp around the edge where the drill has entered the workpiece.

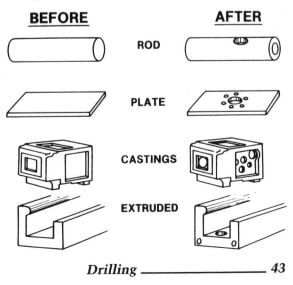

BEFORE   AFTER

ROD

PLATE

CASTINGS

EXTRUDED

Setup and Equipment

The drill press normally consists of a drill head, table, column, and base. The drill head contains a drive motor, a spindle, a feed mechanism, and a toolholding device (the chuck). A drill bit is gripped firmly in the chuck, and the work is positioned on the table. The drill is advanced into the workpiece by the feed mechanism. Regular shaped workpieces are held in a workholding device (the vise). Irregularly shaped workpieces are held in special fixtures.

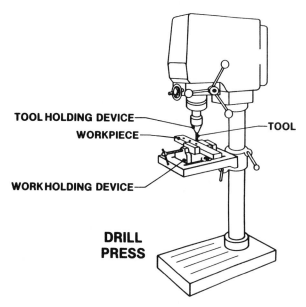

Typical Tools and Geometry Produced

There is a wide variety of drills in use today, including conventional twist drills, counterbores, subland, or multidiameter drills, countersinks, spade drills, and many others for special functions. Each of these drills is designed to impart a desired feature on the workpiece. It is even possible to drill square and hexagon-shaped holes using special equipment and tools.

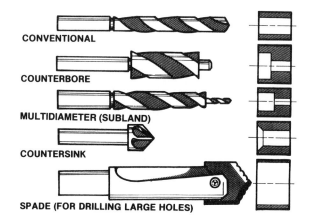

CONVENTIONAL

COUNTERBORE

MULTIDIAMETER (SUBLAND)

COUNTERSINK

SPADE (FOR DRILLING LARGE HOLES)

Workpiece Geometry

Shown are some typical geometries that may be produced by drilling. The majority of holes drilled are between 0.125 in. and 1.50 in. in diameter. However, with special tooling, it is possible to obtain holes 0.007 in. to 10 in. in diameter. The length-to-diameter ratio is usually between 1:1 and 1:5. However, it is possible to have a length-to-diameter ratio as high as 1:20.

HOLE | COUNTERSINK | COUNTERBORE | COMBINATION

DRILLING CAPABILITIES

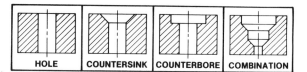

TYPICAL RANGE   FEASIBLE RANGE

Toolholding Methods

Drills are classified as either straight shank or taper shank. Straight shank drills are often held in a keyless chuck for quick load/unload time. Tapered shank drills are held in the female Morris taper in the end of the machine tool spindle. Removal of the drill is accomplished by driving a tapered drift through a hole in the spindle and against the end of the drill's tang.

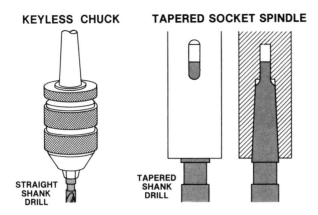

KEYLESS CHUCK   TAPERED SOCKET SPINDLE

STRAIGHT
SHANK
DRILL

TAPERED
SHANK
DRILL

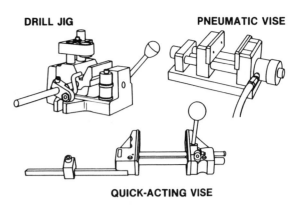

DRILL JIG   PNEUMATIC VISE

QUICK-ACTING VISE

Tool Style

There are many types of drills available in a wide range of sizes and shapes. The countersink/ center drill is used to center holes accurately. The twist drill is most commonly used and produces a majority of the holes drilled. Other types of drills such as the subland, spade, and indexable are considered specialty drills.

DESCRIPTION	STYLE	APPLICATION
COUNTERSINK/ CENTER DRILL		ACCURATE CENTERING OF HOLES
TWIST DRILL		MOST COMMONLY USED
SUBLAND DRILL		TWO OR MORE DIA-METERS DRILLED SIMULTANEOUSLY
SPADE DRILL		LARGE DIAMETER HOLES (1 IN. OR LARGER)
INDEXABLE DRILL		EXCELLENT CHIP CONTROL, BALANCED CUTTING FORCES

Workholding Methods

General purpose vises such as those shown may be used to hold the workpiece during drilling. Frequently, jigs and fixtures are designed for a specific workpiece and operation. Some factors to consider when designing jigs and fixtures are number of parts to be drilled, rigidity, strength, accuracy of location, workpiece clamping method, chip control, and ease of operation.

Tolerances and Surface Finish

Shown are the tolerances that may be obtained with a twist drill with and without the use of a bushing and/or center drilling. For greater accuracy, center drilling should be performed prior to drilling, or a drill bushing should be used. Surface finishes may range from 32 to 500 microinches, with the range usually between 63 and 250 microinches. Finish cuts will generate surfaces near 32 microinches, and roughing will be near 500 microinches.

TOLERANCES						
DRILL DIAMETER (IN)	1/8 to 1/4		1/4 to 3/4		3/4 to 1-1/2	
CONDITION	OVER-SIZE	LOCA-TION	OVER-SIZE	LOCA-TION	OVER-SIZE	LOCA-TION
NO CENTER-DRILLED HOLE—NO BUSHING	.003	±.007	.006	±.008	.008	±.009
CENTER-DRILLED HOLE—NO BUSHING	.003	±.004	.003	±.004	.004	±.005
DRILL BUSHING	.002	±.002	.003	±.002	.004	±.003

SURFACE FINISH									
PROCESS	MICROINCHES (A.A.)								
DRILLING	2	4	8	16	32	63	125	250	500

■ TYPICAL RANGE   ▨ FEASIBLE RANGE

Effects on Work Material Properties

Drilling may affect the mechanical properties of the workpiece by creating low residual stresses around the hole opening and a very thin layer of highly stressed and disturbed material on the newly formed surface. This causes the workpiece to become more susceptible to corrosion at the stressed surface.

Work material properties	Effects of drilling
Mechanical	* Low residual stresses left around hole opening * Very thin layer of highly stressed and disturbed material created on newly formed surface
Physical	* Little effect
Chemical	* May increase susceptibility to corrosion at stressed surface

Tool Geometry

Shown are the three most common angles on a twist drill: the point, helix, and lip relief angles. The standard point angle of 118° is most common, and the standard helix angle is 24°. Point angles between 90° and 118° are used on brittle material, whereas angles between 118° and 135° are used on ductile and hard materials. High helix angles are used for high removal rates. The lip relief angle decreases as the diameter of the drill increases.

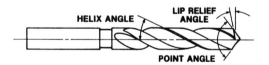

MATERIAL	POINT ANGLE (°)	HELIX ANGLE (°)	LIP RELIEF ANGLE (°)
ALUMINUM	90 to 135	32 to 48	12 to 26
BRASS	90 to 118	0 to 20	12 to 26
CAST IRON	90 to 118	24 to 32	7 to 20
MILD STEEL	118 to 135	24 to 32	7 to 24
STAINLESS STEEL	118 to 135	24 to 32	7 to 24
PLASTICS	60 to 90	0 to 20	12 to 26

Process Conditions

Shown are the possible cutting speeds and feed rates for a high speed steel drill using cutting fluids. Some factors that influence the speed and feed rates are workpiece composition and hardness, tool material, depth of hole, and use of cutting fluids.

Typical speeds and feeds					
Workpiece material	Cutting speed (sfpm)	Drill diameter (in.)			
		0.25	0.5	0.75	1.0
		Feed (ipr)			
Aluminum	650 to 1300	0.012	0.016	0.024	0.032
Brass	200 to 250	0.008	0.010	0.016	0.020
Cast iron*	80 to 110	0.005	0.007	0.010	0.012
Mild steel	110 to 120	0.008	0.010	0.016	0.020
Stainless steel	20 to 50	0.005	0.007	0.010	0.012
Plastics	65 to 250	0.004	0.006	0.008	0.010

* Cast iron is usually drilled dry.

Typical Workpiece Materials

Soft materials such as aluminum and plastics are rated good to excellent, while cast iron and mild steel have a fair to good machinability rating. Stainless steel has a poor to fair rating because of its toughness and its tendency to work harden when machined.

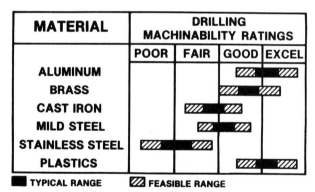

Tool Materials

High speed steel is most commonly used because it provides strength, toughness, and high temperature hardness at a comparatively low cost. Carbide tipped drills are used for special applications, such as drilling abrasive materials of low tensile strength or heat resistant alloys. Solid carbide drills are used where extreme rigidity, accuracy, and long tool life are required.

Tool materials	Applications
High speed steel	* Most commonly used tool * Strength, toughness and high-temperature hardness at comparatively low cost

Carbide, tipped and solid	* Commonly used for drilling cast irons, aluminum, reinforced plastic and steel harder than 495 HB
	Tipped:
	  * Cost less than solid carbide
	  * Abrasive metals of low tensile strength
	Solid:
	  * Used for extreme rigidity and drilling accuracy
	  * Extended tool life

Work material	Cutting fluid	Application
Aluminum	Soluble oil, mineral oil	Spray
Brass	Mineral oil, soluble oil	Flood
Cast iron	Cold air, none	Air jet
Mild steel	Soluble oil, sulfurized oil	Flood
Stainless steel	Soluble oil, sulfurized mineral oil	Flood
Plastics	Mineral oil, soluble oil 100:1, cold air, none	Flood, spray, air jet

Factors Affecting Process Results

Tolerance and surface finish depend upon the following:

* Tool geometry
* Cutting speed and feed rate
* Rigidity of tool, workpiece, and machine
* Alignment of machine components and fixtures
* Cutting fluid
* Composition and hardness of the workpiece
* Accuracy of point angle, lip clearance, and lip length

Lubrication and Cooling

Cutting fluids for drilling include mineral, synthetic, and water-soluble oils. Application of these fluids is usually done by flooding the workpiece or by applying a spray mist. The main function of cutting fluids is to cool the tool, which increases tool life and makes higher cutting feeds and speeds possible. Some secondary benefits are chip removal and lubrication, which contribute to better workpiece surface finish.

Time Calculations

Understanding the drilling setup is important in calculating the time needed to drill, retract, and position. Below are shown the formulas used to calculate cutting time, retract time, positioning time, rpm, and feed rate. These calculations combined comprise the actual machining time per piece.

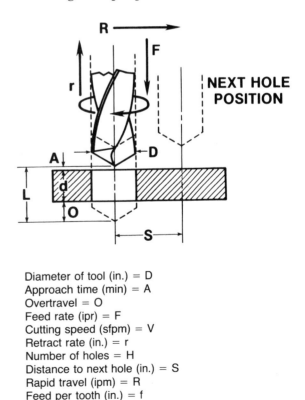

Diameter of tool (in.) = D
Approach time (min) = A
Overtravel = O
Feed rate (ipr) = F
Cutting speed (sfpm) = V
Retract rate (in.) = r
Number of holes = H
Distance to next hole (in.) = S
Rapid travel (ipm) = R
Feed per tooth (in.) = f
Number of teeth = N
Length of cut (in.) = L
Depth of hole (in.) = d

$$\text{Cutting time} = \frac{L}{F}$$

$$\text{Retract time} = \frac{L}{r}$$

$$\text{Positioning time} = H\left(\frac{S}{R}\right)$$

$$\text{rpm} = \frac{4 \times V}{D} \text{ (approx.)}$$

$$\text{Feed rate} = f \times N \times \text{rpm}$$

Power Requirements

Shown is the formula to calculate the machine horsepower, using unit power and material removal rate. For example, if mild steel is drilled with an HB hardness of 330 to 370, and $10\,\text{in.}^3$ of material are removed per minute, then 14 hp drill press is needed (1.4 unit power $\times\ 10\,\text{in.}^3/\text{min} = 14\,\text{hp}$). Unit power is based on HSS drills, feed of 0.002 ipr to 0.008 ipr, and 80% efficiency.

Machine hp = unit power × removal rate (in.³/min)

Material	Hardness (HB)	Unit power*
Aluminum	30 to 150	0.16
Brass	50 to 145	0.48
	145 to 240	0.8
Cast iron	110 to 190	1.0
	190 to 320	1.6
Mild steel	85 to 200	1.0
	330 to 370	1.4
	485 to 560	2.1
Stainless steel	135 to 275	1.1
	275 to 430	1.2
Plastics	N/A	0.05 est.

* Unit power based on: ● HSS drills ● feed of 0.002 to 0.008 ipr ● 80% efficiency.
est. = estimate.

Cost Elements

Cost elements include the following:

* Setup time
* Load/unload time
* Idle time
* Cutting time
* Tool costs
* Direct labor rate
* Overhead rate
* Amortization of equipment and tooling

Safety Factors

The following risks should be taken into consideration:

* Personal
  - Rotating tool
  - Hot, sharp chips
  - Eye and skin irritation from cutting fluids

End Milling

End milling is a multipoint cutting process in which material is removed from a workpiece by a rotating tool. The material is usually removed by both the end and the periphery of the tool. Generally, the cutter rotates about an axis *perpendicular* to the surface. On occasion, a single-point tool, such as a fly cutter, may be used.

Process Characteristics

* Uses a rotating cutter to produce a machined surface and creates small, discontinuous chips
* Uses vertical and horizontal milling machines
* Removes materials with the face and/or periphery of the cutter
* Uses a wide variety of tools, including square end mills, ball end mills, shell end mills, and t-slot mills
* Produces slots, angles, pockets, radii, and many other workpiece geometries

Process Schematic

In end milling, the tool rotates rapidly, and the workpiece is moved relative to the tool, or the tool is moved relative to the workpiece. The teeth on the end and the periphery of the tool cut the material.

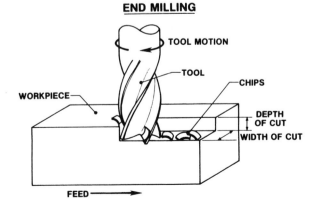

Workpiece Geometry

The original workpiece may have almost any configuration. Depending on the milling machine used, a variety of grooves, slots, and pockets may be milled in the workpiece. Milling with the end of a cutter may leave spiral feed marks.

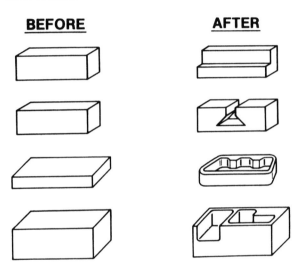

Setup and Equipment

A vertical milling machine is commonly used in end milling, although a horizontal machine may also be used. The tool is mounted in a chuck or collet and the workpiece is secured by a workholding device on the bed of the milling machine.

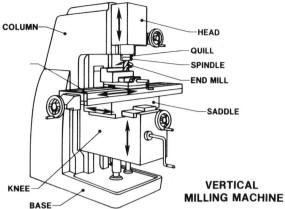

COLUMN
HEAD
QUILL
SPINDLE
END MILL
SADDLE
KNEE
BASE
VERTICAL MILLING MACHINE

Typical Tools and Geometry Produced

Tools used in end milling include square end cutters, ball end cutters, t-slot cutters, and shell mills. Square end cutters mill square slots, pockets, and workpiece edges. Ball end cutters mill radiused slots or fillets. T-slot cutters, as the name implies, are used to cut T-shaped slots. Shell end cutters are used primarily for milling large, flat surfaces and angles.

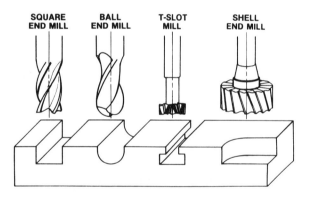

SQUARE END MILL   BALL END MILL   T-SLOT MILL   SHELL END MILL

Geometrical Possibilities

A variety of tools creates many workpiece shapes. Square and round bottom slots, planes and angles, t-slots, dovetail slots, or any combination of these can be produced by end milling. Shown are the typical and feasible ranges for width and depth of cut.

SQUARE SLOT	ROUND BOTTOM SLOT	PLANES/ANGLES
T-SLOT	DOVETAIL SLOT	STEPS

END MILLING CAPABILITIES

	.1	1	10	100
WIDTH OF CUT (IN)				

	.001	.01	.1	1	10
DEPTH OF CUT (IN)					

■ TYPICAL RANGE   ▨ FEASIBLE RANGE

Tolerances and Surface Finish

In production work, end milling machines are capable of holding a tolerance of ±0.001 in., whereas in precision work, tolerances of ±0.0005 in. are possible. Surface finishes can feasibly range from 16 to 500 microinches, but commonly range from 32 to 125 microinches.

TOLERANCES		
END MILLING	TYPICAL	FEASIBLE
DEPTH	±0.001	±0.0005

SURFACE FINISH	
PROCESS	MICROINCHES (A.A.)
END MILLING	2 4 8 16 32 63 125 250 500

■ TYPICAL RANGE   ▨ FEASIBLE RANGE

Tool Style

One of the attractive features of end milling is the great variety of tools available. Square end mills, face mills, t-slot mills, and dovetail mills are a few of the commonly used tools.

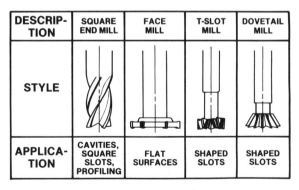

DESCRIP-TION	SQUARE END MILL	FACE MILL	T-SLOT MILL	DOVETAIL MILL
STYLE				
APPLICA-TION	CAVITIES, SQUARE SLOTS, PROFILING	FLAT SURFACES	SHAPED SLOTS	SHAPED SLOTS

Workholding Methods

Two common workholding devices are vises and strap clamps. The vise exerts horizontal pressure on the workpiece securely, and the strap clamp exerts vertical pressure. Pneumatic vises are quick and easy to activate so that workpieces may be quickly loaded and unloaded. The direction of the holding and cutting forces must be considered to insure that the workpiece does not move while being machined.

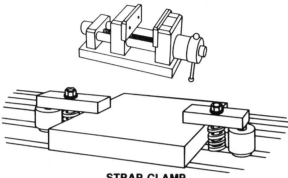

PNEUMATIC VISE CLAMP

STRAP CLAMP

Effects on Work Material Properties

The mechanical properties of a workpiece are affected adversely by a dull tool or a tool with a built-up edge. End milling causes little change in the physical properties of a workpiece. The chemical structure of the workpiece *is* affected if excessive heat is generated when milling alloy steels. Secondly, a rapid, severe heating and cooling of the workpiece may cause a martensitic surface layer to be produced.

Work material properties	Effects of end milling
Mechanical	* A built-up edge on cutter causes a rough workpiece surface * Dull tools cause severe surface damage and high residual stresses
Physical	* Little effect
Chemical	* An untempered martensitic layer (0.001 in.) may be produced when milling heat-treated alloy steels

Typical Workpiece Materials

Fairly soft or brittle materials, such as plastics, aluminum, brass, mild steel, or cast iron, have good to excellent machinability ratings, whereas tough and work-hardened materials, such as stainless steel, have poor to fair machinability ratings.

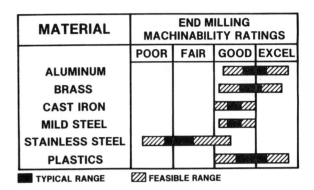

MATERIAL	END MILLING MACHINABILITY RATINGS			
	POOR	FAIR	GOOD	EXCEL
ALUMINUM			▨	▨
BRASS			▨	▨
CAST IRON			▨	
MILD STEEL			▨	
STAINLESS STEEL	▨	▨		
PLASTICS			▨	▨

■ TYPICAL RANGE   ▨ FEASIBLE RANGE

Tool Materials

Common tool materials are high speed steels, carbides, ceramics, and diamonds. High speed steels are useful for special tool shapes and low production. Cutters with carbide inserts are used because of their ability to perform at high speeds. Ceramic inserts are good for high speed applications and high production. Diamond end mills are suitable for nonferrous and non-metallic materials.

Tool materials	Applications
High speed steel	* Special tool shapes * Low production
Carbides (inserts)	* Most commonly used cutters * High production
Ceramics (inserts)	* High speed machining * High production * Uninterrupted cuts
Diamonds (inserts)	* High surface qualities, fine tolerances * Nonferrous or nonmetallic material

Factors Affecting Process Results

The tolerances and surface finishes produced depend on tool geometry and sharpness; cutter and feed rate; rigidity of machine, tool, and workpiece; alignment of machine components and fixtures; and cutting fluids.

Tolerance and surface finish depend upon the following:

* Tool geometry and sharpness
* Cutting speed and feed rate
* Rigidity of tool, workpiece, and machine
* Alignment of machine components and fixtures
* Cutting fluid

Tool Geometry

The four critical angles on the cutter are the end cutting edge angle, axial relief angle, radial relief angle, and radial rake angle. Shown are graphical representations of each of these angles and values commonly used for optimum machining.

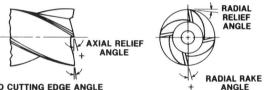

Workpiece material	Radial rake angle (°)	Cutting edge angle (°)	Axial relief angle (°)	Radial relief angle (°)
Aluminum	20 to 35	7 to 12	3 to 5	10 to 12
Brass	10 to 12	7 to 12	3 to 5	5 to 10
Cast iron	−5 to −10	5 to 10	4 to 7	4 to 7
Mild steel	10 to 15	5 to 10	5 to 7	3 to 7
Stainless steel	5 to 12	5	8 to 10	8 to 10
Plastics	18	15	6	6

Process Conditions

The cutter and feed speeds depend on the material being machined. Softer materials, such as plastics and aluminum, may be machined at fairly high cutting speeds with good feed rates. Harder and tougher materials, such as cast iron and stainless steel, require lower cutting speeds and decreased feed rates. Process con-

ditions for peripheral cutting using a 1/2-in. carbide end mill and cutting 0.02 in. to 0.06 in. deep are shown below.

Typical speeds and feeds

Workpiece material	Cutting speed (sfpm)	Feed rate (ipt)
Aluminum	700 to 1300	0.006 to 0.010
Brass	450 to 900	0.005 to 0.010
Cast iron	200 to 300	0.003 to 0.010
Mild steel	350 to 550	0.005 to 0.008
Stainless steel	200 to 350	0.004 to 0.007
Plastics*	400 to 1000	0.005 to 0.016

* High speed tool steel.

Lubrication and Cooling

Cutting fluids used in milling include mineral oil, fatty oil, water-soluble oil, sulfurized mineral oil, and chemical and synthetic oil. Spraying or flooding the tool and workpiece is common.

Work material	Cutting fluid	Application
Aluminum	None, mineral oil, fatty oil	Spray, flood
Brass	Mineral oil, specialty fluid	Spray, flood
Cast iron	Soluble oil, chemical and synthetic oil, none	Spray, flood
Mild steel	Chemical and synthetic oil, soluble oil	Spray, flood
Stainless steel	Sulfurized mineral oil, fatty soluble oil, chemical and synthetic oil	Spray, flood
Plastics	Mineral oil, soluble oil, cold air, none	Spray, flood, air jet

Power Requirements

Unit power is based on the horsepower required to remove one cubic inch of material per minute. Generally, the power required is proportional to the hardness of the material being machined. Less horsepower is required to remove one cubic inch per minute of plastic or aluminum than is required to remove one cubic inch per minute of mild steel or stainless steel.

Machine hp = unit power × removal rate (in.³/min)

Material	Hardness (HB)	Unit power*
Aluminum	30 to 150	0.3
Brass	50 to 145	0.6
	145 to 240	1.0
Cast iron	110 to 190	0.6
	190 to 320	1.1
Mild steel	85 to 200	1.1
	330 to 370	1.5
	485 to 560	2.1
Stainless steel	135 to 275	1.4
	275 to 430	1.5
Plastics	N/A	0.05 est.

* Unit power based on: ● HSS and carbide tools ● feed of 0.005 to 0.012 ipt ● 80% efficiency.

Cost Elements

Cost elements include the following:

* Setup time
* Load/Unload time
* Idle time
* Cutting time
* Tool costs
* Direct labor rate
* Overhead rate
* Amortization of equipment and tooling

Time Calculations

The variables needed for calculating milling time and positioning time are cutter diameter, depth of cut, length of cut, workpiece length, cutter velocity, rapid traverse distance, rapid traverse rate, cutter feed rate, feed per tooth rate, overtravel, number of teeth in cutter, and approach.

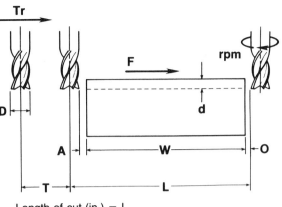

Length of cut (in.) = L
Diameter of cutter (in.) = D
Depth of cut (in.) = d
Length of workpiece (in.) = W
Approach = A
Rapid traverse distance = T
Rapid traverse rate (ipm) = Tr
Number of teeth in cutter = N
Cutter feed rate (ipm) = F
Cutting speed (sfpm) = V
Feed per tooth (in.) = f
Overtravel = O

$$\text{Milling time} = \frac{L}{F}$$

$$\text{Traverse time} = \frac{Td}{Tr}$$

$$\text{rpm} = \frac{4 \times V}{D} \text{ (approx.)}$$

$$\text{Feed rate} = f \times N \times \text{rpm}$$

Safety Factors

The following risks should be taken into consideration:

* Personal
  – Rotating tool
  – Hot and sharp chips
  – Eye and skin irritation from cutting fluids

Fine Blanking

Fine blanking is a controlled shearing process in which a tightly clamped workpiece is forced through a fixed die opening to produce accurate workpieces with a fine finish and straight edges. A V-shaped ring around the perimeter of the workpiece presses into the stock to control material flow.

Process Characteristics

* Die clearance is approximately 1% of stock thickness
* Produces clean, smooth edges
* Hole sizes and spacing can equal stock thickness
* Material thicknesses of 0.0006 in. to 0.60 in. for steel, brass, aluminum, etc.
* Produces minimal surface distortion
* Punch does not enter die
* Uses a V-ring that is embedded in the stock to control fracture

Process Schematic

Fine-blanking dies use very small clearances (near zero percent of sheet metal thickness). A special triple-action press is used to operate the blanking die. High squeezing pressures and a stinger V-shaped impingement ring are used to create a square and fully burnished blank edge. In essence, the part is extruded out of the metal strip without fracture or die breakage.

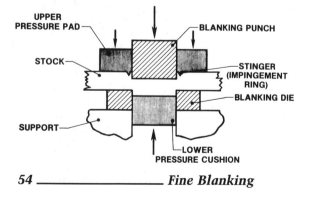

Workpiece Geometry

By machining critical areas, parts can be processed from thicker sheets; then fine blanking finishes the part. Today, fine-blanked parts are used in many areas of industry for such things as gears, cams, and other precision parts.

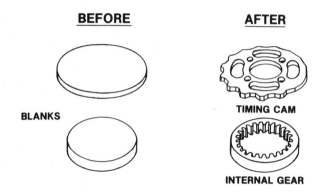

Setup and Equipment

A triple-action hydraulic or mechanical press is used for fine blanking. An outer slide holds the stock firmly against the die rings and forces a V-shaped impingement ring into the metal surrounding the outline of the part. The stock is stripped from the punch during the upstroke of the slides. The press must supply sufficient shear pressure, V-ring pressure, and counter-pressure to cause material to pinch around the die and produce smooth, accurate parts.

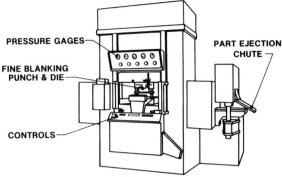

TRIPLE-ACTION PRESS

Typical Tools and Geometry Produced

Most fine-blanked parts have intricate shapes and close tolerances. One of the main advantages of fine blanking is the smooth, square edge that is produced. Typical blanking thicknesses are 0.06 in. to 0.4 in., and feasible thicknesses are 0.006 in. to 0.9 in.

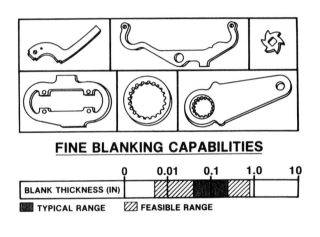

FINE BLANKING CAPABILITIES

	0	0.01	0.1	1.0	10
BLANK THICKNESS (IN)					

▓ TYPICAL RANGE ▨ FEASIBLE RANGE

Tool Geometry

Tool components include the ejector, die, stripper, and blanking punch. Extremely small punch-to-die clearance is involved. The tools must be sturdy and well supported to prevent deflections and tool failure.

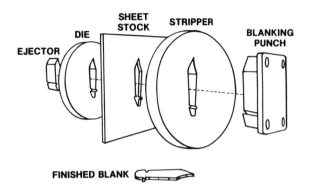

FINISHED BLANK

Tolerances and Surface Finish

For most fine-blanking operations, tolerances are typically held within ±0.001 in. and tolerances of approximately ±0.0005 in. are feasible. Surface finish may range from 8 to 75 microinches, with the range typically between 18 and 40 microinches.

TOLERANCES		
FINE BLANKING	**TYPICAL**	**FEASIBLE**
	±.001	±.0005

SURFACE FINISH									
PROCESS	**MICROINCHES (A.A)**								
FINE BLANKING	4	8	16	32	63	125	250	500	1000

▓ TYPICAL RANGE ▨ FEASIBLE RANGE

Tool Style

Shown here is a fine-blanking die set with a sliding punch used on a triple-action press. Other components include the pressure plate, die plate, inner form punch, retaining plates, backup plates, and die-set bolster plates. Fine-blanking dies are quite complex but can produce very accurate finished parts that require few subsequent operations.

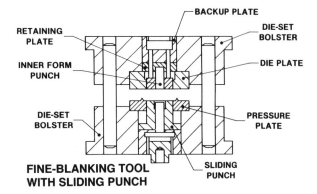

FINE-BLANKING TOOL WITH SLIDING PUNCH

Work material properties	Effects of fine blanking
Mechanical	* Workhardening on edge of blanked part * Smooth edge is formed with a small burr * Increase in tensile strength in the shear zone
Physical	* Little effect
Chemical	* Little effect

Workholding Methods

A conventional blanking tool does not have the same impingement ring that is common in fine blanking. A much smaller clearance also distinguishes fine from conventional blanking. The impingement ring forms a raised ridge on the die plate and pressure plate at a prescribed distance and height (A, B, C) from the cutting edge. The impingement ring confines the material outside the shear zone during the blanking operation. If a thin (1/8 in. or less) tough material is being worked, then only one ring on the die plate is needed.

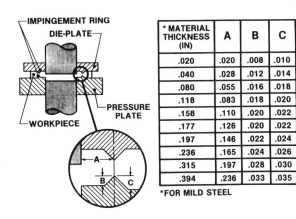

* MATERIAL THICKNESS (IN)	A	B	C
.020	.020	.008	.010
.040	.028	.012	.014
.080	.055	.016	.018
.118	.083	.018	.020
.158	.110	.020	.022
.177	.126	.020	.022
.197	.146	.022	.024
.236	.165	.024	.026
.315	.197	.028	.030
.394	.236	.033	.035

*FOR MILD STEEL

Effects on Work Material Properties

Effects include workhardening on the edge of the blanked part and a smooth edge with a small burr. Also, there is an increase in tensile strength in the shear zone.

Typical Workpiece Materials

Shown are some typical workpiece materials and associated fine-blanking ratings. Aluminum and brass have a fair to good rating. For aluminum alloys, tensile strengths should be below 16 tons/in.^{2}, and brass alloys should have no lead content. Bronze has a good to excellent rating for soft alloys. Steel has an excellent rating for alloys, having a tensile strength between 40 and 80 kips/in.^{2} (ksi), and stainless steels have a fair to good rating.

MATERIAL	FINE BLANKING RATINGS			
	POOR	FAIR	GOOD	EXCEL
ALUMINUM		▨▦▨		
BRASS		▨▦▨		
BRONZE			▨▦▨	
MILD STEEL				▨▦▨
STAINLESS STEEL		▨▦▨		

▓ TYPICAL RANGE   ▨ FEASIBLE RANGE

Tool Materials

Some of the more common tool materials used for fine blanking are molybdenum alloys, high carbon, high chromium, oil hardened tool steels, and carbides. Molybdenum is good for piercing punches and small inner form punches. The high carbon, high chromium tool steels are best for die and pressure plates, and piercing and inner form punches. Oil hardened tool steels are best for retaining pressure and backup plates. Carbides are very expensive and are used for very high production.

Tool materials	Applications
Molybdenum alloys	* Piercing punches * Small inner form punches
High carbon, high chromium tool steels	* Die plates * Piercing punches * Inner form punches * Pressure plates
Oil hardened tool steels	* Retaining plates * Pressure plates * Backup plates
Carbides	* Piercing punches * Dies

Factors Affecting Process Results

Tolerance and surface finish depend upon the following:

* Punch and die clearance
* Tool design
* Material thickness
* Configuration of workpiece
* Structural composition of material
* Tensile strength

Tool Clearance

In general, a greater clearance means higher production rates. For this reason the greatest possible punch-die clearance should be applied consistent with product quality. Outer clearance is normally 1% of the material thickness to be worked. Inner clearance depends on the diameter or length and width of the workpiece, along with the material thickness.

Production Rates

Presses may vary from as much as 600 cycles per minute to as little as 20 cycles per minute, depending on size and construction.

Approximate number of cycles per minute			
Workpiece thickness (in.)	Workpiece material		
	Mild steel	Aluminum	Stainless steel
0.05	480	560	320
0.10	240	280	160
0.20	120	140	80
0.40	60	70	40
0.80	30	35	20

Force Requirements

Force requirements have been tabulated for fine-blanking working workpieces with a 5-in. perimeter.

Material thickness (in.)			
Low carbon steel	Aluminum	Stainless steel	Required blanking force (lbs)
0.06	0.07	0.04	15,000
0.12	0.13	0.06	30,000
0.25	0.25	0.19	60,000
0.37	0.40	0.25	90,000
0.50	0.63	0.38	120,000

* For a workpiece with a 5-in. perimeter.

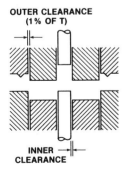

OUTER CLEARANCE (1% OF T)

INNER CLEARANCE

MATERIAL* THICKNESS (IN)	INNER CLEARANCE (% OF T)		
	$\phi < T$	$T \leq \phi < 5T$	$\phi > 5T$
.020	2.5	2.0	1.0
.040	2.5	2.0	1.0
.080	2.5	1.0	0.5
.118	2.0	1.0	0.5
.158	1.7	0.75	0.5
.240	1.7	0.5	0.5
.394	1.5	0.5	0.5
.590	1.0	0.5	0.5

* MILD STEEL

T = MATERIAL THICKNESS
ϕ = DIAMETER, OR LENGTH AND WIDTH OF FORM

Time Calculations

Operational elements for fine blanking include feed time, closing time, blanking time, opening time, and ejection time. Feed time is a function of the distance between the centers of fine-blanked parts and the feed rate.

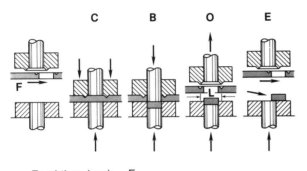

Feed time (sec) = F
Closing time (sec) = C
Blanking time (sec) = B
Opening time (sec) = O
Ejection time (sec) = E
Feed rate (in./sec) = R
Distance between center of blanked parts (in.) = L

$$\text{Feed time} = \frac{L}{R}$$

$$\text{Cycle time} = F + C + B + O + E$$

Cost Elements

Cost elements include the following:

* Setup time
* Load/unload time
* Idle time
* Blanking time
* Tool change time
* Direct labor rate
* Overhead rate
* Amortization of equipment and tooling

Safety Factors

The following risks should be taken into consideration:

* Personal
    - Contact with moving parts
    - Noise

Gear Hobbing

Gear hobbing is a multipoint machining process in which gear teeth are progressively generated by a series of cuts with a helical cutting tool (hob). Both the hob and the workpiece revolve constantly as the hob is fed across the face width of the gear blank.

Process Characteristics

* Is a gear generating process that uses a helical hob cutter
* Cutters and blanks rotate in a timed relationship
* Maintains a proportional feed rate between the gear blank and the hob
* Cuts several teeth on a progressive basis
* Is used for high production runs

Process Schematic

With gear hobbing, the cutting action is continuous, and as many as four or five teeth may be cut at the same time. A hob removes metal as it "generates" the tooth form. The straight-sided teeth on the hob cut into the gear blank one after the other, each tooth being in a slightly different position. Speeds between gear blank and hob are proportional.

GEAR HOBBING

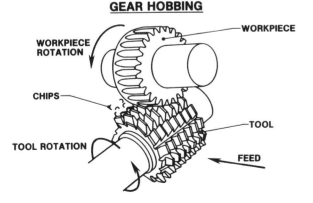

Workpiece Geometry

Shown are typical workpieces before and after gear hobbing. Hobbing is a practical method for cutting teeth in spur gears, worm gears, helical gears, and many special forms. This process is rapid and economical. More gears are cut by hobbing than by any other process.

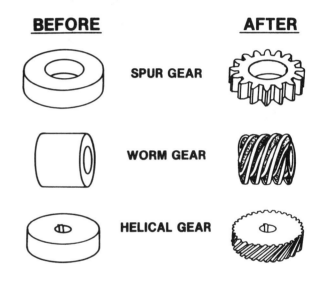

Setup and Equipment

Gear hobbing machines come in a wide range of sizes. They normally consist of a chuck and tailstock to hold the workpiece, a spindle to house the hob cutter, and a drive motor. A wide variety of tooth forms can be hobbed in production for applications ranging from tiny instrument pinions to marine drive gears as large as 10 ft in diameter.

Typical Tools and Geometry Produced

Shown is a standard gear hob for cutting gears. A hob can be thought of as one long rack tooth that has been wrapped around a cylinder in the form of a helix and gashed at intervals to provide a number of cutting edges. Special hob cutters are also available to cut such forms as sprockets and splines.

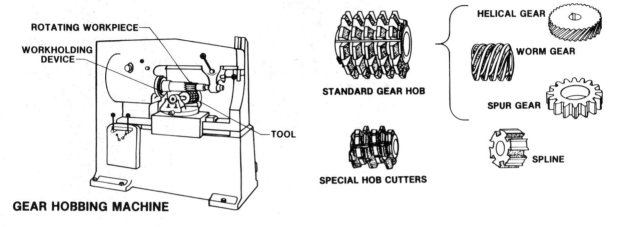

ROTATING WORKPIECE

WORKHOLDING DEVICE

TOOL

GEAR HOBBING MACHINE

STANDARD GEAR HOB

SPECIAL HOB CUTTERS

HELICAL GEAR

WORM GEAR

SPUR GEAR

SPLINE

Gear Milling

Gear milling is a multipoint machining process in which individual tooth spacings are created by a rotating multiedge cutter having a cross-section similar to that of the generated teeth (involute). After cutting each space, the gear is returned to its original position, and the gear blank is indexed for the next cut.

Process Characteristics

* Uses a rotating form cutter
* Gear blanks are indexed after each cut
* Gear teeth are produced individually
* Requires deburring
* Is a low production process
* Is normally used only when generating processes (i.e., hobbing, shaping) are unavailable

Process Schematic

The schematic shows teeth in a spur gear being cut by peripheral milling with a form cutter. The gear blank is gradually fed against the rotating cutter. At the completion of the cut, the gear blank is rapidly returned to its original position. The blank is indexed a suitable amount, and another tooth space is cut.

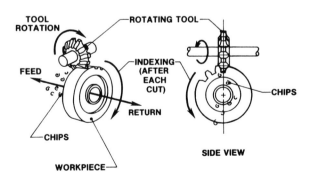

Workpiece Geometry

Shown are typical workpieces before and after being cut by gear milling. The rack is usually held in a plain vise instead of being used with an indexing head. In practice, gear milling is usually confined to replacement gears or to small-lot production. Although high quality gears can be produced by milling, the accuracy of tooth spacing is limited by the inherent inaccuracy of the indexing head.

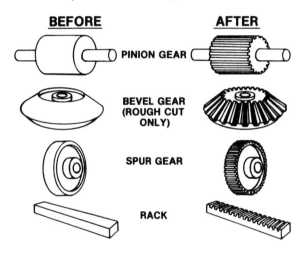

Setup and Equipment

Shown here is a horizontal milling machine. A gear mill cutter is mounted on a cutter arbor. This arbor is then held in the spindle and extends through the arbor support. The tail-stock and indexing head are secured on the machine table as close to the column as the workpiece will permit. A gear blank is mounted on the workpiece arbor and held between the tail-stock and the indexing head.

Typical Tools and Geometry Produced

In form-cutting gears on a milling machine, standard cutters can be used. Unlike gear generating tools, gear milling cutters cannot be used for all numbers of teeth. Instead, a standard gear milling cutter is designed for cutting varying numbers of teeth within a certain range. There is also a variety of special cutters available for uses such as cutting worm gears, splines, and sprockets.

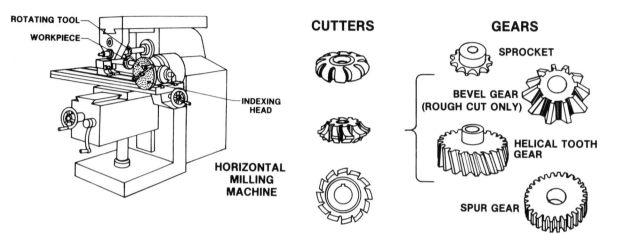

ROTATING TOOL

WORKPIECE

INDEXING HEAD

HORIZONTAL MILLING MACHINE

CUTTERS

GEARS

SPROCKET

BEVEL GEAR (ROUGH CUT ONLY)

HELICAL TOOTH GEAR

SPUR GEAR

Gear Shaping

Gear shaping is a multipoint machining process for generating teeth by a reciprocating cutter. Gear shaping is one of the most versatile of all gear cutting operations.

Process Characteristics

* Uses a gear-shaped cutter that is reciprocated and rotated, in relationship to a blank, to produce the gear teeth
* Cutters rotate in timed relationship with the workpiece
* Produces internal gears, external gears, and integral gear-pinion arrangements

Process Schematic

Gear shaping is a generating process in which a toothed disk cutter reciprocates in axial strokes as it rotates. The workpiece rotates on a second spindle (arbor). The workpiece rotation is synchronized with the cutter. The workpiece gradually feeds into the cutter as the workpiece rotates. If a two-step process is used, all tooth spaces are partially cut before the finishing step.

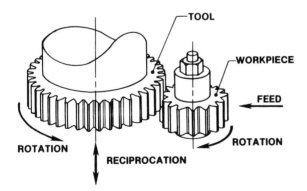

Workpiece Geometry

Shown are typical workpieces before and after gear shaping. Either straight or helical gears can be cut readily. The gear shaping setup can be easily changed to cut internal, integral gears and special shaped parts. (Gear shaping cannot be used to cut teeth in bevel gears.) Because tooling costs are relatively low, shaping is practical for any quantity of production.

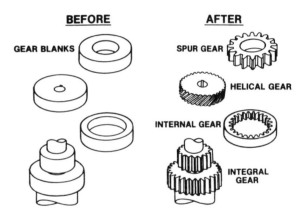

Setup and Equipment

The gear shaping machine normally consists of a base, column spindle, and arbor. The gear cutter is mounted on the spindle, and the gear blank is mounted on the arbor. The cutter reciprocates up and down while the work is gradually fed into the cutter. At the end of each cutting stroke, the spindle is retracted slightly to provide clearance between the work and the tool on the return stroke (refer to Process Schematic for detailed illustration).

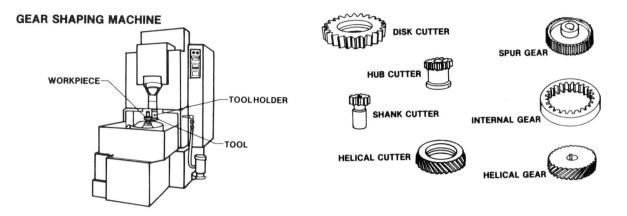

GEAR SHAPING MACHINE

WORKPIECE

TOOL HOLDER

TOOL

DISK CUTTER

SPUR GEAR

HUB CUTTER

SHANK CUTTER

INTERNAL GEAR

HELICAL CUTTER

HELICAL GEAR

Typical Tools and Geometry Produced

The types of cutters used generally fall into four groups: disk, hub, shank, and helical cutters. The cutters are essentially gears that have been relieved to form cutting teeth. This method of gear cutting is based on the principle that any two gears will mesh if they are of the same pitch, proper helix angle, and proper tooth depth and thickness.

Honing

In honing, a simultaneous rotating and reciprocating motion between an abrasive stone and the workpiece causes the removal of a small amount of material. The resulting workpiece surface is one of high precision and uniform finish.

Process Characteristics

* Uses a bonded abrasive stone for removing small amounts of material from metallic and nonmetallic surfaces
* Is a two-motion process (rotation and reciprocation)
* Stones are self-sharpening
* Minimizes heat and distortion
* Lubricants are essential
* Provides size and microinch finish control with crosshatch surface pattern

Process Schematic

Shown is an internal cylinder being honed by a typical honing machine. The abrasive honing stone can be moved outward by the expanding tool holder to remove the desired amount of material. The amount of material usually removed is less than 0.005 in. Surface speeds in honing vary from 50 fpm to 300 fpm (feet per minute). Lubricants flush chips away and maintain a uniform cutting action.

HONING

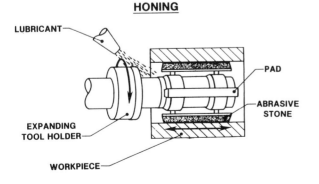

Workpiece Geometry

Honing is a sizing and finishing operation. It is also used to remove undesired waviness caused by grinding or machining operations. The majority of vertical honing operations are done on the internal surface of cylinders, such as automobile cylinder walls. When abrasive particles become dull, they break off of the stone, allowing other sharp grains to contact the work surface. General honing only improves the surface quality and not the main geometry.

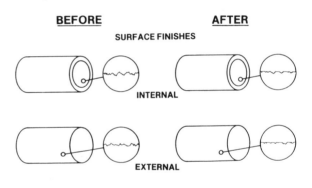

Setup and Equipment

Honing machines come with horizontal or vertical spindles. On the vertical honing machine shown, a honing tool is mounted in the rotating spindle. The workpiece is slipped over the honing tool and supported on the workrest. As the honing stones are rotated they are moved up and down relative to the workpiece to produce accurately sized bores.

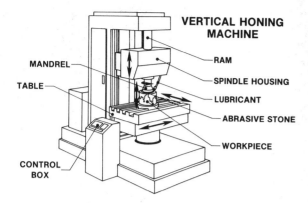

VERTICAL HONING MACHINE

MANDREL

TABLE

RAM

SPINDLE HOUSING

LUBRICANT

ABRASIVE STONE

WORKPIECE

CONTROL BOX

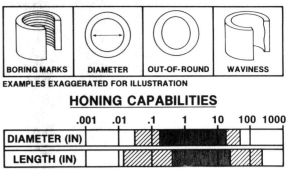

BORING MARKS | DIAMETER | OUT-OF-ROUND | WAVINESS

EXAMPLES EXAGGERATED FOR ILLUSTRATION

HONING CAPABILITIES

	.001	.01	.1	1	10	100	1000
DIAMETER (IN)							
LENGTH (IN)							

■ TYPICAL RANGE   ▨ FEASIBLE RANGE

Typical Tools and Geometry Produced

Almost all internal and external honing is done with stones made of fine bonded abrasives. The abrasive grains range in size from 70 to 600 grit and produce a uniform surface with a distinctive crosshatched appearance.

Tolerances and Surface Finish

Typical tolerances for both diameter and roundness are ±0.0005 in. It is possible, with finer stones and more accurately controlled conditions, to hone tolerances as close as ±25 millionths of an inch. Surface finishes for honing are typically between 4 and 16 microinches. However, surface finishes as fine as 1 microinch and as rough as 63 microinches are possible.

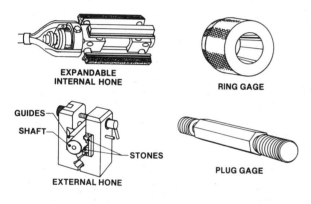

EXPANDABLE INTERNAL HONE

RING GAGE

GUIDES

SHAFT

STONES

EXTERNAL HONE

PLUG GAGE

Geometrical Possibilities

Geometrical conditions that can be remedied by honing include the removal of boring marks, sizing to a specified diameter, correcting for a limited amount of out-of-roundness, and eliminating waviness. Examples are exaggerated to illustrate the point. With honing, generally only 0.001 to 0.010 in. of material can be honed away. Workpiece sizes typically range from 3/8 in. to 30 in. in length and from 1/2 in. to 24 in. in diameter. However, it is possible to hone workpieces with diameters between 1/16 in. and 60 in. and 1/64 in. and 400 in. in length.

TOLERANCES		
HONING	**TYPICAL**	**FEASIBLE**
DIAMETER (IN)	±0.0005	±0.000025
ROUNDNESS (IN)	±0.0005	±0.000025

SURFACE FINISH											
PROCESS	**MICROINCHES (A.A.)**										
HONING	.5	1	2	4	8	16	32	63	125	250	

■ TYPICAL RANGE   ▨ FEASIBLE RANGE

Tool Style

Honing tools are categorized by how they are used (internal or external operations) and by the type of stroke used for the tools (automatic, power, or manual). Generally, multistone tools are used for power stroking larger diameters, whereas the single-stone tools are used for relatively smaller diameters. Manual stroking is the primary honing method.

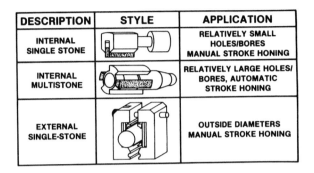

DESCRIPTION	STYLE	APPLICATION
INTERNAL SINGLE STONE		RELATIVELY SMALL HOLES/BORES MANUAL STROKE HONING
INTERNAL MULTISTONE		RELATIVELY LARGE HOLES/BORES, AUTOMATIC STROKE HONING
EXTERNAL SINGLE-STONE		OUTSIDE DIAMETERS MANUAL STROKE HONING

Workholding Methods

For manual stroking of small, lightweight parts, the operator actually holds the workpiece and strokes it over the rotating honing tool. The work rest helps prevent the workpiece from being twisted out of the operator's hand. For heavy workpieces, a floating hanger is used simply to support the bulk of the weight. In the power stroking method, the workpiece is either clamped in a vise or mounted in a fixture while either the fixture or the machine spindle reciprocates.

MANUAL STROKING

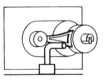

WORK REST

HANGER

POWER STROKING

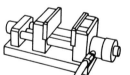

CLAMPING DEVICE

FIXTURE

Effects on Work Material Properties

Effects on work material properties are very minor. In some cases, however, the low honing pressures and speed may produce a slight effect on the workpiece. Mechanical properties include finer surface finishes with only fair finishes on ductile materials due to *smear*. The controlled crosshatch finish produced by honing allows for good lubrication carrying abilities. There is little or no effect on the material's chemical characteristics because little heat is generated.

Work material properties	Effects of honing
Mechanical	* Only fair surface finish on ductile materials
Physical	* Little effect
Chemical	* Little effect

Typical Workpiece Materials

The machinability ratings for honing some of the softer materials, such as aluminum, brass, and synthetics, are only fair to good because of material buildup on the abrasive stones. Mild steels and cast iron have fair to excellent ratings, whereas tool steels or stainless steels have good to excellent ratings.

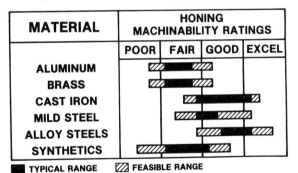

MATERIAL	HONING MACHINABILITY RATINGS			
	POOR	FAIR	GOOD	EXCEL
ALUMINUM				
BRASS				
CAST IRON				
MILD STEEL				
ALLOY STEELS				
SYNTHETICS				

■ TYPICAL RANGE   ▨ FEASIBLE RANGE

Abrasive Materials

Aluminum oxide and silicon carbide are the most common abrasives used in honing stones. Aluminum oxide is typically used on ferrous metals, except cast iron. Silicon carbide is used on cast iron as well as all other nonferrous materials. Borazon is used on tool steels, chromium-plated steels, and ceramics, whereas diamond is used on carbides, ceramics, and high-alloyed steels. Typical grit sizes and common binders are shown in the table.

Abrasive materials	Grit size	Application
Aluminum oxide	60–320	All ferrous materials except cast iron
Silicon carbide	80–600	Cast iron and nonferrous materials
Borazon	100–1000	Tool steels, ceramics, chromium-plated
Diamond	150–2000	Carbides, ceramics, high-alloyed steels

Common binders include: vitrified ceramic, resinoid, cork, carbon, and metal.

Factors Affecting Process Results

The main factors that influence tolerances and surface finishes are abrasive grit size, bonding material, tool rotation speed and reciprocation speed, lubrication, tool pressure, workpiece geometry, and workpiece material type.

Tolerances and finishes from honing depend upon the following:

* Size of grit and hardness of bond
* Speed (rotation and reciprocation)
* Lubrication
* Pressure
* Workpiece geometry and material

Tool Geometry

The tool geometry is limited to the honing of internal bores and external cylinders. Both internal and external honers are available with single or multiple stones. Multiple-stone tools are used for larger diameters and for power stroking, whereas single-stone tools are generally used for smaller diameters using manual stroking. The chart lists the typical number of stones and shoes for a particular style and diameter of a honing operation.

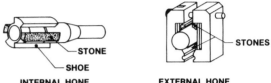

OPERATION	DIAMETER OF HONE (IN)	NO. OF STONES	NO. OF SHOES
MANUAL STROKE	1/8 to 1	1	2
	1 to 4	3	3
POWER STROKE	4 to 12	3 to 5	3 to 5
	12 to 60	10 to 15	10 to 15

Process Conditions

Recommended speeds for spindle rotation vary from 45 rpm to 350 rpm, depending on stone diameter and work material type. Speeds are generally higher for the relatively harder materials, such as cast iron and alloy steels, because harder materials do not build up on the stone.

Workpiece material	Honing speed (sfpm)	
	Rotation (rpm)	Reciprocation
Aluminum	50 to 210	9 to 75
Brass	50 to 250	9 to 90
Cast iron	65 to 330	11 to 120
Mild steel	45 to 300	8 to 120
Alloy steels	60 to 350	15 to 150
Synthetics	40 to 100	8 to 50

Honing pressures typically range from 150 psi to 400 psi. Crosshatch angle is between 25° and 35°.

Lubrication and Cooling

Water-soluble and oil-based fluids are used to lubricate and cool nonferrous workpieces, and mineral oil compounds are used for most ferrous workpieces. Cast iron workpieces require a kerosene-based mineral oil, whereas synthetic materials require an oil-based lubricant that is water-soluble. All lubricants are usually flooded over the tool and the workpiece. The main purposes of honing fluids are to remove chips and abrasive particles, to prevent the stone from loading and glazing, and to maintain a constant workpiece temperature, thus preventing dimensional inaccuracies due to thermal distortion.

Work material	Honing fluids	Application
Aluminum	Oil-based water solubles	Flooded
Brass	Water-soluble compounds	Flooded
Cast iron	Kerosene-based mineral oil	Flooded
Mild steel	Mineral oil compounds	Flooded
Tool steel	Mineral oil compounds	Flooded
Synthetics	Oil-based water solubles	Flooded

Power Requirements

Power requirements for honing are similar to many of the other machining processes. Because hardness of a material is one of the primary determinants of power requirements, it is listed in the chart. Unit power requirements are based on the number of stones used, whether proper stone pressure is maintained, use of the correct honing fluid, and the proper combination of rotation and reciprocation speeds.

Machine hp = unit power × removal rate (in.³/min)

Material	Hardness (HB)	Unit power*
Aluminum	30 to 150	2
Brass	50 to 145	4
	145 to 240	6.5
Cast iron	110 to 190	4
	190 to 320	9
Mild steel	85 to 200	7
	330 to 370	9
	485 to 560	12
Tool steel	135 to 275	8.5
	275 to 430	9
Synthetics	—	0.4 est.

* Unit power based on: • 180 grit aluminum oxide • rotation speed 100 sfpm • pressure 200 psi.

Cost Elements

Cost elements include the following:

* Honing time/piece
* Tool change time
* Setup time
* Direct labor rate
* Overhead rate
* Tool materials consumed (stone, mandrel, shoes, fluid)
* Amortization of equipment and tooling costs

Time Calculation

Honing time per piece is a function of the volume of material to be removed and removal rate.

Diameter to be honed (in.) = D
Length to be honed (in.) = L
Material removal rate (in.³/min) = R
Depth to be removed (in.) = d
Material to be removed (in.³) = Q

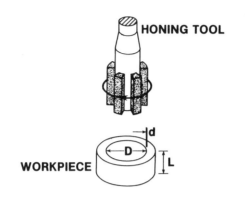

$$Q = \pi \times D \times L \times d$$

$$\text{Honing time} = \frac{Q}{R}$$

Safety Factors

The following hazards should be taken into consideration:

* Personal
  - Rotating tools
  - Reciprocating workpiece/tool
  - Eye and skin irritation from honing fluids
* Environmental
  - Waste particles and fluid
  - Inflammable honing fluids

Horizontal Boring

Horizontal boring is a single-point cutting operation used to produce an accurate internal cylindrical surface by enlarging an existing opening in a workpiece. The workpiece moves parallel to the axis of rotation of the cutting tool.

Process Characteristics

* Produces accurate internal cylindrical surfaces
* Uses a single-point cutting tool
* Enlarges and straightens existing holes
* Has a tool motion parallel to the axis of rotation
* Leaves helical internal feed marks

Process Schematic

In horizontal boring, an existing hole is enlarged and straightened by advancing one or more rotating single-point cutters horizontally into a stationary workpiece. As the boring bar is advanced horizontally into the workpiece, material is removed in the form of chips. The workpiece may also be fed into the cutter. The bored hole is always concentric with the axis of rotation of the cutting tool.

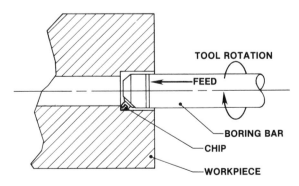

Workpiece Geometry

Horizontal boring is adaptable to many types of workpieces. Workpiece sizes range from medium to large. Other machining operations such as milling, drilling, reaming, and tapping may be performed with the same setup, thereby saving time by avoiding extra setups or routing to other machines.

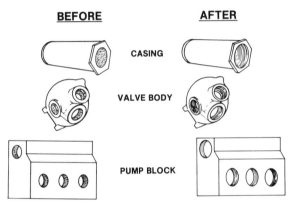

Setup and Equipment

Horizontal boring mills are generally large machine tools designed for heavy workpieces. The table type machine shown here has a rotating work table that can feed horizontally, parallel and at right angles to the spindle axis. The headstock moves vertically on the column, and the spindle has a horizontal feed motion.

HORIZONTAL BORING MACHINE

WORKPIECE

WORKHOLDING DEVICE

TOOL

Typical Tools and Geometry Produced

Tools commonly used for horizontal boring include boring bars with micrometer adjustable cartridges and blade-type boring bars. In addition to these and other standard insert-type boring bars, multiple-edged, multiple-diameter bars are used for high speed production applications. Parts having more than one inside diameter along the same axis require using these special bars.

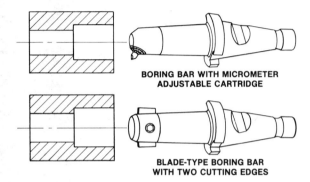

BORING BAR WITH MICROMETER ADJUSTABLE CARTRIDGE

BLADE-TYPE BORING BAR WITH TWO CUTTING EDGES

Internal Grinding

Internal grinding is an abrasive machining process by which small chips of material are removed from an internal surface or hole by means of a plain or formed grinding wheel. The workpiece may be held in either a chucking-type or centerless-type workholding device.

Process Characteristics

* Is applicable on internal surfaces or holes
* Produces close tolerances and smooth surfaces
* Can produce nonsymmetrical parts by using auxilliary holding devices
* Is most efficient for use on rings, sleeves, etc.
* Requires coolant
* Is primarily a finishing process

Process Schematic

The workpiece is rotated in a counterclockwise direction, and the grinding wheel is rotated clockwise. The workpiece material is removed in the form of small chips by the abrasive particles in the grinding wheel.

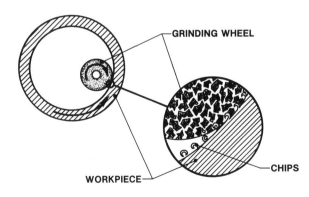

Workpiece Geometry

Shown are some workpiece geometries that can be machined. Close dimensional tolerances as well as smooth surface finishes can be achieved.

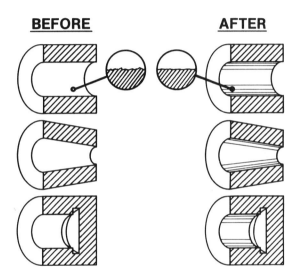

Setup and Equipment

Chucking-type internal grinding machines consist of a grinding wheel, a chuck that holds the workpiece, a headstock, and a tailstock. Machines are capable of performing several different types of operations.

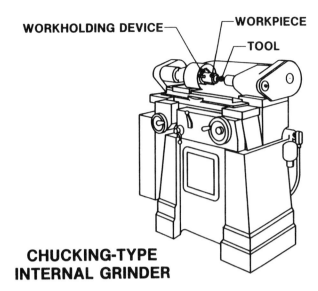

WORKHOLDING DEVICE — WORKPIECE — TOOL

CHUCKING-TYPE INTERNAL GRINDER

Typical Tools and Geometry Produced

Most applications use a plain (nonformed) grinding wheel that imparts no special contours to the workpiece. With a plain grinding wheel, straight, tapered, and blind geometries may be produced. When blind holes are being ground, a recess at the end of the hole is often incorporated in the design to reduce a tendency for tapering as the grinding wheel wears. To grind a formed surface, a formed grinding wheel is required.

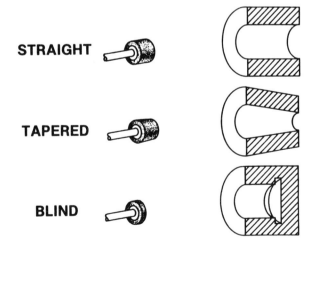

STRAIGHT

TAPERED

BLIND

Geometrical Possibilities

Shown are typical shapes produced when internal grinding with plain and formed grinding wheels. Both symmetrical and nonsymmetrical workpieces may be internally ground. The pattern produced on the workpiece is a "mirror image" of the grinding wheel shape. Typical workpiece sizes range from 0.1 in. to 20 in. in diameter and 0.5 in. to 16 in. in length. However, it is possible to grind workpieces that are between 0.02 in. and 80 in. in diameter and 0.05 in. and 100 in. in length.

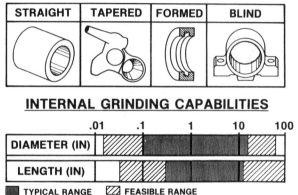

STRAIGHT	TAPERED	FORMED	BLIND

INTERNAL GRINDING CAPABILITIES

	.01	.1	1	10	100
DIAMETER (IN)					
LENGTH (IN)					

▓ TYPICAL RANGE   ▨ FEASIBLE RANGE

Tolerances and Surface Finish

For most applications, tolerances are held within ±0.001 in. for both diameter and roundness. For precision applications, tolerances for both diameter and roundness as low as ±0.00002 in. may be achieved. Surface finish may range from 2 to 125 microinches, with the typical range between 8 and 32 microinches.

TOLERANCES		
INTERNAL GRINDING	TYPICAL	FEASIBLE
DIAMETER	±.001	±.00002
ROUNDNESS	±.001	±.00002

SURFACE FINISH	
PROCESS	MICROINCHES (A.A.)
INTERNAL GRINDING	2 4 8 16 32 63 125 250 500

▓ TYPICAL RANGE   ▨ FEASIBLE RANGE

Tool Style

Internal grinding tools are available in a wide range of shapes. Plain grinding wheels are most common and are used to grind straight and tapered geometries. Special geometries may be produced in the workpiece by using a wide variety of formed grinding wheels.

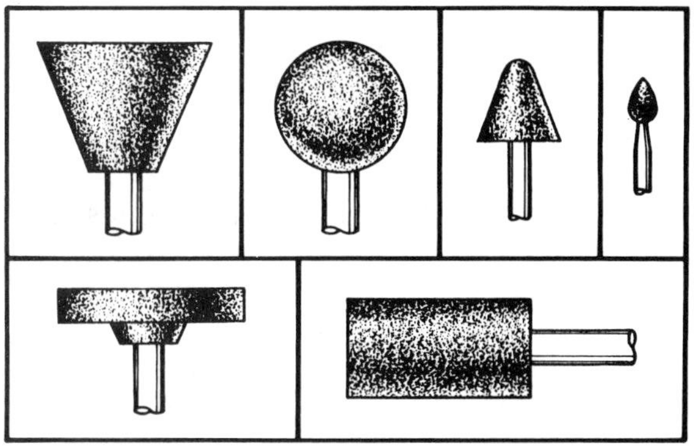

Workholding Methods

The conventional chuck and centerless grinding arrangement are two common workholding methods. With the conventional-type chuck, the workpiece is secured to the workhead by clamps. This method also allows nonsymmetrical workpieces to be ground. With the centerless arrangement, only cylindrical workpieces may be ground. These are supported by the pressure wheel, support wheel, and regulating wheel.

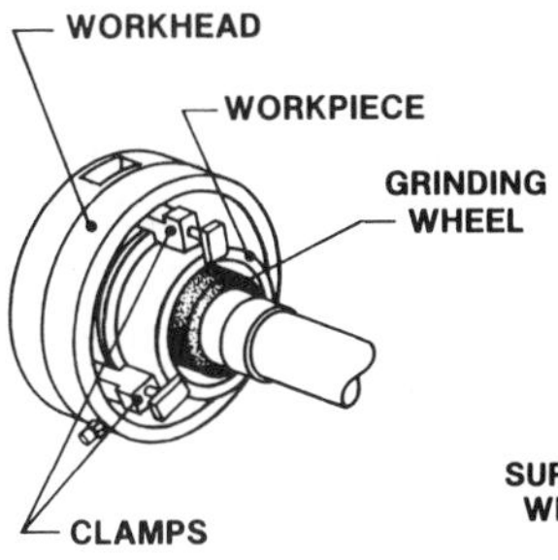
WORKHEAD
WORKPIECE
GRINDING WHEEL
CLAMPS

CONVENTIONAL TYPE

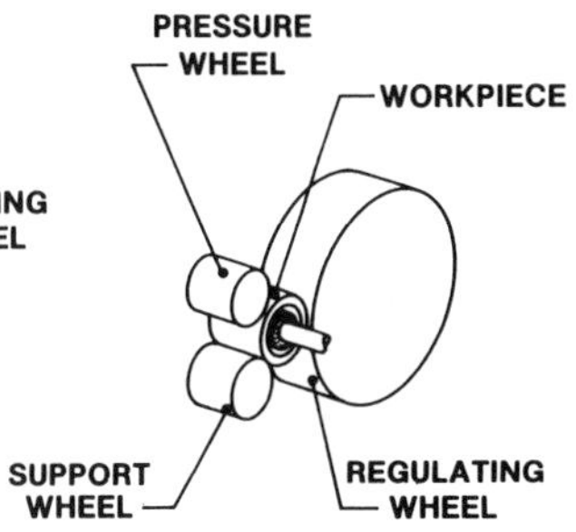
PRESSURE WHEEL
WORKPIECE
SUPPORT WHEEL
REGULATING WHEEL

CENTERLESS TYPE

Effects on Work Material Properties

High grinding temperatures create residual stresses along with a thin martensitic layer. This thin layer of martensite may reduce material fatigue strength due to microcracks. If the grinding temperature is elevated above the curie temperature, a loss of magnetic properties may occur in ferromagnetic materials. Internal grinding may also reduce corrosion resistance.

Work material properties	Effects of internal grinding
Mechanical	* Residual surface stresses * A thin martensitic layer may form on the part surface * Fatigue strength may be reduced
Physical	* Magnetic properties may be lost on ferromagnetic materials
Chemical	* May increase susceptibility to corrosion

Typical Workpiece Materials

Aluminum, brass, and plastics have poor to fair ratings, whereas cast iron and mild steel have good machinability characteristics. On the other hand, stainless steel has a poor to fair rating because of its toughness and tendency to work harden.

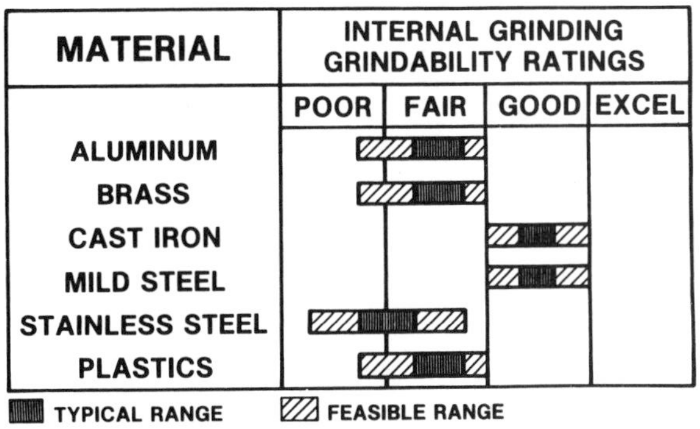
MATERIAL
INTERNAL GRINDING GRINDABILITY RATINGS
POOR
FAIR
GOOD
EXCEL
ALUMINUM
BRASS
CAST IRON
MILD STEEL
STAINLESS STEEL
PLASTICS
TYPICAL RANGE
FEASIBLE RANGE

Abrasive Materials

Aluminum oxide, silicon carbide, diamond, and cubic boron nitride (CBN) are four commonly used abrasive materials for internal grinding wheels. Of these materials, aluminum oxide is the most common. Because of cost, diamond and CBN grinding wheels are generally made with a core of less expensive material surrounded by a layer of diamond or CBN. Diamond and CBN wheels are very hard and are capable of economically grinding materials, such as ceramics and carbide, that cannot be ground by aluminum oxide or silicon carbide wheels.

Abrasive materials	Applications
Aluminum oxide (A)	* Best for most steels and steel alloys * Low cost * High volume applications
Silicon carbide (C)	* Best for cast iron, nonferrous metals, and nonmetallic materials * Low cost
Diamond (D, MD, SD)	* Can be natural or man-made * Best for most carbides and some nonmetallic materials * Very high cost
Cubic boron nitride (B)	* Superior for high speed steel * Long life * Cool cutting * Very high cost

Factors Affecting Process Results

Tolerance and surface finish depend upon the following:

* Workpiece material
* Grinding wheel material
* Cutting speed and feed rate
* Proper grit, grade, and bond of wheel
* Cutting fluid
* Rigidity of tool, workpiece, and machine

Tool Geometry

This figure shows the standard marking system used for grinding wheels. The marking

identifies five characteristics of a grinding wheel.

1. The abrasive material is identified by a letter (A—aluminum oxide, C—silicon carbide, etc.)
2. Grain size is indicated by a number from 8 (coarse) to 600 (very fine)
3. The grade is denoted by a letter from A (hard wheel) to Z (soft wheel)
4. The spacing between abrasive grains is labeled by a number from 1 (dense spacing) to 16 (open spacing)
5. Six common grinding wheel bonds are vitrified (V), resinoid (B), silicate (S), shellac (E), rubber (R), and oxychloride (O)

The manufacturer may add additional non-standardized identification letters or numbers.

STANDARD MARKING SYSTEM

A 36 L 6 V

ABRASIVE MATERIAL — GRAIN SIZE — GRADE — STRUCTURE — BOND

ABRASIVE MATERIAL	ALUMINUM OXIDE(A), SILICON CARBIDE(C)
GRAIN SIZE	COARSE ⟶ FINE 8 to 600 46 to 80 MOST COMMON
GRADE (BOND STRENGTH)	K–P SOFT WORKPIECES I–N HARD WORKPIECES J–N MOST COMMON
STRUCTURE (GRAIN SPACING)	DENSE ⟶ OPEN 1 to 16
BOND MATERIAL	VITRIFIED(V), RESINOID(B), SILICATE(S), SHELLAC(E), RUBBER(R), OXYCHLORIDE(O)

Process Conditions

Wheel speed for internal grinding may range from 5000 to 6500 surface feet per minute. Work speed varies from 75 to 200 inches per minute (50 to 150 ipm for plastics). The type of wheel used depends on the workpiece material, hardness, and the desired surface finish.

Workpiece material	Hardness (HB)	Infeed (in./pass)	Wheel ID No. ANSI	
			Rough	Finish
Aluminum	30 to 150	0.003 to 0.0002	C 46 J 9 V	C 46 J 9 V
Brass	40 to 200	0.002 to 0.0002	C 46 J 9 V	C 46 J 9 V
Cast iron, ductile	520 max	0.002 to 0.0002	C 54 J 7 V A 60 J 7 V	C 54 J 7 V A 60 J 7 V
Mild steel	500 max	0.0005 to 0.0002	A 54 M 7 V	A 54 M 7 V
Stainless steel	150 to 200	0.0005 to 0.0002	A 60 K 7 V	A 60 K 7 V
Thermo-plastic	—	0.001 to 0.0005	A 60 K 9 V	A 60 K 9 V

Lubrication and Cooling

The use of grinding fluids is essential to cool the wheel and workpiece, lubricate their interface, and to remove chips. Among the commonly used grinding fluids are water-soluble chemical fluids, water-soluble oils, synthetic oils, and petroleum-based oils. It is important that the grinding fluid be applied directly to the cutting area to insure that the fluid is not carried away by the fanlike action of the rapidly rotating grinding wheel.

Work material	Cutting fluid	Application
Aluminum	Light duty oil	Flood
Brass	Light duty oil	Flood
Cast iron	Heavy duty emulsifiable oil, light duty chemical and synthetic oil	Flood
Mild steel	Heavy duty water-soluble oil	Flood
Stainless steel	Heavy duty emulsifiable oil, heavy duty chemical and synthetic oil	Flood
Plastics	Water soluble oil, dry, heavy duty emulsifiable oil, light duty chemical and synthetic oil	Flood

Power Requirements

The metal removal rate in cubic inches per minute and the efficiency of the grinding wheel drive mechanism determine power requirements. The following graph may be used to compute the required horsepower quickly.

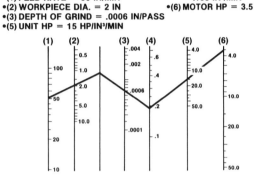

BASED ON:
• 8620 STEEL, 170 HB
• (1) FEED RATE = 50 IN/MIN
• (2) WORKPIECE DIA. = 2 IN
• (3) DEPTH OF GRIND = .0006 IN/PASS
• (5) UNIT HP = 15 HP/IN³/MIN

THEN:
• (4) METAL REMOVAL RATE = .188 IN³/MIN
• (6) MOTOR HP = 3.5

Time Calculations

Grinding time for traverse grinding is a function of the length of the part to be ground, feed rate, and the number of passes required. The feed rate is primarily a function of the wheel width and desired surface.

TRAVERSE GRINDING

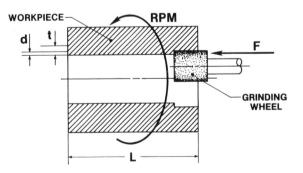

Length of cut (in.) = L
Feed rate (ipm) = F
Approach time (min) = A
Spark-out time (min) = s
Width of grinding wheel (in.) = W
Depth of cut per pass (in.) = d
Thickness of material
to be removed (in.) = t
Number of passes required = N

$$\text{Grinding time} = N \times \frac{L}{F}$$

$$\text{Number of passes} = \frac{t}{d}$$

The cutting time for internal plunge grinding is a function of the length of cut (which in this case is the thickness of material to be removed) and the feed rate. Feed rates for plunge grinding may vary from 0.0001 in./rev to 0.001 in./rev (inches per revolution) of the workpiece. Approach and spark-out time are also part of the total time.

$$\text{Cutting time} = \frac{t}{F}$$

$$\text{Total time} = A + \frac{t}{F} + s$$

Cost Elements

Cost elements include the following:

* Setup time
* Load/unload time
* Idle time
* Grinding time
* Tool change time
* Tool costs
* Direct labor rate
* Overhead rate
* Amortization of equipment and tooling

PLUNGE GRINDING

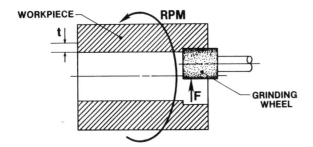

WORKPIECE — RPM

t

F GRINDING WHEEL

Safety Factors

The following risks should be taken into consideration:

* Personal
  - Rotating workpiece and grinding wheel
  - Eye irritation from fine chips
  - Eye and skin irritation from cutting fluids
  - Grinding wheel disintegration
* Environmental
  - Cutting fluid disposal
  - Grinding sludge disposal

Jig Boring

Jig boring is a single-point cutting operation used to produce a precise internal cylindrical surface by enlarging an existing opening in a workpiece. The tool is rotated and advanced parallel to the axis of the hole in the workpiece.

Process Characteristics

* Enlarges and straightens existing holes
* Tool is rotated and advanced parallel to the axis of the hole in the workpiece
* Bored hole location and dimensions are usually highly precise
* Can maintain accurate relationships between multiple holes or surfaces

Process Schematic

In jig boring, an existing hole is enlarged and straightened by advancing a rotating, single-edge cutter vertically into a stationary workpiece. Holes with highly precise dimensions and locations are created using this process. Tolerances can be held readily within ±0.0005 in.

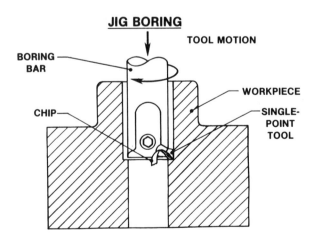

Workpiece Geometry

Jig boring is typically used to produce parts that require a high level of dimensional accuracy such as jigs, tools, and fixtures. A wide variety of component parts are commonly manufactured using this process.

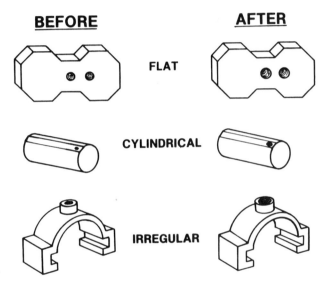

Setup and Equipment

The jig boring machine consists primarily of a base, precision work table, column, and spindle housing. The spindle, spindle bearings, and work table positioning devices are constructed with very high accuracy. Jig boring machines are designed to establish the location of holes in surfaces to the highest level of accuracy possible with a machine tool.

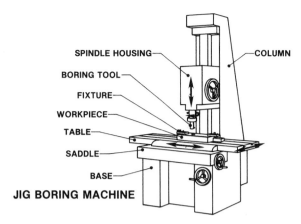

SPINDLE HOUSING
BORING TOOL
FIXTURE
WORKPIECE
TABLE
SADDLE
BASE
COLUMN

JIG BORING MACHINE

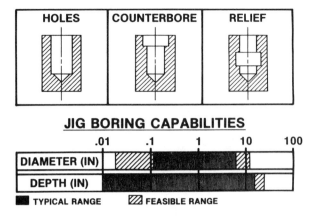

| HOLES | COUNTERBORE | RELIEF |

JIG BORING CAPABILITIES

	.01	.1	1	10	100
DIAMETER (IN)					
DEPTH (IN)					

■ TYPICAL RANGE   ▨ FEASIBLE RANGE

Typical Tools and Geometry Produced

Some of the most commonly used tools for jig boring are solid boring bars with replaceable inserts or with solid boring tool bits and boring bars with a micrometer adjustable cartridge. Solid boring bars are rigid, which makes them especially useful for boring deep holes. Single-point tool bits are generally used for boring small holes, but with the necessary attachments, they may be adapted for larger holes.

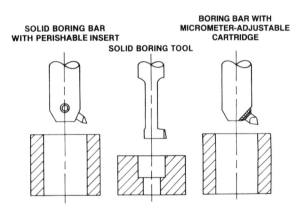

SOLID BORING BAR
WITH PERISHABLE INSERT

SOLID BORING TOOL

BORING BAR WITH
MICROMETER-ADJUSTABLE
CARTRIDGE

Geometrical Possibilities

Shown are some of the shapes that can be produced with jig boring tools. Precision holes are accurately located for tooling work. Typically, diameters range from 0.125 in. to 8 in., and depths are up to 20 in. However, it is possible to jig bore workpieces that range between 0.030 in. to 12 in. in diameter and 18 in. and 36 in. in depth.

Tolerances and Surface Finish

For most jig boring applications, tolerances are held within ±0.001 in. for diameter and ±0.003 in. for depth. For precision applications, tolerances can be held within ±0.0002 in. for diameter and ±0.0005 in. for depth. Surface finish may range from 4 to 125 microinches, with the typical range between 16 and 63 microinches.

TOLERANCES		
JIG BORING	**TYPICAL**	**FEASIBLE**
DIAMETER	±.001	±.0002
DEPTH	±.003	±.0005
LOCATION	±.001	±.0002

SURFACE FINISH									
PROCESS	**MICROINCHES (A.A.)**								
JIG BORING	2	4	8	16	32	63	125	250	500

■ TYPICAL RANGE   ▨ FEASIBLE RANGE

Tool Style

Jig boring tools are available in a wide range of sizes and shapes depending upon their application. Shown are three common styles used for jig boring. Adjustable insert boring bars utilize different types of inserts, such as diamond, round, square, and triangular. Offset heads are often used for formed relief and counterboring. Solid boring bars are used for straight boring on relatively small holes.

DESCRIPTION	STYLE	APPLICATION
ADJUSTABLE INSERT		STRAIGHT AND TAPERED BORING
OFFSET HEAD		RELIEF AND COUNTERBORING
SOLID BORING BAR		STRAIGHT BORING

Workholding Methods

Shown are two frequently used workholding devices: a rotary magnetic chuck and a strap clamp. When the rotary magnetic chuck is used, the workpiece must be a ferrous metal to be held securely. When the strap clamp is used, the workpiece is positioned on the table, and the straps are bolted down on top of the workpiece. Often, "parallels" are used to support the workpiece when through-bored holes are required.

MAGNETIC ROTATING TABLE

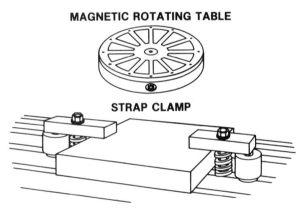

STRAP CLAMP

Effects on Work Material Properties

Jig boring has little effect on either physical or chemical properties. However, residual surface stresses may create microcracking.

Work material properties	Effects of jig boring
Mechanical	* Creates residual surface stresses
Physical	* Little effect
Chemical	* Little effect

Typical Workpiece Materials

Aluminum, brass, and plastics have good to excellent ratings, whereas cast iron and mild steel have good machinability characteristics. Stainless steel has a poor to fair rating because of its toughness and its tendency to work harden when machined.

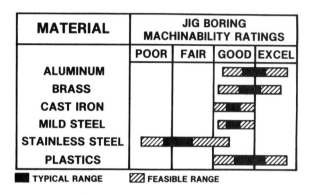

MATERIAL	JIG BORING MACHINABILITY RATINGS			
	POOR	FAIR	GOOD	EXCEL
ALUMINUM			▨	▨
BRASS			▨	▨
CAST IRON			▨ ▨	
MILD STEEL			▨ ▨	
STAINLESS STEEL	▨	▨		
PLASTICS			▨	▨

■ TYPICAL RANGE   ▨ FEASIBLE RANGE

Tool Materials

High speed tools made of steel have been used for some time but are being replaced in many applications by carbide, ceramic, and diamond tooling. Carbide inserts are easily replaceable and are long lasting. Ceramic tools can resist high temperatures, so high speed machining is possible. Carbides and ceramics often fracture if cuts are interrupted. Diamond tools produce superior surface finish and are used for nonferrous and nonmetallic materials.

Tool materials	Applications
High speed steel	* Special tool shapes * Low production * Best for interrupted cuts
Carbides (inserts)	* Most commonly used tool * High production
Ceramics (inserts)	* High speed machining * Avoid interrupted cuts * Avoid positive rake angles
Diamonds (inserts)	* Abrasive machining * High surface qualities, fine tolerances * Nonferrous or nonmetallic materials

Factors Affecting Process Results

Tolerance and surface finish depend upon the following:

* Tool geometry
* Cutting speed and feed rate
* Rigidity of tool, workpiece, and machine
* Alignment of machine components and fixtures
* Cutting fluid
* Chip removal
* Environmental conditions, such as humidity, temperature, vibration, and cleanliness

Tool Geometry

Boring tool geometry is designated by angles on the tool. Shown are the four angles that are most often varied on the tool. Generally, the back and side rake angles are higher for soft materials. Negative rake angles are often used on hard materials. Relief angles must be increased when holes are less than 0.5 in. in diameter.

WORKPIECE MATERIAL	BACK RAKE ANGLE (°)	SIDE RAKE ANGLE (°)	END RELIEF ANGLE (°)	SIDE RELIEF ANGLE (°)
ALUMINUM	0 to 15	5 to 15	8 to 13	5 to 8
BRASS	0 to 10	5 to 20	8 to 13	5 to 8
CAST IRON	-5 to 5	-5 to 5	5 to 10	5 to 8
MILD STEEL	-10 to 0	0 to 15	5 to 10	2 to 3
STAINLESS STEEL	3 to 10	0 to 15	5 to 10	2 to 3
PLASTICS	-5 to 5	0 to 15	20 to 30	15 to 20

Process Conditions

Shown are the suggested ranges for depth of cut, feed rate, and cutting speed using cutting fluid. Generally, the cutting speeds are high for nonferrous materials, such as aluminum and brass, and low for ferrous materials, such as cast iron and stainless steel. Depth of cut, feed rate, and cutting speed will vary depending upon the type of cut and the geometry of the workpiece.

Carbide tool with cutting fluid

Conditions	Rough	Finish
Depth of cut (in.)	0.015 to 0.094	0.005 to 0.015
Feed rate (ipr)	0.005 to 0.015	0.002 to 0.005

Workpiece material	Cutting speed (sfpm)	
	Rough	Finish
Aluminum	450 to 700	700 to 1000
Brass	600 to 700	700 to 800
Cast iron	350 to 450	450 to 600
Mild steel	550 to 700	700 to 1200
Stainless steel	300 to 400	400 to 500
Plastics	400 to 650	650 to 1000

Lubrication and Cooling

This chart shows the typical cutting fluids that are used for a variety of work materials. Cutting fluids cool the tool, which reduces tool wear and makes higher cutting speeds and feed rates possible. Secondary functions include lubrication, flushing chips, rust prevention, and preventing the tool from building up an edge that would mar the surface finish.

Work material	Cutting fluid	Application
Aluminum	None, water-soluble oil, synthetic oil, kerosene	Brush, spray
Brass	None, water-soluble oil, synthetic oil	Brush
Cast iron	None	–
Steel, all types	None, water-soluble oil, synthetic oil, sulfurized oil	Brush
Plastics	None, emulsifiable oil, synthetic oil	Brush, spray

Power Requirements

The formula used to calculate the machine horsepower is a function of unit power and material removal rate. For example, if mild steel is bored with a HB hardness of 85 to 200, and 10 in.^{3} of material is removed per minute, then an 11 hp engine lathe is needed (1.1 unit power × 10 in.^{3}/min = 11 hp). Unit power is based on HSS and carbide tools, feeds of 0.005 ipr to 0.020 ipr, and 80% efficiency.

Machine hp = unit power × removal rate (in.³/min)

Material	Hardness (HB)	Unit power*
Aluminum	30 to 150	0.3
Brass	50 to 145	0.6
	145 to 240	1.0
Cast iron	110 to 190	0.7
	190 to 320	1.4
Mild steel	85 to 200	1.1
	330 to 370	1.4
	485 to 560	2.0
Stainless steel	135 to 275	1.3
	275 to 430	1.4
Plastics	N/A	0.05 est.

* Unit power based on: ● HSS and carbide tools ● feed of 0.005 ipr to 0.020 ipr ● 80% efficiency.

Cost Elements

Cost elements include the following:

* Setup time
* Load/unload time
* Idle time
* Cutting time
* Tool change time
* Tool costs
* Direct labor rate
* Overhead rate
* Amortization of equipment and tooling

Time Calculations

Understanding the jig boring setup is important in calculating the time needed to turn, face, retract, and traverse. The elements involved in jig boring are depth of bore, approach distance, feed rate, traverse distance, traverse rate, retract distance, and retract rate.

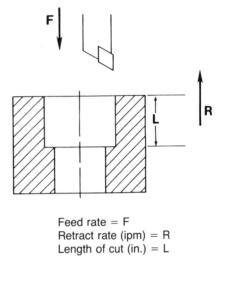

Feed rate = F
Retract rate (ipm) = R
Length of cut (in.) = L

$$\text{Boring time} = \frac{\text{length of cut}}{\text{feed rate}} = \frac{L}{F}$$

$$\text{Retract time} = \frac{L}{R}$$

Safety Factors

The following risks should be taken into consideration:

* Personal
  - Rotating tool
  - Hot and sharp chips
  - Eye and skin irritation from cutting fluids

Lancing

Lancing is a combined shearing and bending operation where a portion of the periphery of a hole is cut into the workpiece and the remainder is bent to the desired shape. No material is removed from the workpiece by this process.

Process Characteristics

* Cuts a portion of the periphery of the hole, and the remainder is bent to the desired shape
* Removes no metal from the workpiece
* Can use a single cut to facilitate the making of special features

Process Schematic

Shown is a piece of sheet metal with a vent being formed. In this operation, the punch shears one side of the workpiece and bends the other side. The cut does not have a closed contour and does not release a blank or a piece of scrap.

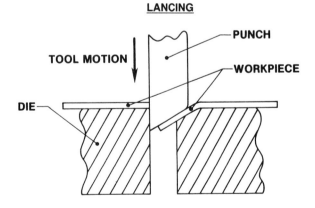

Workpiece Geometry

In addition to producing louvers, tabs, and vents, lancing is used to free metal for subsequent forming and to cut partial contours for blanked parts.

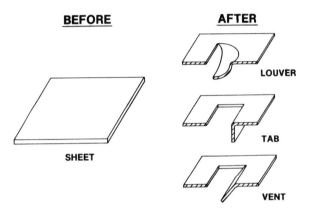

Setup and Equipment

Lancing is commonly done on a mechanical press. The punch is mounted to the ram and the die is mounted to the bolster plate.

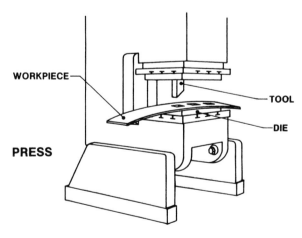

Typical Tools and Geometry Produced

Lancing is frequently combined with bending to form louvers, tabs, and vent openings. Punches and dies for this operation are made of tool steel.

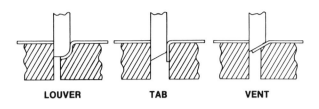

LOUVER   TAB   VENT

Lapping

Lapping is an abrasive process in which a rotating *lap*, charged with a loose abrasive slurry, removes very small amounts of material from flat metallic or nonmetallic surfaces. Low speed and low pressure result in finely finished surfaces of extreme flatness.

Process Characteristics

* Contact points between tool and workpiece are constantly changed through relative motions
* Rotating tools store the abrasive particles
* Is a low speed, low pressure abrading operation
* Results in a fine surface finish and extreme flatness

Process Schematic

A rotating lap is coated with an abrasive slurry (abrasive grains in an oil media). A weighted workpiece is placed on top of the lap, and, as the lap rotates, the abrasive grains cut very small chips from the workpiece surface. The workpiece is usually placed within a control ring to restrict its movement.

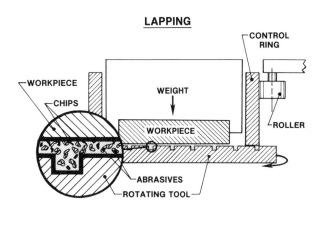

LAPPING

Workpiece Geometry

Many types of metals or nonmetals can be successfully lapped. Shown are an aluminum aircraft housing, a flat brass gear, and a steel pump housing. The average surface finish produced is usually between 2 and 16 microinches.

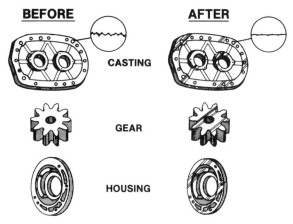

Setup and Equipment

The lapping machine usually consists of a rotating lap, which can hold several workpieces at one time. The workpieces are contained within control rings that prevent them from moving off the lap. Because pressure on the workpiece is essential, weights or pressure cylinders are used.

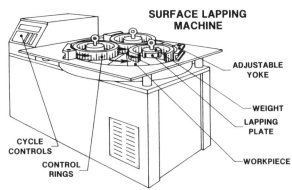

SURFACE LAPPING MACHINE

Typical Tools and Geometry Produced

Almost any flat workpiece can be lapped. The tooling is very simple and consists mainly of a lap charged with abrasives, a rotating control ring, and a weight to hold the workpiece.

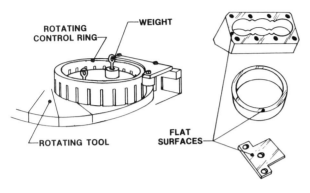

Geometrical Possibilities

Three of the more common shapes that can be lapped are cylindrical, spherical, and flat surfaces. Workpiece size ranges from around 2 in. to as long as 80 in. in length, depending on equipment and workpiece diameter. The diameter ranges from 0.25 in. to 20 in. Diameter is limited only by equipment.

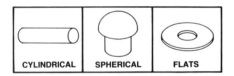

LAPPING CAPABILITIES

	0.001	0.01	0.1	1	10	100
LENGTH (IN)						
DIAMETER (IN)						

■ TYPICAL RANGE   ▨ FEASIBLE RANGE

Tolerances and Surface Finish

The surface finishes achieved by lapping are some of the best available in terms of flatness, smoothness, and freedom from stress.

TOLERANCES (IN)		
LAPPING	TYPICAL	POSSIBLE
DIAMETER	±0.00005	±0.000005
ROUNDNESS	±0.000025	±0.000005
STRAIGHTNESS	±0.000025	±0.000025
PARALLELISM	±0.00005	±0.000025

SURFACE FINISH	
PROCESS	MICROINCHES (A.A.)
LAPPING	.5 1 2 4 8 16 32 63 125 250

■ TYPICAL RANGE   ▨ FEASIBLE RANGE

Tool Style

The two plates shown at the top of the figure below are used for lapping two parallel surfaces. The workpiece is sandwiched between the lap faces and then placed in the lapping machine. The special lapping ring shown below is used for lapping round or tubular workpieces. The workpiece is rotated, and the lapping ring is moved along the workpiece surface.

DESCRIPTION	STYLE	APPLICATION
LAPPING PARALLEL SURFACES		FORM MATING PARTS FORM FLAT SURFACES
RING LAPPING		PROVIDES EXTREMELY CLOSE TOLERANCES CORRECTS OUT OF ROUNDNESS

Workholding Methods

The workpieces are placed between the plates of a holder and held by the small retainer disk in the center. Workpieces are also placed at a slight angle off the true axis to promote uniform cutting.

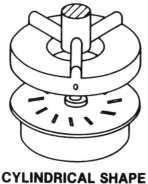

**CYLINDRICAL SHAPE
WORKHOLDER**

Effects on Work Material Properties

The effects of lapping on work material properties are minimal. One benefit of lapping is that it relieves surface stress. A mechanical property that causes problems in soft materials is that the abrasives tend to embed in the surface of the workpiece. There is little effect on physical and chemical properties.

Work material properties	Effects of lapping
Mechanical	* Relief of surface stress * Surface inclusions in soft materials
Physical Chemical	* Little effect * Little effect

Typical Workpiece Materials

The machinability ratings are best for hard materials such as cast iron, mild steel, stainless steel, and tool steel, but satisfactory results can be obtained for materials such as aluminum and brass.

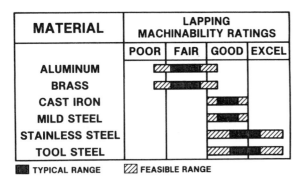

Abrasive Materials

Given here are the different types of abrasive materials and grit size for various applications. The grit sizes range from 100 parts per inch for silicon carbide to as small as 1 micron for ferric oxide.

Abrasive Materials	Grit size	Applications
Silicon carbide	100 to 400 600 to 1000	* Tool-room lapping * Roughing hard steels
Corundum	400 to 800	* Roughing and finishing soft steels
Garnet	600 to 800	* Finishing brass and bronze
Ferric oxide	1 micron	* Polishing soft metals
Aluminum oxide	400 to 900	* Finishing of soft steel, nonferrous metals
Diamond	0.5 to 20 microns	* Hardened materials

Factors Affecting Process Results

The main factors that influence tolerances and surface finishes are size of grit, type of lapping material, rotational speed, lapping motion, pressure, workpiece material, and workpiece hardness. Secondary factors are number of parts per setup, number of lapping operations, and cleanliness of the environment.

Tolerances and finishes from lapping depend upon the following:

* Size of grit
* Type of lapping material
* Speed and lapping motion
* Pressure
* Workpiece material
* Hardness of workpiece material
* Number of parts per setup
* Number of lapping operations
* Cleanliness of the environment

Lapping Grit Size

Abrasive grit for lapping ranges from 200 parts per inch to as small as 1000 parts per inch. Generally, harder materials allow for the use of finer grit, which produces better surface finishes.

Workpiece material	Grit size	Surface finish (microinches)
Cast iron	240 to 600	16 to 10
Copper	300 to 500	16 to 12
Lead	200 to 400	20 to 15
Mild steel	300 to 500	16 to 12
Stainless steel	400 to 600	14 to 10
Hard steels	600 to 1000	10 to 2

Process Conditions

When lapping between plates, the lapping pressure is around 1 psi to 3 psi for softer materials and 3 psi to 10 psi for harder materials. Lapping speeds range from 300 sfpm to 800 sfpm.

Lapping between plates

Conditions	Hard metals	Soft metals
Lapping pressure	3 to 10 psi	1 to 3 psi

Workpiece material	Lapping speed (sfpm)
Aluminum	500 to 800
Brass	500 to 800
Cast iron	300 to 800
Mild steel	400 to 800
Stainless steel	300 to 800

The type of fluid used for lapping depends on the type of lap and the workpiece material. When the workpiece is aluminum, brass, or copper, a water-soluble compound is used. When the workpiece is cast iron, hardened steel, or tool steel, a weak oil- or grease-based fluid is often used.

Work material	Lapping fluid
Aluminum	Water-soluble compounds
Brass	Water-soluble compounds
Cast iron	Oil- or grease-base fluids
Stainless steel	Oil- or grease-base fluids
Mild steel	Oil- or grease-base fluids

Power Requirements

The power requirements for lapping are small but depend on the type of workpiece material, the pressure applied, and the lapping speed. Power requirements range from 0.3 hp for aluminum to around 0.9 hp for stainless steel, when lapping workpieces with a surface area of 25 in.^{2}.

Workpieces with a surface area of 25 in.^{2}

Workpiece material	Pressure (lbs)	Lapping (sfpm)	Power required (hp)
Aluminum	2 to 4	500 to 800	0.3
Brass	3 to 5	500 to 800	0.5
Cast iron	4 to 7	300 to 800	0.8
Mild steel	4 to 8	400 to 800	0.8
Stainless steel	5 to 9	300 to 800	0.9

Cost Elements

Cost elements include the following:

* Lapping time/piece
* Tool changing time
* Setup time
* Direct labor rate
* Overhead rate
* Amortization of equipment and tooling

Time Calculation

To calculate the amount of material to be removed, one of two formulas is used, depending on workpiece shape. The material to be removed is then divided by the removal rate to obtain lapping time.

Removal rate (in.^{2}/min) = R
Amount to be removed (in.^{3}) = Q
Thickness to be removed (in.) = d
Workpiece length (in.) = L
Workpiece diameter (in.) = D
Workpiece width (in.) = W

$$Q \text{ (rounds)} = \pi \times D \times L \times d$$

$$Q \text{ (flats)} = L \times W \times d$$

$$\text{Lapping time} = \frac{Q}{R}$$

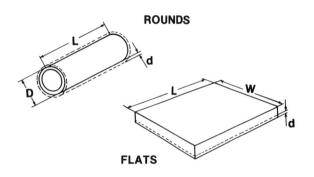

ROUNDS

FLATS

Safety Factors

The following risks should be taken into consideration:

* Personal
  - Skin irritation
  - Rotating machinery

Lathe Boring

Lathe boring is a cutting operation that uses a single-point cutting tool to produce conical and cylindrical surfaces by enlarging an existing opening in a workpiece. The cutting tool moves parallel to the axis of rotation. To produce a taper, the cutting tool moves at an angle to the axis of rotation.

Process Characteristics

* Uses a single-point cutting tool
* Enlarges, straightens, and improves position of existing holes
* Produces accurate internal conical and cylindrical surfaces
* Tool motion is usually parallel to the axis of rotation for cylindrical surfaces
* Produces internal helical feed marks

Process Schematic

This process usually requires that the workpiece be held in a chuck and rotated. As the workpiece is rotated, a boring bar with an insert attached to the tip of the bar is fed into an existing hole. When the cutting tool engages the workpiece, a chip is formed. Depending on the type of tool used, the material, and the feed rate, the chip may be continuous or segmented. The surface produced is called a *bore*.

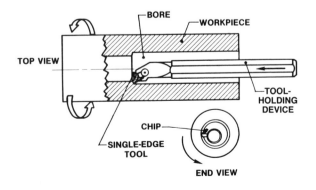

Workpiece Geometry

The geometry produced by lathe boring is usually of two types: straight holes and tapered holes. Several diameters can also be added to each shape hole if required. To produce a taper, the tool may be fed at an angle to the axis of rotation or both feed and axial motions may be concurrent. Straight holes and counterbores are produced by moving the tool parallel to the axis of workpiece rotation.

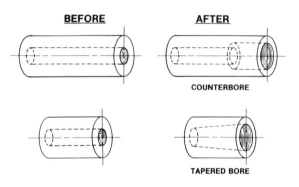

Setup and Equipment

Boring on a lathe requires no special equipment, except the boring bar. All other equipment is standard. The boring bar shown in these illustrations is an insert-type boring bar. The holder can easily and quickly be changed if another type is desired. The workpiece is usually held in a chuck on one end. A tailstock is not normally used for boring but may be used to support a drill when creating an initial hole.

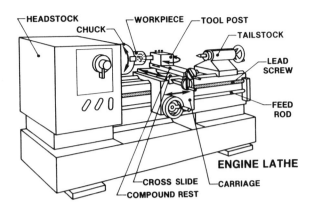

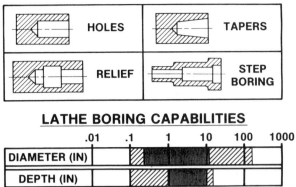

LATHE BORING CAPABILITIES

	.01	.1	1	10	100	1000
DIAMETER (IN)						
DEPTH (IN)						

■ TYPICAL RANGE ▨ FEASIBLE RANGE

Typical Tools and Geometry Produced

Several types of inserts can be used, such as high speed steel, brazed carbide, carbide, or ceramics.

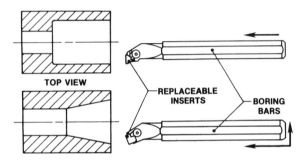

Geometrical Possibilities

Most often, straight or tapered holes are machined. Less frequent operations would include relief and step boring. Typically, diameters range from 3/8 in. to 14 in. and depths range from 1 in. to 10 in. However, it is possible to bore holes that range between 0.125 in. and 120 in. in diameter and between 0.125 in. and 24 in. in depth.

Tolerances and Surface Finish

For most lathe boring applications, tolerances are held within ±0.002 in. for deep holes. For precision applications, tolerances can be held within ±0.0005 in. only for shallow holes. Surface finish may range from 8 to 250 microinches, with a typical range between 32 and 125 microinches.

TOLERANCES		
LATHE BORING	TYPICAL	FEASIBLE
DIAMETER	±.002	±.0005
DEPTH	±.002	±.0005

SURFACE FINISH									
PROCESS	MICROINCHES (A.A.)								
LATHE BORING	2	4	8	16	32	63	125	250	500

■ TYPICAL RANGE ▨ FEASIBLE RANGE

Tool Style

Lathe boring tools are available in a wide range of sizes and shapes depending upon their application. Shown are three common styles used for lathe boring. Adjustable insert boring bars utilize different types of inserts, such as diamond, round, square, and triangular. Offset heads are often used for formed relief boring. Multiple cutting edges can be utilized for rough, finish, chase, and step boring.

DESCRIPTION	STYLE	APPLICATION
ADJUSTABLE INSERTS		STRAIGHT AND TAPERED BORING
OFFSET HEAD		RELIEF BORING
MULTIPLE HEAD		STEP BORING

Workholding Methods

The three most commonly used workholding devices are the three-jaw chuck, the four-jaw chuck, and the face plate. The three-jaw chuck is used to hold round workpieces because the work is automatically centered. The four-jaw chuck is used to hold irregular shapes because of its independent action on each jaw. The face plate is also used for irregular shapes that need to be through-bored.

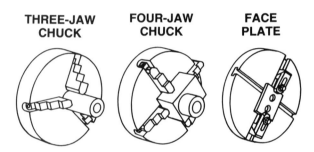

THREE-JAW CHUCK	FOUR-JAW CHUCK	FACE PLATE

Effects on Work Material Properties

Residual surface stresses may create microcracking. Dull tools may cause workhardening of some materials, and high temperatures may cause tempering of hardened materials. Lathe boring has little effect on either the physical or chemical properties.

Work material properties	Effects of lathe boring
Mechanical	* Creates residual surface stresses * May create microcracks * May cause workhardening * May cause tempering in hardened materials
Physical Chemical	* Little effect * Little effect

Typical Workpiece Material

Aluminum, brass, and plastics have good to excellent ratings, whereas gray cast iron and mild steel have good machinability characteristics. Stainless steel has a poor to fair rating because of its toughness and its tendency to work harden when machined.

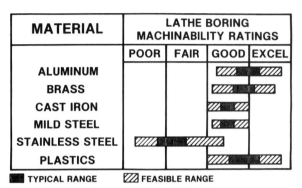

MATERIAL	LATHE BORING MACHINABILITY RATINGS			
	POOR	FAIR	GOOD	EXCEL
ALUMINUM			▨▨	▨
BRASS			▨	▨
CAST IRON			▨	
MILD STEEL			▨	
STAINLESS STEEL	▨	▨		
PLASTICS			▨	▨

■ TYPICAL RANGE   ▨ FEASIBLE RANGE

Tool Materials

High speed steel tools have been used for some time but are being replaced by carbide, ceramic, and diamond tooling. Carbide inserts are easily replaceable and are long lasting. Ceramic tools can resist high temperatures, so high speed machining is possible. Carbides and ceramics often fracture if cuts are interrupted. Diamond tools produce superior surface finish and are used for nonferrous and nonmetallic materials.

Tool materials	Applications
High speed steel	* Special tool shapes * Low production * Best for interrupted cuts
Carbides (inserts)	* Most commonly used tool * High production
Ceramics (inserts)	* High speed machining * Avoid interrupted cuts * Avoid positive rake angles
Diamonds (inserts)	* Abrasive machining * High surface qualities, fine tolerances * Nonferrous or nonmetallic materials

Factors Affecting Process Results

Tolerance and surface finish depend upon:

* Tool geometry
* Cutting speed and feed rate
* Rigidity of tool, workpiece, and machine
* Alignment of machine components and fixtures
* Cutting fluid
* Chip removal
* Environmental conditions, such as humidity, temperature, vibration, and cleanliness

Tool Geometry

Boring tool geometry is designated by angles on the tool. Generally, the back and side rake angles are higher for soft materials. Negative rake angles are often used on hard materials. Relief angles must be increased for small diameter holes to prevent friction.

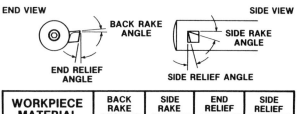

WORKPIECE MATERIAL	BACK RAKE ANGLE (°)	SIDE RAKE ANGLE (°)	END RELIEF ANGLE (°)	SIDE RELIEF ANGLE (°)
ALUMINUM	0 to 15	5 to 15	8 to 13	5 to 8
BRASS	0 to 10	5 to 20	8 to 13	5 to 8
CAST IRON	-5 to 5	-5 to 5	5 to 10	5 to 8
MILD STEEL	-10 to 0	0 to 15	5 to 10	2 to 3
STAINLESS STEEL	3 to 10	0 to 15	5 to 10	2 to 3
PLASTICS	-5 to 5	0 to 15	20 to 30	15 to 20

Process Conditions

Shown are the suggested ranges for depth of cut, feed rate, and cutting speed using cutting fluid. Generally, cutting speed is high for soft materials, such as aluminum and brass, and is low for hard materials, such as cast iron and stainless steel. Depth of cut, feed rate, and cutting speed will vary depending on the type of cut and on the geometry of the workpiece.

Carbide tool with cutting fluid

Conditions	Rough	Finish
Depth of cut (in.)	0.094 to 0.187	0.015 to 0.094
Feed rate (ipr)	0.015 to 0.030	0.005 to 0.015

Workpiece material	Cutting speed (sfpm)	
	Rough	Finish
Aluminum	300 to 450	450 to 700
Brass	500 to 600	600 to 700
Cast iron	250 to 350	350 to 450
Mild steel	400 to 550	550 to 700
Stainless steel	250 to 300	300 to 400
Plastics	250 to 400	400 to 650

Lubrication and Cooling

The chart shows the typical cutting fluids used for a variety of work materials. Cutting fluids cool the tool, which reduces tool wear and makes higher cutting speeds and feed rates possible. Secondary functions include lubrication, flushing chips away, rust prevention, and controlling edge buildup on the tool.

Work material	Cutting fluid	Application
Aluminum	None, water-soluble oil, synthetic oil, kerosene	Flood, spray
Brass	None, water-soluble oil, synthetic oil	Flood
Cast iron	None	–
Steel, all types	None, water-soluble oil, synthetic oil, sulfurized oil	Flood
Plastics	None, emulsifiable oil, synthetic oil	Flood, spray

Power Requirements

Shown is the formula used to calculate the machine horsepower, using unit power and material removal rate. For example, if brass is bored with an HB hardness of 145 to 240 and 10 in.^{3} of material is removed per minute, then a 10 hp engine lathe is needed (1.0 unit power $\times$ 10 in.^{3}/min = 10 hp). Unit power is based on HSS and carbide tools, feed of 0.005 ipr to 0.020 ipr, and 80% efficiency.

Machine hp = unit power × removal rate (in.^{3}/min)

Material	Hardness (HB)	Unit power*
Aluminum	30 to 150	0.3
Brass	50 to 145	0.6
	145 to 240	1.0
Cast iron	110 to 190	0.7
	190 to 320	1.4
Mild steel	85 to 200	1.1
	330 to 370	1.4
	485 to 560	2.0
Stainless steel	135 to 275	1.3
	275 to 430	1.4
Plastics	N/A	0.05 est.

*Unit power based on: • HSS and carbide tools • feed of 0.005 ipr to 0.020 ipr • 80% efficiency.

Cost Elements

* Setup time
* Load/unload time
* Idle time
* Cutting time
* Tool changing time
* Tool costs
* Direct labor rate
* Overhead rate
* Amortization of equipment and tooling

Time Calculations

Speed, feed rate, retract rate, and length of cut are major variables for economic machining.

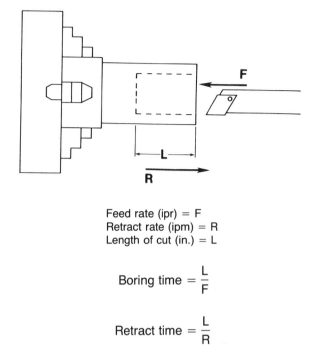

Feed rate (ipr) = F
Retract rate (ipm) = R
Length of cut (in.) = L

$$\text{Boring time} = \frac{L}{F}$$

$$\text{Retract time} = \frac{L}{R}$$

Productivity Tip

Choice of equipment *and* tooling depends on certain factors, such as size and configuration of the workpiece, equipment capacity (speed, feed, and horsepower), production quantity, dimensional accuracy, and surface finish.

Where possible, it is best to avoid boring holes that have a flat bottom because of the possibility of damaging the end of the boring bar. Relief not only protects the boring bar but also provides space for trapped chips. Also, interrupted cuts are difficult to perform with accuracy, so where possible, cuts should be finished.

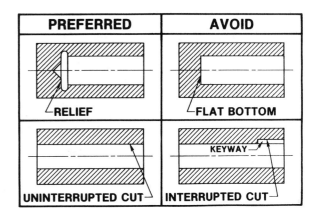

Safety Factors

The following risks should be taken into consideration:

* Personal
  - Rotating workpiece
  - Hot and sharp chips
  - Eye and skin irritation from cutting fluids

Nibbling

Nibbling is a shearing process that utilizes a series of overlapping cuts to make complex shapes from sheet metal.

Process Characteristics

* Sheet metal thickness usually does not exceed 0.25 in.
* Utilizes a press with an oblong or rectangular punch and die or shear blade
* Machines usually stroke 300 to 900 times per minute
* Produces complex shapes from sheet metal material

Process Schematic

In one form of nibbling, an oblong punch and die is used. The workpiece is fed into the reciprocating punch, which produces a narrow kerf along the periphery of the desired part. Then in a nibbling machine, shear blades cut metal with a scissor-type action. Nibbling can produce virtually any geometry in sheet metal parts.

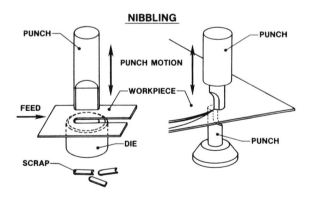

Workpiece Geometry

In this process both curved and straight lines can be produced on either the part interior or exterior. To produce interior surfaces, a hole must be punched or drilled beforehand. Most parts need filing or grinding after the nibbling process to make them smooth and accurate.

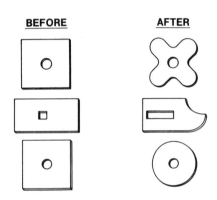

Setup and Equipment

Nibbling machines are usually simple and versatile, so that any desired shape can be nibbled from sheet metal material. Nibbling machines usually stroke 300 to 900 times per minute.

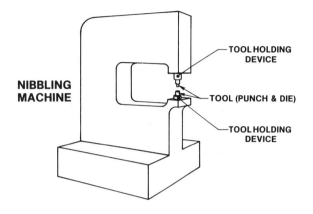

Typical Tools and Geometry Produced

A workpiece parted by the use of an oblong or rectangular die set has a narrow kerf and produces long and narrow pieces of scrap. This type of die set produces minimal waste and allows for longer travel between strokes than with round punches. Shear blades produce virtually no scrap when the sheet metal is sheared.

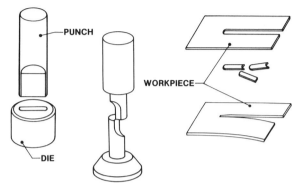

Notching

Notching is a shearing operation by which metal scrap is removed from the outside edge of a workpiece by multiple shear blades set at right angles to each other. Notching can be used to provide relief from wrinkling before drawing or forging. It is a manually operated, low production process.

Process Characteristics

* Removes metal from the edges of the workpiece
* Can produce different angled notches by adjusting the position of the workpiece
* Directly produces re-entrant cuts not possible by shearing
* Can facilitate bending or drawing to achieve final geometry

Process Schematic

Notching is used to remove metal from the perimeter of a workpiece. The knives force the workpiece down past the table, shearing the metal at all points of contact.

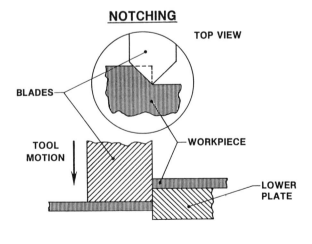

Workpiece Geometry

Shown are typical workpieces before and after notching. Notches in a part can facilitate bending or drawing to achieve final workpiece geometry.

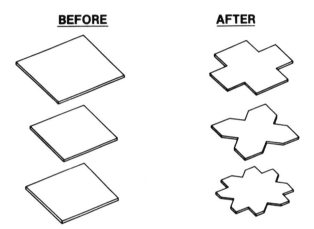

Setup and Equipment

The notching machine normally consists of a work locating fence, table, and hydraulic ram. The blades are bolted to the ram at right angles to each other. The workpiece is pushed against the work locating fence, and then the hydraulic ram closes over the workpiece, shearing it at the point of contact.

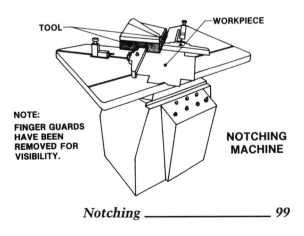

Typical Tools and Geometry Produced

The tool used for notching has two blades, usually mounted at right angles to each other. The blades are made from hardened tool steel. Various cuts can be made by altering the position of the work locating fences.

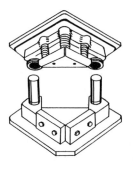

NOTCHING DIE

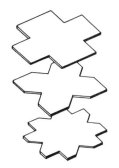

NOTCHED SHEET METAL PARTS

Parting/Grooving

Parting is a machining operation that uses a single-point cutoff tool to sever a complete section of a workpiece from the raw stock. Grooving is similar to parting except that grooves are cut to a specified depth in one pass by a form tool. Grooving does not sever a complete section from the workpiece. In both parting and grooving, a tool moves radially with respect to a rotating workpiece.

Process Characteristics

* Advances a single-point cutting tool radially into a rotating workpiece
* Severs the workpiece from the stock
* Tools must be the same shape as the desired groove form
* Removes material in the form of chips

Process Schematic

During the parting operation, the workpiece is rotated, and the cutting tool is advanced radially into the workpiece until complete separation occurs. During the grooving operation, the form tool is advanced radially into the rotating workpiece until the desired groove depth is obtained. Grooving can be achieved on internal as well as external surfaces.

Workpiece Geometry

Shown here are typical examples of workpiece geometry before and after parting or grooving operations.

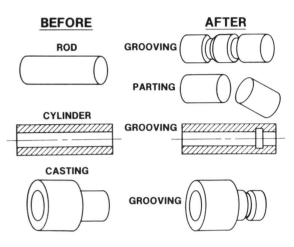

Setup and Equipment

Parting and grooving operations are performed on engine lathes, screw machines, turret lathes, etc. The workpiece is held and rotated by a chuck attached to the headstock spindle. The parting or grooving tool is held by a tool post that is attached to the carriage.

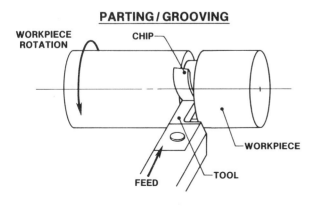

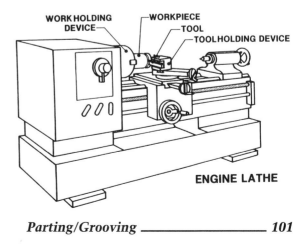

Typical Tools and Geometry Produced

Shown are some typical tools. The shape of a resulting surface is determined by the shape and size of the groove cutting tool. Successive pieces of turned, faced, or bored work are frequently separated or cut away by a cutoff tool. Usually, a special toolholder is needed to hold the high speed steel or carbide tool firmly.

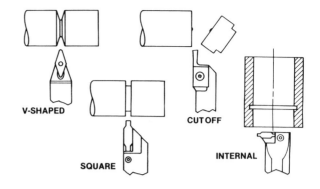

V-SHAPED

CUTOFF

SQUARE

INTERNAL

Perforating

Perforating is a punching process in which a desired pattern of holes is cut into the workpiece by means of multiple punches and dies.

Process Characteristics

* Utilizes multiple punches and dies
* Hole quality is controlled by the punch and die clearance
* Punches shear the waste from the workpiece as they enter the dies
* Workpieces have burnished and sheared areas on the sidewalls of the holes
* Produces burrs on the bottom surface
* Produces a pattern of punched holes in similar or continuous rows

Process Schematic

Shown are two of the punches and dies in a die set. There may be many in a given die set. It is important that proper clearance be maintained between the punch and die, so that there is a clean break on the sheet metal. As the punch enters the die, a blank, or waste, is pushed through the bottom of the die, leaving a perforated workpiece.

PERFORATING

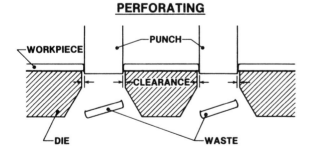

Workpiece Geometry

Shown are some typical sheet metal parts before and after perforating. Patterns similar to these might be punched for several different reasons. Perforating allows light, gas, or liquid to pass, has decorative purposes, or enhances sound deadening.

BEFORE   **AFTER**

SHEET

PERFORATED SHEETS:

Setup and Equipment

Perforating can be performed on many types of presses. Shown is a hydraulic press with a multiple punch and die set. The design creates the desired pattern and location of holes of the die. The workpiece may be advanced to produce another series of holes close to the first set.

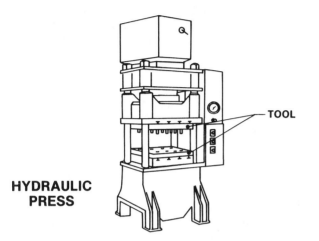

HYDRAULIC PRESS

TOOL

Typical Tools and Geometry Produced

Shown are three different types of punches: square, round, and oblong. Often a pattern of several similar punches is used to obtain the desired pattern of holes.

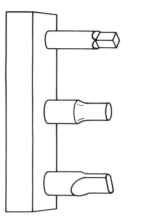 

DIE BLOCK WITH PUNCHES   **METAL STRIP**

Precision Boring

Precision boring is a single-point cutting operation used to produce a precise internal cylindrical surface by enlarging an existing opening in a workpiece. The workpiece moves *parallel* to the axis of rotation of the cutting tool.

Process Characteristics

* Uses a single-point cutting tool
* Enlarges and straightens existing holes
* Produces accurate internal cylindrical surfaces
* Utilizes carbide, ceramic, and diamond tools

Process Schematic

In precision boring, an existing hole is enlarged and straightened by advancing the workpiece into a rotating single-point cutter. The material is removed in the form of small chips. The cutter may also be fed into the workpiece. Tolerances can be held within 0.0001 in.

PRECISION BORING

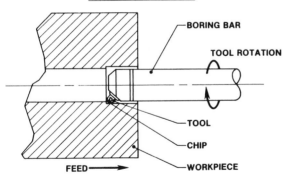

Workpiece Geometry

Precision boring is typically used in the production of parts that require accurate finishing on internal bearing surfaces. A wide variety of machines are available in single- and multiple-spindle types. The wide variety of machines make possible other precision operations such as turning, facing, grooving, chamfering, and contouring.

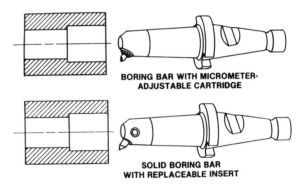

BORING BAR WITH MICROMETER-ADJUSTABLE CARTRIDGE

SOLID BORING BAR WITH REPLACEABLE INSERT

Setup and Equipment

Precision boring machines are available in a wide range of sizes and types. The single-spindle machine shown here has a rotating toolholding device (boring bar) with a single-point cutting tool. The boring operation is performed by advancing the workpiece horizontally into the rotating tool.

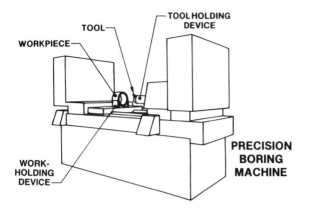

WORKPIECE — TOOL — TOOL HOLDING DEVICE

WORK-HOLDING DEVICE

PRECISION BORING MACHINE

Typical Tools and Geometry Produced

One of the most commonly used toolholders has an adjustable cartridge for holding the insert. Another common toolholder is the solid boring bar type with a replaceable insert. Carbide tools are used most often, but ceramic and diamond tools are also used.

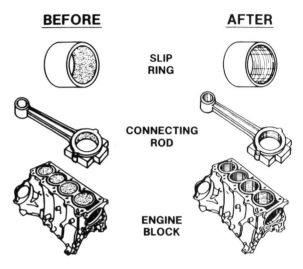

BEFORE   **AFTER**

SLIP RING

CONNECTING ROD

ENGINE BLOCK

Punching

Punching is a shearing process in which a scrap slug is separated from the workpiece when the punch enters the die. The sidewall of the resulting hole displays a burnished area, roll-over, and die break.

Process Characteristics

* Is the most economical method of making holes in sheet or strip metal for medium to high production
* Can produce various shaped holes
* Punches and dies are normally made of conventional tool steel or carbides
* Produces a burnished area, roll-over, and die break on the sidewall of the resulting hole

Process Schematic

Punching involves forcing a hardened steel punch through a workpiece to pierce a hole matching the diameter of the punch. The scrap material normally drops through a die and bolster plate into a container.

PUNCHING

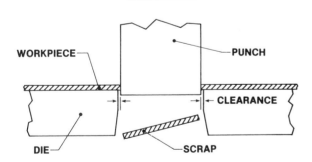

Workpiece Geometry

The material being punched is usually in sheets but may also be in rolls. Holes of various sizes and shapes can be made quickly.

BEFORE   **AFTER**

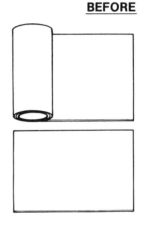
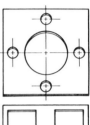

Setup and Equipment

Major components of this mechanical press are the frame, motor, ram, die posts, bolster, and bed. The punch is mounted into the ram, and the die is mounted to the bolster plate. The scrap material drops through as the workpiece is advanced for the next hole.

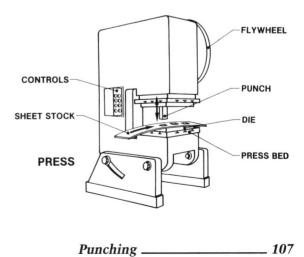

Typical Tools and Geometry Produced

Three typical punches are shown. The clearance between the punch and die is determined by the thickness and type of material and by the specified tolerance. Multiple punches are frequently mounted together to produce a complete part in one press stroke.

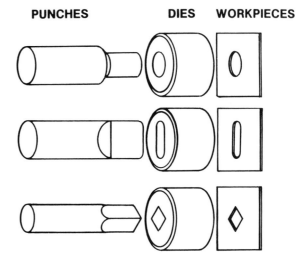

Reaming

Reaming is a cutting process in which an existing hole is enlarged and accurately sized by means of a multifluted cutting tool. As the reamer and workpiece are advanced toward each other, chips are produced by shaving thin sections from the existing hole. A reamer is an accurate tool and is designed for removing only a small amount of material.

Process Characteristics

* Uses a multifluted tool
* Must have an existing hole in the workpiece
* May utilize either tool or workpiece rotation
* Produces close-tolerance holes
* Removes very small amounts of material

Process Schematic

Reaming involves relative axial and rotational motions between the reamer and the workpiece. The reamer can rotate and advance into the workpiece or the workpiece can rotate as the reamer advances into it. Small chips are removed by flowing along flutes in the reamer. The amount of material removed by the reamer depends on the material being cut and the hole size.

REAMING

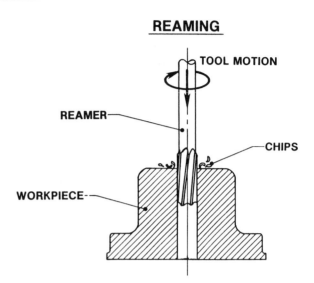

Workpiece Geometry

Shown are the typical workpieces before and after reaming. Reaming may be used in drilled, bored, cast, pierced, or molded holes. Most holes reamed are between 1/8 in. and 1-1/4 in. in diameter. However, specially designed reamers are obtainable for smaller or larger holes.

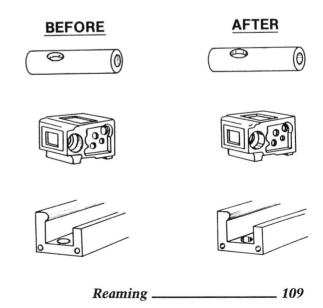

Setup and Equipment

Reaming is often done on a drill press, but it may also be done on lathes, machining centers, and the like. The reamer holder "floats," allowing the reamer to follow the existing hole. The workpiece is commonly held firmly in place by some type of vise, chuck, or fixture as the reamer is advanced into the workpiece.

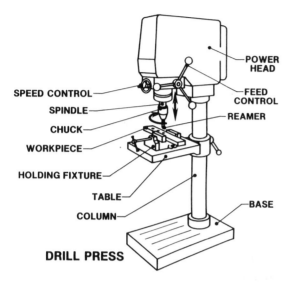

DRILL PRESS

Typical Tools and Geometry Produced

There are a wide variety of reamers in use today, including hand reamers, chucking reamers, shell reamers, taper reamers, expansion reamers, adjustable reamers, and special purpose reamers. Selection of the reamer is usually governed by the following considerations:

1. Composition and hardness of the workpiece
2. Hole diameter
3. Hole configuration
4. Hole length
5. Amount of stock removed
6. Required accuracy and finish
7. Required production
8. Cost

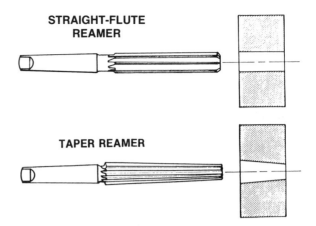

STRAIGHT-FLUTE REAMER

TAPER REAMER

Geometrical Possibilities

Shown are some typical workpiece geometries that can be reamed. Most holes reamed are between 0.125 in. and 1.250 in. in diameter. However, specially designed reamers are available for holes as small as 0.005 in. and as large as 6 in. in diameter. The depth of hole successfully reamed depends on reamer diameter, method of holding and driving the reamer, and required dimensional accuracy. Typically, holes less than 6 in. in depth are reamed. However, holes as deep as 17 in. may be reamed.

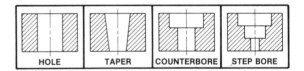

| HOLE | TAPER | COUNTERBORE | STEP BORE |

REAMING CAPABILITIES

TYPICAL RANGE   FEASIBLE RANGE

Workholding Methods

General purpose vises and jigs, such as those shown, may be used to hold the workpiece during reaming. Drill jigs and fixtures are frequently designed to guide a tool for a specific workpiece and operation. Some factors to consider when designing jigs and fixtures are rigidity, strength, location and clamping of the workpiece, chip control, and ease of operation.

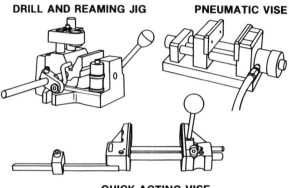

DRILL AND REAMING JIG   PNEUMATIC VISE

QUICK-ACTING VISE

Tool Style

Reamers are available in a wide range of sizes and shapes. A few of these are the straight-flute chucking reamer (for general purpose reaming), the tapered reamer (for tapered holes), the helical-flute shell reamer (for lightweight finishing operations), the multidiameter reamer (for multidiameter holes), and the insert-blade adjustable reamer (for a large range of hole sizes). Reamers are often provided with spiral flutes to reduce the tendency to chatter and to control chip removal.

DESCRIPTION	STYLE	APPLICATION
STRAIGHT-FLUTE REAMER		GENERAL PURPOSE, SUITED FOR REAMING MOST METALS
TAPER REAMER		TAPERED HOLES
HELICAL-FLUTE SHELL REAMER		HARDER MATERIALS LARGE DIAMETER— SHALLOW DEPTH
MULTIDIAMETER REAMER		MULTIDIAMETER HOLES AND HOLE ALIGNMENT
INSERT-BLADE ADJUSTABLE REAMER		ADJUSTABLE OVER LARGE RANGE OF HOLE SIZES

Toolholding Methods

Reamers may be classified as either straight shank or taper shank. Straight-shank reamers are often held in a keyless chuck for quick changing. Taper-shank reamers fit into the Morse taper sleeve of the machine spindle and may be removed by driving a *drift* between the end of the reamer (tang) and the spindle.

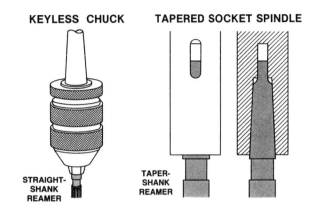

KEYLESS CHUCK   TAPERED SOCKET SPINDLE

STRAIGHT-SHANK REAMER

TAPER-SHANK REAMER

Tolerances and Surface Finish

For most reaming operations, tolerances are held within 0.001 in. to 0.003 in. For precision applications, tolerances can be held to less than 0.001 in. Surface finishes for reaming may range from 16 to 125 microinches, with a typical range between 32 and 63 microinches. Finishes depend on workpiece material and hardness, condition of cutting edges, workpiece feed, and cutting speed.

TOLERANCES	
TYPICAL	.001 TO .003
FEASIBLE	LESS THAN .001

SURFACE FINISH									
PROCESS	MICROINCHES (A.A.)								
	2	4	8	16	32	63	125	250	500
REAMING									

■ TYPICAL RANGE   ▨ FEASIBLE RANGE

Effects on Work Material Properties

Reaming affects the mechanical properties of the workpiece by removing a very thin layer of highly stressed and disturbed material from the surface of a drilled hole, producing a good surface finish. Reaming has no noticeable effect upon the chemical and physical properties of the workpiece.

Work material properties	Effects of reaming
Mechanical	* Removes a very thin layer of highly stressed and disturbed material from the surface of a drilled hole, producing a good surface finish
Physical	* No noticeable effect
Chemical	* No noticeable effect

Tool Geometry

Shown are the two most common parameters that vary on a high speed steel (HSS) reamer. The margin width increases as the diameter of the hole increases, and the margin relief angle decreases as the diameter of the hole increases. Also shown are the chamfer angle and chamfer relief angle, which are 45° and 7° to 12°, respectively.

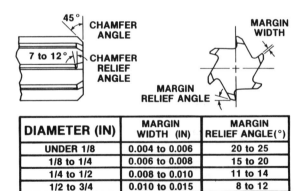

DIAMETER (IN)	MARGIN WIDTH (IN)	MARGIN RELIEF ANGLE(°)
UNDER 1/8	0.004 to 0.006	20 to 25
1/8 to 1/4	0.006 to 0.008	15 to 20
1/4 to 1/2	0.008 to 0.010	11 to 14
1/2 to 3/4	0.010 to 0.015	8 to 12
3/4 to 1	0.012 to 0.017	7 to 10
1 to 1-1/2	0.014 to 0.018	5 to 8
1-1/2 to 2	0.016 to 0.022	5 to 8
OVER 2	0.018 to 0.025	5 to 8

Process Conditions

Shown are the typical cutting speeds and feed rates for reaming with a HSS reamer using cutting fluids. Generally speaking, reaming speeds are 65% to 70% lower, and feeds are two to three times higher than when drilling the corresponding diameter. Material, reamer design, rigidity of setup, and depth of hole all influence reaming speeds and feed rates.

Workpiece material	Cutting speed (sfpm)	Reamer diameter (in.)			
		0.25	0.50	0.75	1.0
		Feed (in./rev)			
Aluminum	150 to 300	0.008	0.012	0.016	0.020
Brass	150 to 180	0.008	0.012	0.016	0.020
Cast iron*	40 to 60	0.007	0.010	0.013	0.016
Mild steel	50 to 60	0.007	0.011	0.015	0.018
Stainless steel	20 to 45	0.005	0.007	0.009	0.012
Plastics	40 to 100	0.005	0.007	0.009	0.012

Typical speeds and feeds

* Cast iron is usually reamed dry.

Typical Workpiece Materials

Aluminum and brass have good to excellent machinability ratings. Cast iron, mild steel, and plastics have good machinability ratings. Stainless steel has a poor to fair rating because of its toughness and its tendency to work harden when machined, especially with the light cuts used in reaming.

MATERIAL	REAMING MACHINABILITY RATINGS			
	POOR	FAIR	GOOD	EXCEL
ALUMINUM			▨██	█▨
BRASS			▨██	█▨
CAST IRON			▨█	
MILD STEEL			▨█▨	
STAINLESS STEEL	▨█	█▨		
PLASTICS			▨██	▨

██ TYPICAL RANGE   ▨▨ FEASIBLE RANGE

Tool Materials

High speed steel is most commonly used because it is relatively inexpensive and works well on most materials. Carbide reamers outlast HSS 10 to 1 when reaming steel and are usually used to ream steel between 370 HB and 475 HB. Reamers should have a hardness of RC65 or higher to obtain maximum surface finish and tool life.

Tool materials	Applications
High speed steel	* Most commonly used * Used on hard or abrasive metals
Carbide	* More expensive than HSS * Will outlast HSS 10 to 1 when reaming steel * Used to ream steel between 370 and 475 HB
High cobalt alloy steel	* Used to ream titanium and stainless steel * More expensive than HSS or carbide

Factors Affecting Process Results

Tolerance and surface finish depend upon the following:

* Tool geometry
* Cutting speed and feed rate
* Rigidity of tool, workpiece, and machine
* Alignment of machine components and fixtures
* Cutting fluid
* Composition and hardness of the workpiece
* Accuracy of the reamer

Lubrication and Cooling

Cutting fluids for reaming include mineral, synthetic, and water-soluble oils. These fluids are usually applied by flooding or spraying. The main function of cutting fluids is to cool the tool, which increases tool life and makes higher cutting feeds and speeds possible. Some secondary benefits are chip removal and lubrication, which contribute to better workpiece surface finish.

Work material	Cutting fluid	Application
Aluminum	Soluble oil, kerosene, synthetic fluid	Flood
Brass	None, soluble oil	Flood
Cast iron	Cold air, none	Air jet
Mild steel	Soluble oil, sulfurized oil	Flood
Stainless steel	Soluble oil, sulfurized oil	Flood
Plastics	None, mineral oil, synthetic oil	Flood, spray

Power Requirements

Shown is the formula to calculate the machine horsepower using power and material removal rate. For example, if mild steel is reamed with an HB hardness of 330 to 370, and 10 in.^3 of material is removed per minute, then a 0.14 hp motor is needed (1.4 unit power $\times$ 0.10 in.^{3}/min. = 0.14 hp). Unit power is based on HSS reamers, feed of 0.002 ipr to 0.008 ipr, and 80% efficiency.

Machine hp = unit power × removal rate (in.^{3}/min)

Material	Hardness (HB)	Unit power
Aluminum	30 to 150	0.16
Brass	50 to 145	0.48
	145 to 240	0.8
Cast iron	110 to 190	1.0
	190 to 320	1.6
Mild steel	85 to 200	1.0
	330 to 370	1.4
	485 to 560	2.1
Stainless	135 to 275	1.1
steel	275 to 430	1.2
Plastics	N/A	0.05 est.

* Unit power based on: • HSS reamers • feed of 0.002 ipr to 0.008 ipr • 80% efficiency.

Time Calculations

Understanding the reaming setup is important in calculating the time needed to position, ream, and retract. The elements involved in reaming are diameter of the tool, approach, hole depth, overtravel, feed rate, cutting speed, retract rate, number of holes, distance to next hole position, positioning rate, number of teeth on cutter, and the feed per tooth, and length of cut.

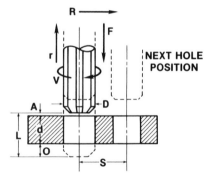

Diameter of tool (in.) = D
Approach (in.) = A
Overtravel (in.) = O
Feed rate (ipr) = F
Cutting speed (sfpm) = V
Retract rate (ipm) = r
Number of holes = H
Distance to next hole position (in.) = S
Number of teeth = N
Feed per tooth (in.) = f
Positioning rate (ipm) = R
Length of cut (in.) = L
Depth of hole (in.) = d

$$\text{Positioning time} = H \times \frac{S}{R}$$

$$\text{Cutting time} = \frac{L}{F}$$

$$\text{Feed rate} = f \times N \times rpm$$

$$\text{Retract time} = \frac{L}{r}$$

$$rpm = \frac{4 \times V}{D} \text{ (approx.)}$$

Cost Elements

Cost elements include the following:

* Setup time
* Load/unload time
* Idle time
* Cutting time
* Direct labor rate
* Overhead rate
* Amortization of equipment and tooling costs

Productivity Tip

Where possible, provision should be made for the reamer to pass through the workpiece to avoid the need to ream a blind hole. When this is unavoidable, the depth of cut should be controlled to prevent bottoming and possible damage to the reamer. (If a reamer hits the bottom of a blind hole, it could break.) It is advisable to avoid making holes that are not perpendicular to the surface, and to avoid operations requiring multidiameter reamers.

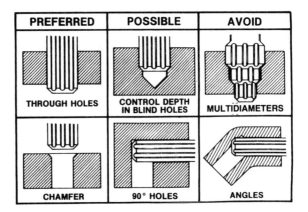

PREFERRED	POSSIBLE	AVOID
THROUGH HOLES	CONTROL DEPTH IN BLIND HOLES	MULTIDIAMETERS
CHAMFER	90° HOLES	ANGLES

Safety Factors

The following risks should be taken into consideration:

* Personal
  - Rotating tool
  - Hot and sharp chips
  - Eye and skin irritation from cutting fluids

Reciprocating Filing

Reciprocating filing is a multipoint cutting process in which a workpiece is fed into a reciprocating file, and chips are removed on each downstroke by the cutting teeth. The cutting teeth are arranged in succession along the file surface. Usually, the workpiece is fed manually into the file teeth.

Process Characteristics

* Uses a multipoint cutting tool blade
* Feeds a reciprocating blade into a stationary workpiece
* Cutting action is in one direction
* Produces a narrow kerf
* Often produces coarse tool marks and burrs on the workpiece

Process Schematic

The metal-removing action in filing is basically the same as in sawing: chips are removed by cutting teeth that are arranged in succession along the same plane on the surface of the file. As the file reciprocates, the workpiece is advanced into the file. The cutting action takes place on each downstroke.

RECIPROCATING FILING

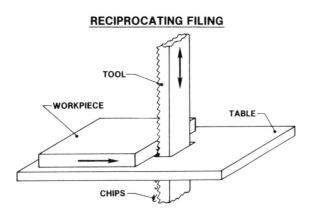

Workpiece Geometry

Shown are some of the operations performed by reciprocating filing. The type of file required depends upon the material used and the desired form. Quite accurate work can be done, but because of the reciprocating action, approximately 50% of the operating time is nonproductive.

BEFORE

AFTER

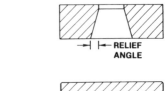

RELIEF ANGLE

BURRS

Setup and Equipment

A reciprocating file machine normally consists of a base, a table, and a drive motor. The file extends upward through the worktable and is held in a support arm. The file rides against a roller guide at its upper end, and cutting occurs on the downstroke, so the cutting force tends to hold the work against the table. The table may be tilted to the desired angle.

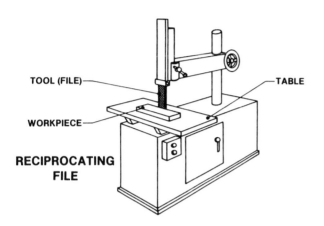

TOOL (FILE)

WORKPIECE

TABLE

RECIPROCATING FILE

Typical Tools and Geometry Produced

There are a wide variety of files in use today including square files, round files, half-round files, flat files, and other special function files.

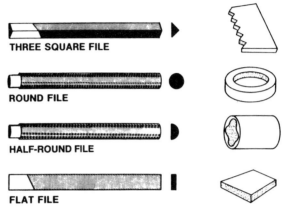

THREE SQUARE FILE

ROUND FILE

HALF-ROUND FILE

FLAT FILE

Reciprocating Sawing

Reciprocating sawing is a multipoint cutting process in which a reciprocating blade is advanced against a stationary workpiece to sever parts.

Process Characteristics

* Uses a multipoint cutting tool blade
* Feeds a reciprocating blade into a stationary workpiece
* Cutting action is in one direction
* Produces a narrow kerf
* Often produces coarse tool marks and burrs on the workpiece
* Has poor dimensional accuracy and repeatability

Process Schematic

The workpiece is held securely in a workholder while the tool reciprocates and slowly feeds into the work. Actual cutting takes place in only one direction. The saw blade is lifted slightly on the return stroke, reducing wear on the saw teeth. Sections of considerable size can be severed from the workpiece with a small amount of material loss in the form of chips.

Workpiece Geometry

Shown are typical workpieces before and after sawing. Reciprocating power hacksaws are used primarily for cutting metal of various sizes, kinds, and shapes to length. Care must be taken in setting up the work, including clamping workpieces in the holding device, for efficient operation. At least two teeth must be in contact with the workpiece at all times, or they may be stripped out.

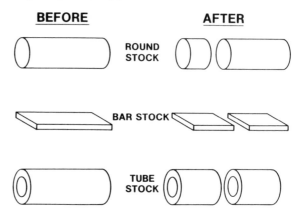

RECIPROCATING SAWING

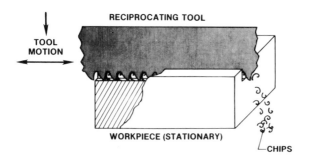

Setup and Equipment

The reciprocating hacksaw normally consists of a frame, a base, a workholder, and a tool. The reciprocating motion is provided by hydraulics or a crankshaft mechanism. Lubrication is recommended for all operations to cool the tool and to wash away small chips that accumulate between the teeth. Principle disadvantages of reciprocating sawing are limited accuracy and cutting being limited to one direction.

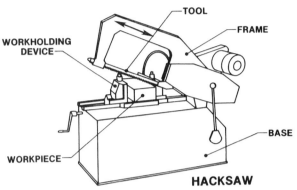

Typical Tools and Geometry Produced

The three most common tooth forms include the *precision tooth form*, which gives accurate cuts with a smooth finish; the *claw tooth form*, which gives additional chip clearance for fast cuts in soft, thick materials; and the *buttress tooth form*, which provides larger chip loads and faster cutting.

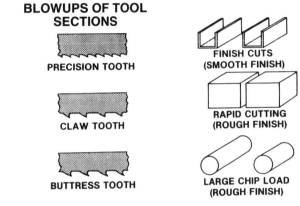

Routing

In routing, specially designed, high speed cutters trim and shape a variety of machinable materials.

Process Characteristics

* Is typically limited to machining soft metals and rigid nonmetals
* Uses small, specially designed cutters
* Cutter is powered by high speed, electrical, or pneumatic motors
* Cutter path is controlled by hand, with the aid of templates and routing fixtures, or by machine control
* Can be used to gouge, contour, sculpt, and shape recessed cavities

Process Schematic

This process schematic illustrates the routing tool as it cuts the workpiece. Note that the guide bushing runs along the template pattern to insure an accurate cutter path. All routing cutters rotate clockwise and are specially designed to prevent the chips from clogging the cutting edge.

ROUTING

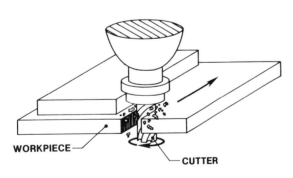

Workpiece Geometry

Routing is typically a shaping process used to produce a finished edge or shape. Materials that are difficult to work with, such as fiberglass, kevlar, and graphite, can be shaped and finished very neatly. Not only can finished edges be produced, but cutaways, holes, and contours also can be shaped using a routing process.

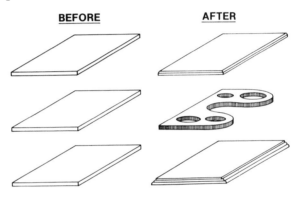

Setup and Equipment

The setup and equipment for routing primarily consist of an air/electric driven router, a cutting tool, and a guide template. The routing machine shown is a radial-arm model. A power lever is manipulated by the operator to control the rpm of the cutter.

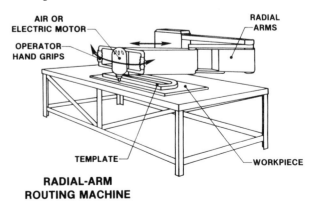

RADIAL-ARM ROUTING MACHINE

Typical Tools and Geometry Produced

Shown are three general classifications of routers. The most commonly used tool is the fluted cutter. It may be single, double, or multifluted. Fluted cutters are used primarily for edging and trimming. Profile cutters are used for shaping and trimming. Helical cutters are used on easily machined materials, such as aluminum, for drilling, shaping, and trimming. Diamond insert cutters and rotary files are used for abrasive materials such as fiberglass or other composites.

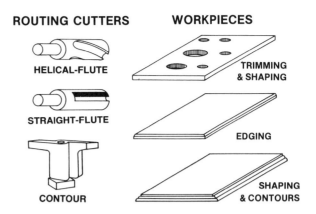

ROUTING CUTTERS   WORKPIECES

HELICAL-FLUTE

STRAIGHT-FLUTE

CONTOUR

TRIMMING & SHAPING

EDGING

SHAPING & CONTOURS

Sandblasting

Sandblasting is a cleaning process in which dry sand particles are propelled through a nozzle by compressed air to remove contaminants or burrs from the surface of a workpiece. Other media that may be used include glass beads and silicon carbide.

Process Characteristics

* Uses sand particles as medium
* Propels the sand particles by compressed air
* Uses a nozzle to direct the sand particles
* Can be adapted for portable use
* Produces finely textured matte surfaces

Process Schematic

Sand is propelled through a nozzle by a pres surized stream of air. The high velocity sand particles abrasively remove scale, rust, and other contaminants. Sandblasting can also be used to remove burrs and produce decorative finishes.

SANDBLASTING

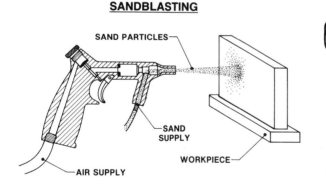

Workpiece Geometry

Because this is a cleaning process, basic workpiece geometry is unaltered. Sandblasting is rarely used for nonmetallic workpieces. Care must be taken when sandblasting ductile materials because the surface may be severely damaged. Sandblasted parts have finely textured, clean matte surfaces that are suitable for painting or other coating procedures.

BEFORE   **AFTER**

 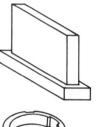

Setup and Equipment

Sandblasting equipment normally consists of a hand-held nozzle that is used to direct the sand particles toward the workpiece. A mixing chamber is used to mix sand with air to transport it to the nozzle where a high velocity air stream propels the sand particles toward the workpiece.

Typical Tools and Geometry Produced

Sandblasting nozzles come in a variety of shapes and sizes. Several kinds of materials are used to make nozzles. Boron carbide is a popular nozzle material because it resists abrasive wear. When parts are so large that they cannot be transported to sandblasting machines, portable units can be used.

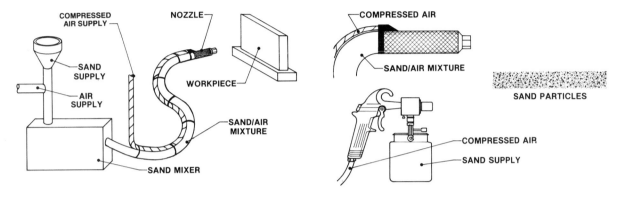

Shaping/Planing

Shaping is a material removal process in which a single-point cutting tool reciprocates across the face of a stationary workpiece to produce a plane or sculpted surface. Planing is a material removal process in which the workpiece reciprocates against a stationary single-point cutting tool producing a plane surface. Shaping/planing operations are rapidly being replaced by milling.

Process characteristics

* Uses single-point cutting tools
* Involves a reciprocating motion between the tool and workpiece
* Produces plane or sculpted surfaces
* Leaves parallel feed marks

Process Schematic

In shaping, the tool is moved into position with the workpiece. The reciprocating, single-point cutting tool moves in a straight line across the workpiece as the workpiece is fed incrementally across the line of motion to produce a flat surface. For shaped pieces, the tool reciprocates across the stationary workpiece. Planing motion is opposite that of shaping.

Workpiece Geometry

Shown are typical workpieces before and after shaping or planing. Planing can be used to produce horizontal, vertical, or inclined flat surfaces on workpieces usually too large for shaping. Shaping is used not only for flat surfaces, but also for external or internal surfaces (either horizontal or inclined). Curved and irregular surfaces can also be produced by using special attachments.

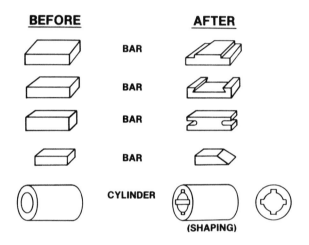

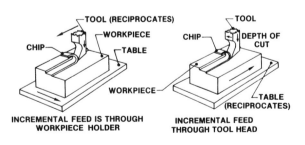

Setup and Equipment

Horizontal shapers, which are common, produce flat, angular, and contoured surfaces. In shaping, the workholding device has a very heavy movable jaw to withstand cutting forces. The size of the planer is determined by the largest workpiece that can be machined on it. Usually, several clamping and supporting devices are used to hold large workpieces on the planer.

Typical Tools and Geometry Produced

Shaping/planing tools are usually carbide-tipped or made of high speed steel. Cutting tools resemble those used in facing and turning, except for slight differences in tool angles. The main advantages of single-point cutting tools over multipoint tools are that they can be made and sharpened more easily. Internal shapes can be made with a special extension tool.

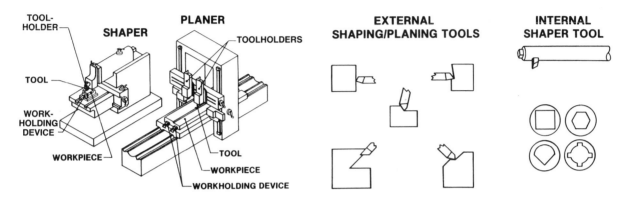

Shearing

Shearing is a process for straight-line cutting of flat stock. It is performed by forcing an upper blade and a lower blade past each other with a desired offset. Usually, one blade is stationary.

Process Characteristics

* Is used for straight-line cuts on flat sheet stock
* Takes place between the upper and lower shear blades
* Produces burred and slightly deformed edges
* The blades can be mounted at an angle to reduce the necessary force required so that the relative length of material being cut at any given instant is small

Process Schematic

The workpiece is positioned between an upper vertical blade and a lower stationary blade. The upper blade forces the workpiece against the lower blade, shearing the metal at the point of contact. Shearing actually causes the workpiece to fracture, creating a slightly deformed and burred edge.

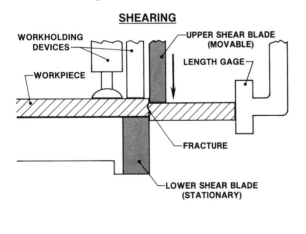

Workpiece Geometry

The first example shows how metal can be cut to a required length. The other three examples show different angles that can be sheared (the dashed lines indicate material removed).

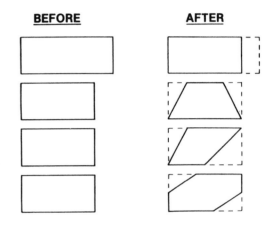

Setup and Equipment

A shear basically consists of a shear table, shear blades (upper and lower), workpiece holding devices, and gaging devices. The gage insures consistent lengths. The hold-down devices hold the workpiece in position, in addition to preventing workpiece edge buckling during cutting.

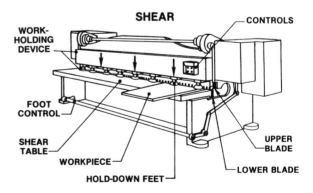

Typical Tools and Geometry Produced

The upper blade is mounted at an angle or slope with respect to the horizontal lower blade. Only basic straight-line cutting of flat sheet stock can be performed with a shear. Various geometric shapes can be produced, however.

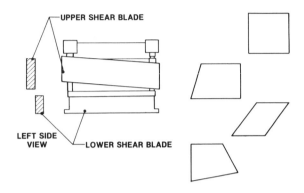

Geometrical Possibilities

Shapes are formed primarily from sheet, plate, bar, and angle stock. A wide variety of geometrical possibilities may be produced from sheet and plate as opposed to bar and angle materials, which are cut only to length. All sheared parts have straight-line geometries and are often used as blanks for a secondary process. Sheet material from 0.1 in. to over 1 in. thick may be sheared in widths up to 1000 in.

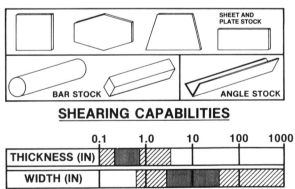

Tolerances and Surface Finish

For most shearing operations, tolerances are typically held within ±0.01 in. for sheet materials and ±0.06 in. for bar stock and angles. Tolerances may be feasibly obtained within ±0.005 in. for sheet materials and ±0.030 in. for bar stock and angles. Surface finishes for sheet, bar, and angle materials may range from 125 to 2000 microinches, with a typical range between 250 and 1000 microinches. If better surfaces are required, a secondary operation is necessary.

TOLERANCES		
SHEARING	TYPICAL	FEASIBLE
SHEET	±.01	±.005
BAR & ANGLE	±.06	±.03

SURFACE FINISH	
PROCESS	MICROINCHES (A.A.)
	8 16 32 63 125 250 500 1000 2000
SHEARING	

▉ TYPICAL RANGE   ▨ FEASIBLE RANGE

Tool Style

There are three types of systems: those for sheet and plate material, those for angles, and those for bar stock. Squaring and bow-tie shear are for sheet and plate materials. Angle and bar shear are for angle and bar stock materials. In general, most shearing systems have both a stationary and movable blade that shears the workpiece.

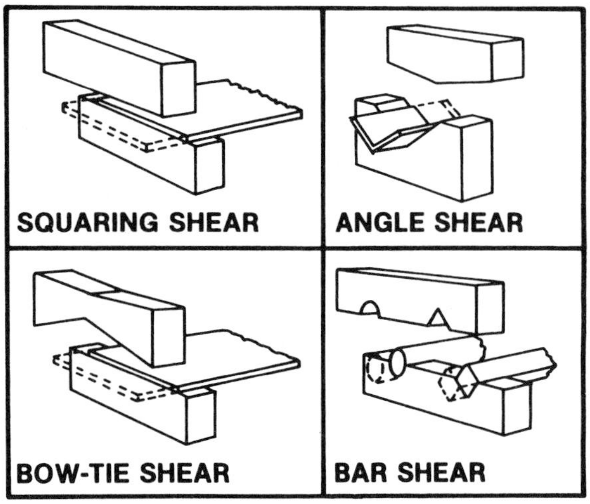
SQUARING SHEAR
ANGLE SHEAR
BOW-TIE SHEAR
BAR SHEAR

Workholding Methods

The workholding methods vary according to the material being sheared. For sheet and plate materials, the workpiece is held between the shear table and the hold-down feet. For bars or rods, special dies like those shown are used. The shearing blade is also designed to match the geometry.

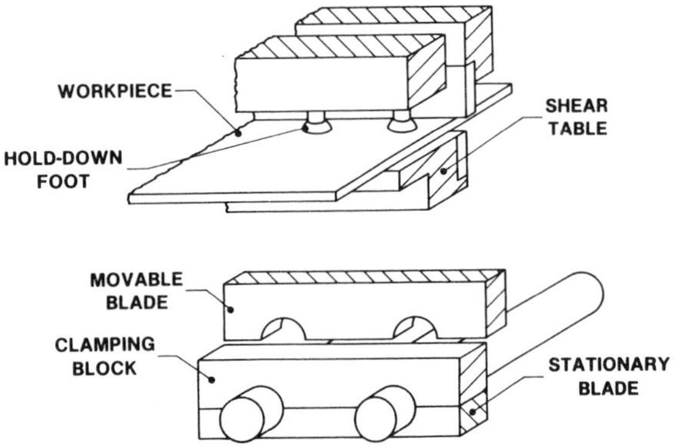
WORKPIECE
SHEAR TABLE
HOLD-DOWN FOOT
MOVABLE BLADE
CLAMPING BLOCK
STATIONARY BLADE

Effects on Work Material Properties

The effects of shearing include workhardening on the edge of the sheared part, along with residual cracks. Also, too much clearance may cause roll-over or heavy burring.

Work material properties	Effects of shearing
Mechanical	* Workhardening on edge of sheared part * Sheared edges may have residual cracks * Excess clearance may cause roll-over or burring
Physical	* Little effect
Chemical	* Little effect

Typical Workpiece Materials

Aluminum has an excellent rating, whereas brass, bronze, and mild steel are rated from good to excellent. Stainless steel has only a fair to good rating because of its ductility and tendency to work harden.

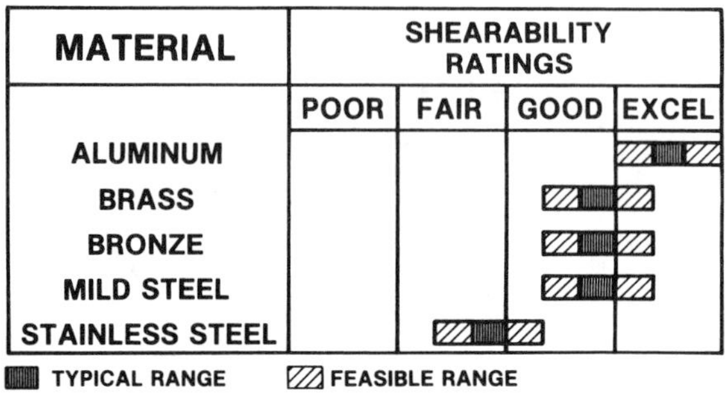
MATERIAL
SHEARABILITY RATINGS
POOR
FAIR
GOOD
EXCEL
ALUMINUM
BRASS
BRONZE
MILD STEEL
STAINLESS STEEL
TYPICAL RANGE
FEASIBLE RANGE

Tool Materials

Low alloy steel, high-carbon steel, high-chromium steel, and shock-resisting steel are common tool steels used in shearing. Low alloy is best for low production of materials up to 1/4 in. thick. High-carbon, high-chromium steel is best for high production of materials up to 1/4 in. thick. Shock-resisting tool steels are best for materials greater than or equal to 1/4 in. in thickness.

Tool materials	Applications
Low alloy steel	* Low production * Materials up to 1/4 in. thick
High-carbon, high-chromium steel	* High production * Materials up to 1/4 in. thick
Shock-resisting steel	* Materials greater than or equal to 1/4 in. thick

Process Capabilities

Major factors that affect tolerances and surface finishes are the following:

* Clearance between shear blades
* Rake angle of upper shear blade
* Material thickness
* Material width
* Material hardness
* Hold-down force

Tool Geometry

The shear angle of the upper blade is the most important factor of the shear and is denoted as the amount of rise in inches per feet. It should be kept as low as possible to reduce the amount of distortion in narrow workpieces.

•AS RAKE AND SHEAR ANGLES INCREASE, CUTTING FORCE DECREASES.

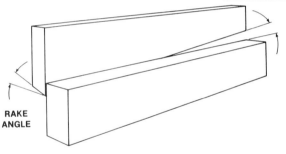

Shearing Capabilities

Shown are the shearing capabilities for equipment used for sheet, angle, and bar materials. It is feasible to shear sheet materials up to 25 ft wide and 2 in. thick. Angle materials may be sheared to 8 × 8 × 1.5 in., but 6 × 6 × 3/4 in. is more typical. It is also feasible to shear bar stock up to 6 in. in diameter, but 2 in. is more typical.

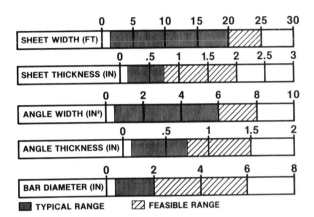

Force Requirements

Force requirements are shown in this table for a 10-in. wide workpiece. With the shear angle increased from 0° to 5°, the forces are reduced by approximately 20%.

Material thickness (in.)			Force requirements (lbs)	
Low-carbon steel	Aluminum	Stainless steel	0° Shear	5° Shear
0.06	0.07	0.04	30,000	24,000
0.12	0.13	0.06	60,000	48,000
0.25	0.25	0.19	125,000	98,000
0.50	0.63	0.38	250,000	196,000
0.75	1.00	0.50	375,000	294,000
1.00	1.50	0.75	500,000	392,000

Time Calculation

Operational elements that influence shearing time include shear blade velocity, thickness of the workpiece, approach distance, and over-travel distance. Also included are retract time, width of stock to be sheared, and the feed rate of the stock.

SHEARING SETUP

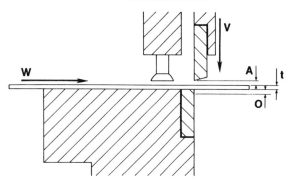

Shear blade velocity (in./sec) = V
Workpiece thickness (in.) = t
Approach distance (in.) = A
Overtravel distance (in.) = O
Retract time (sec) = R
Width of stock to be sheared (in.) = W
Feed rate of stock to be sheared (in./sec) = F

$$\text{Shearing time} = \frac{t + A + O}{V} + R$$

$$\text{Feeding time} = \frac{W}{F}$$

$$\text{Total time} = \frac{t + A + O}{V} + R + \frac{W}{F}$$

Cost Elements

Cost elements include the following:

* Setup time
* Load/unload time
* Cycle time
* Tool maintenance
* Direct labor rate
* Overhead rate
* Amortization of equipment and tooling costs

Safety Factors

The following risks should be taken into consideration:

* Personal
  - Contact with moving parts
  - Sharp edges
  - Noise
* Environmental
  - Vibration

Slitting _____

Slitting is a shearing process used to cut wide coils of material into several coils of narrower width as the material passes lengthwise through circular blades.

Process Characteristics

* Is limited to relatively thin materials (0.001 to 0.125 in.)
* Burrs are normally present to some extent on slit edges
* May be used on ferrous and nonferrous metals, plastics, and paper
* Is a high production, width-control process

Process Schematic

A workpiece is fed horizontally through two offset blades. The shearing forces of the two blades separate the workpiece at the point of contact.

SLITTING

UPPER KNIFE

WORKPIECE

LOWER KNIFE

Workpiece Geometry

A workpiece used in slitting may be either in sheet or in roll form. Slitting is used for such products such as sheet steel, plastics, fabrics, and paper stock.

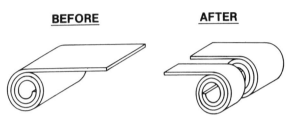

BEFORE　　**AFTER**

Setup and Equipment

Slitter machines are manufactured in a variety of sizes, relative to the material and size of the roll being slit. The basic machine consists of an uncoiler, slitter knives (set at specific widths), and a recoiler.

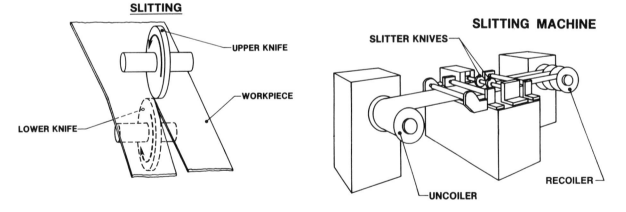

SLITTING MACHINE

SLITTER KNIVES

RECOILER

UNCOILER

Typical Tools and Geometry Produced

The three slitter knives shown are typical of those used on slitter machines. The geometry of the blades is determined by the following:

* Thickness of the workpiece
* Type of material
* Specified tolerances

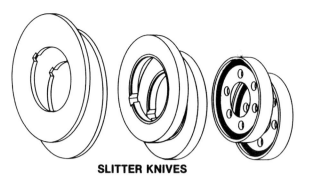

SLITTER KNIVES

Steel-Rule-Die Blanking

Steel-rule-die blanking is a shearing process in which a punch and die set is used to cut a part from stock material. The die consists of a thin strip of steel (formed to the contour of the part) supported on its edge. The punch is a flat surface made of steel, wood, or rubber.

Process Characteristics

* Uses thin strip steel knives as shearing dies
* Is often limited to short production runs
* Is predominantly used for nonmetallic material
* Is used for nonmetallics and thin gage to half hard sheet metals
* Reduces die design and fabrication costs over conventional dies

Process Schematic

Steel-rule-die blanking involves the opposing forces of the steel rule and punch meeting the workpiece. As the press closes, the steel rule penetrates the workpiece downward past the stationary punch, causing the workpiece to shear at the point where the steel rule and punch meet.

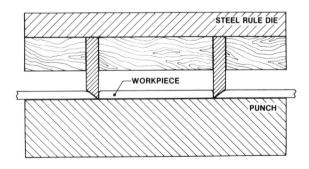

Workpiece Geometry

A steel-rule-die blanked hole may be distinguished from one produced by casting, molding, piercing, or torch cutting, by the presence of a burr and sheared section on the cut edge of the workpiece.

BEFORE

AFTER

GASKET

Setup and Equipment

The major components of this mechanical press are frame, motor, ram, bolster plate, and bed. The upper die block containing the steel rule is mounted on the ram, and the punch is mounted to the bolster plate. A workpiece is fed into the press, where it is secured and blanked in one stroke.

Typical Tools and Geometry Produced

The two typical dies shown below illustrate possible shapes that can be achieved with this process. The blank is usually from 0.0015 in. to 1.375 in. thick. Close tolerances can be held (±0.001 in. on holes and ±0.003 in. on peripheries over 12 in.). Size limits are determined by the press tonnage, and the physical properties and thickness of the workpiece.

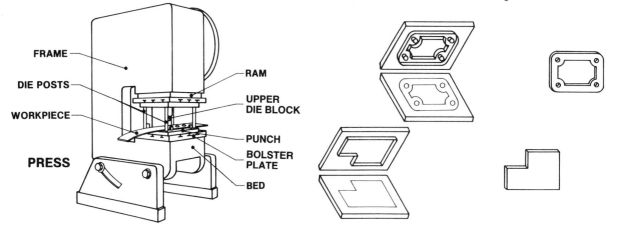

FRAME

DIE POSTS

WORKPIECE

PRESS

RAM

UPPER
DIE BLOCK

PUNCH

BOLSTER
PLATE

BED

Superfinishing

In superfinishing, an abrasive stone, shaped to the contour of the workpiece, removes surface fragmentation or smeared metal from previous finishing processes. Superfinishing uses very low pressure and speeds to produce a controlled low stress surface finish.

Process Characteristics

* Involves no appreciable production of heat to alter metallurgical properties
* Produces a controlled surface finish, typically less than 8 microinches (arithmetic average).
* Utilizes low speeds (50 to 60 sfpm) and low pressures (10 to 40 psi)
* Requires the use of lubricants
* Removes only a thin layer of smear metal, usually less than 0.0005 in.

Process Schematic

Superfinishing requires the bonded abrasive to oscillate over the rotating symmetrical workpiece. As the workpiece rotates, very fine oxidized chips are removed from its surface. Lubricant is required to cool the workpiece and to wash away the chips from the abraded area.

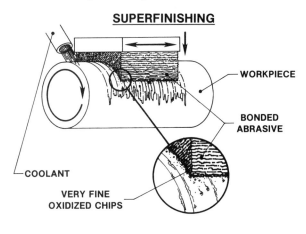

SUPERFINISHING

WORKPIECE
BONDED ABRASIVE
COOLANT
VERY FINE OXIDIZED CHIPS

Workpiece Geometry

This process is not considered a dimension changing process as only an average of 0.0008 in. to 0.0002 in. of stock is removed. Surface finishes range from a mirror finish to 30 microinches or more.

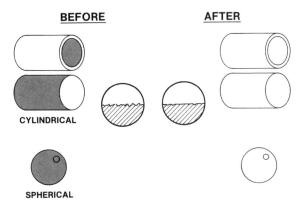

BEFORE   AFTER

CYLINDRICAL

SPHERICAL

Setup and Equipment

The workpiece is held between centers and the tool is attached to a device that oscillates and also traverses the rotating workpiece. The usual superfinishing speed is 50 sfpm to 60 sfpm with a pressure of 10 psi to 40 psi on the stone-contact area. Surfaces larger than 20 in.^{2} can be completed at a rate of 10 in.^{2}/min to 40 in.^{2}/min.

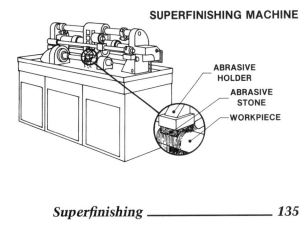

SUPERFINISHING MACHINE

ABRASIVE HOLDER
ABRASIVE STONE
WORKPIECE

Typical Tools and Geometry Produced

The tool style used in superfinishing depends on the shape of the workpiece. For internal diameter objects, a convex stone is preferred. A spherically shaped object requires a cup wheel, and cylindrical objects require a radially formed wheel. Superfinished surfaces are highly wear-resistant, stress-free, and corrosion-resistant.

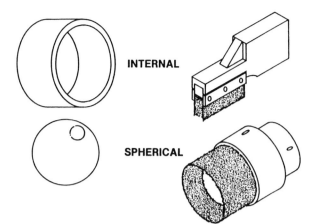

INTERNAL

SPHERICAL

Surface Grinding

Surface grinding is an abrasive machining process in which abrasive particles, bonded into a grinding wheel, remove small amounts of material from a workpiece. The grinding wheel may be shaped to form a variety of workpiece geometries.

Process Characteristics

* Produces very high surface accuracy and smooth surfaces
* Removes very small chips from metallic or nonmetallic materials
* Produces a flat or formed surface
* Uses abrasive grains in the form of a bonded wheel
* Is primarily a finishing process

Process Schematic

A bonded abrasive wheel rotates at a high speed and is vertically fed into a reciprocating workpiece. The abrasive grains cut very small chips from the workpiece interface. Coolant is required to keep the workpiece cool and to wash the chips away from the cutting area.

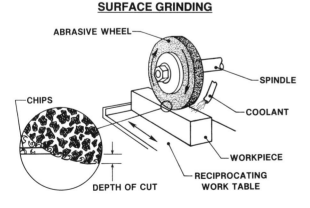

SURFACE GRINDING

Workpiece Geometry

Shown is a typical example of a workpiece that can be smoother and flatter after surface grinding. The surface roughness is usually between 8 and 32 microinches after grinding. If the workpiece requires a finer surface finish, a lapping or honing operation may follow grinding.

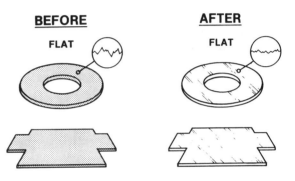

Setup and Equipment

The surface grinder consists of an abrasive wheel, a reciprocating table, and a workholding device. The rotating abrasive wheel is fed slowly into the reciprocating workpiece, which is usually secured by means of a magnetic chuck. If the workpiece is nonmetallic, then adhesive materials or a vacuum chuck may be used to secure the workpiece.

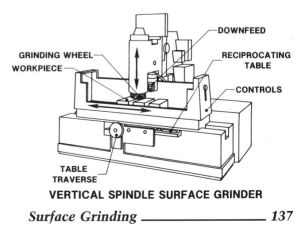

VERTICAL SPINDLE SURFACE GRINDER

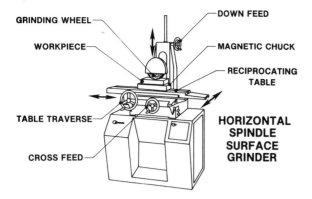

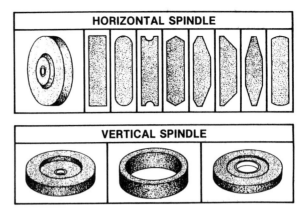

Typical Tools and Geometry Produced

A plain wheel imparts no special contour to the workpiece surface, only a flat surface results. However, a form wheel imparts a geometric shape to the workpiece surface. An abrasive wheel must be prepared true to the desired geometric shape for a formed surface.

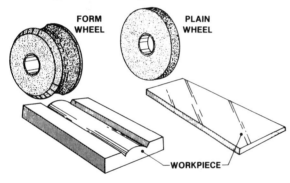

Tool Style

Some of the grinding wheel face contours that have been standardized by the Grinding Wheel Manufacturers' Association are shown. Specialized shapes may be formed on the face of the grinding wheel by a dressing unit, which uses diamond or carbide to "dress" the wheel to the desired shape. The wheel may also be formed by using a crush roll device. The device molds the wheel by exerting great pressure with a shaped roll, breaks away abrasive particles, and produces the desired wheel geometry.

Workholding Methods

Magnetic chucks and vacuum chucks are commonly used to secure a workpiece to the table. A magnetic chuck is useful for holding ferromagnetic workpieces to the table by magnetic forces. A vacuum chuck employs atmospheric pressure to hold nonmagnetic workpieces rigidly on the work table.

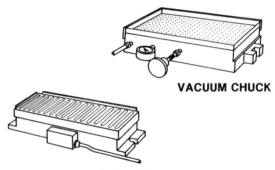

VACUUM CHUCK

ELECTROMAGNETIC CHUCK

Geometrical Possibilities

Geometrical possibilities for surface grinding with straight and formed wheels are illustrated. The geometry produced on the workpiece is a mirror image of the grinding wheel shape. Typical workpiece sizes range from 0.5 in. to 16 in. in width, and 0.5 in. to over 100 in. in length.

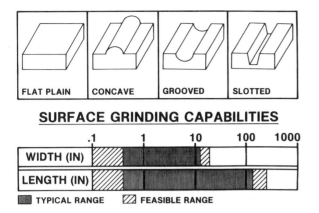

| FLAT PLAIN | CONCAVE | GROOVED | SLOTTED |

SURFACE GRINDING CAPABILITIES

	.1	1	10	100	1000
WIDTH (IN)					
LENGTH (IN)					

▓ TYPICAL RANGE   ▨ FEASIBLE RANGE

Work material properties	Effects of surface grinding
Mechanical	* Residual surface stresses * A thin martensitic layer may form on the part surface * Fatigue strength may be reduced
Physical	* Magnetic properties may be lost on ferromagnetic materials
Chemical	* May increase susceptibility to corrosion

Tolerances and Surface Finish

For general surface grinding applications, tolerances are held within ±0.002 in. for flatness and ±0.003 in. for parallelism. For precision applications, tolerances for both flatness and parallelism as low as ±0.00015 in. can be achieved. Surface finish may range from 2 to 125 microinches, with the typical range between 8 and 32 microinches.

TOLERANCES		
SURFACE GRINDING	TYPICAL	FEASIBLE
FLATNESS	±0.002	±0.00015
PARALLELISM	±0.003	±0.00015

SURFACE FINISH	
PROCESS	MICROINCHES (A.A.)
SURFACE GRINDING	2 4 8 16 32 63 125 250 500

▓ TYPICAL RANGE   ▨ FEASIBLE RANGE

Effects on Work Material Properties

High grinding temperatures create residual stresses along with a thin martensitic layer. This thin layer of martensite may reduce material fatigue strength. If the grinding temperature is elevated above the curie temperature, a loss of magnetic properties may occur in ferromagnetic materials. Surface grinding may also increase susceptibility to corrosion.

Tool Geometry

The figure below shows the standard marking system used for grinding wheels. The marking identifies the following five characteristics of a grinding wheel:

1. The abrasive material is identified by a letter (A—aluminum oxide, C—silicon carbide, etc.)
2. Grain size is indicated by a number from 20 (coarse) to 320 (very fine)
3. The grade is denoted by a letter from A (hard wheel) to Z (soft wheel)
4. The spacing between abrasive grains is labeled by a number from 1 (dense spacing) to 16 (open spacing)
5. Six common grinding wheel bonds are vitrified (V), resinoid (B), silicate (S), shellac (E), rubber (R), and oxychloride (O)

The manufacturer may add additional non-standardized identification letters or numbers.

STANDARD MARKING SYSTEM   A 36 L 5 V

ABRASIVE MATERIAL — GRAIN SIZE — GRADE — STRUCTURE — BOND

FEATURES	HORIZONTAL	VERTICAL
ABRASIVE	ALUMINUM OXIDE (A),	SILICON CARBIDE (C)
GRAIN SIZE	COARSE ⟶ FINE 36 to 320 46 to 60 MOST COMMON	COARSE ⟶ FINE 20 to 54 36 to 54 MOST COMMON
GRADE (BOND STRENGTH)	H–K SOFT WORKPIECES F–J HARD WORKPIECES	H–K SOFT WORKPIECES D–J HARD WORKPIECES
STRUCTURE (GRAIN SPACING)	DENSE ⟶ OPEN 1 to 16	
BOND MATERIAL	VITRIFIED (V), RESINOID (B), SILICATE (S), SHELLAC (E), RUBBER (R), OXYCHLORIDE (O)	

Abrasive Materials

Aluminum oxide, silicon carbide, diamond, and cubic boron nitride (CBN) are four commonly used abrasive materials for surface grinding wheels. Of these materials, aluminum oxide is the most common. Because of cost, diamond and CBN grinding wheels are generally made with a core of less expensive material surrounded by a layer of diamond or CBN. Diamond and CBN wheels are very hard and are capable of economically grinding materials, such as ceramics and carbides, that cannot be ground by aluminum oxide or silicon carbide wheels.

Abrasive materials	Applications
Aluminum oxide (A)	* Best for most steels and steel alloys * Low cost * High volume applications
Silicon carbide (C)	* Best for cast iron, nonferrous metals, and nonmetallic materials * Low cost
Diamond (D, MD, SD)	* Can be natural or man-made * Best for most carbides and some nonmetallic materials * Very high cost
Cubic boron nitride (B)	* Superior for high speed steel * Long life * Cool cutting * Very high cost

Process Conditions

Process conditions with a horizontal spindle machine include wheel speed, table speed, downfeed rate, crossfeed, and wheel identification number for surface grinding with a horizontal spindle and reciprocating table. Wheel speed may range from 4000 to 6500 surface feet per minute (sfpm). The table speed varies from 40 sfpm to 100 sfpm. The table generally moves laterally at a rate of 1/4 to 1/3 of the wheel width per pass. The type of wheel used depends on the workpiece material type and hardness and the desired surface finish.

Workpiece material	Hardness (HB)	Downfeed (in./pass)	Wheel ID No. ANSI Rough	Wheel ID No. ANSI Finish
Aluminum	30 to 150	0.003 to 0.001	C 46 J 7 V	C 46 J 7 V
Brass	340 to 450	0.003 to 0.0005	A 46 J 7 V	A 46 J 7 V
Cast iron, ductile	520 max	0.003 to 0.001	C 36 I 6 V	C 36 I 6 V
Mild steel	500 max	0.003 to 0.001	A 46 J 6 V	A 46 J 6 V
Stainless steel	325 to 450	0.001 to 0.0005	A 46 H 6 V	A 46 H 6 V
Plastics	—	0.005 to 0.0005	C 54 J 7 V	C 54 J 7 V

With a vertical spindle machine, process conditions include wheel speed, table speed, downfeed rate, and wheel identification number for surface grinding with a vertical spindle and rotary table. Wheel speed may range from 3500 sfpm to 6000 sfpm. The table speed varies from 80 sfpm to 350 sfpm. A broad work area is necessary. The type of wheel used depends on the workpiece material type and hardness and the desired surface finish.

Workpiece material	Hardness (HB)	Downfeed (in./pass)	Wheel ID No. ANSI Rough	Wheel ID No. ANSI Finish
Aluminum	30 to 150	0.001 to 0.004	CA 46 G 7 B	CA 60 G 7 B
Brass	40 to 200	0.001 to 0.004	CA 46 G 7 B	CA 60 G 7 B
Cast iron, ductile	520 max	0.001 to 0.005	A 24 G 6 V	A 80 G 6 V
Mild steel	500 max	0.001 to 0.005	A 30 H 6 V	A 80 H 6 V
Stainless steel	325 to 450	0.001 to 0.003	A 30 H 6 V	A 80 H 6 V
Plastics	—	—	—	—

Typical Workpiece Materials

Aluminum, brass, and plastics have poor to fair ratings, because they tend to clog the grinding wheels. Cast iron and mild steel have good machinability characteristics. Generally, stainless steel has a poor to fair rating because of its toughness and tendency to work harden.

MATERIAL	SURFACE GRINDING GRINDABILITY RATINGS			
	POOR	FAIR	GOOD	EXCEL
ALUMINUM		▨▦▨		
BRASS		▨▦▨		
CAST IRON			▨▦▨	
MILD STEEL			▨▦▨	
STAINLESS STEEL	▨▦▨			
PLASTICS		▨▦▨		

▦ TYPICAL RANGE   ▨ FEASIBLE RANGE

Factors Affecting Process Results

Tolerance and surface finish depend upon the following:

* Workpiece material
* Grinding wheel material
* Cutting speed and feed rate
* Proper grit, grade, and bond of wheel
* Cutting fluid
* Rigidity of tool, workpiece, and machine

Power Requirements

The metal removal rate in cubic inches per minute and the efficiency of the grinding wheel drive mechanism determine power requirements. The graph shown can be used to compute quickly grinding increases, metal removal rate increases, and the required horsepower increases.

BASED ON:
•GRAY IRON, 180 HB
•(1) FEED RATE = 665 IN/MIN
•(2) WORKPIECE WIDTH = 2 IN
•(3) DEPTH OF GRIND = .0006 IN/PASS
•(5) UNIT HP = 15 HP/IN³/MIN

THEN:
•(4) METAL REMOVAL RATE
   = .8 IN³/MIN
•(6) MOTOR HP = 15

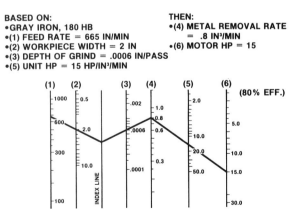

Lubrication and Cooling

Common lubricants include water-soluble chemical fluids, water-soluble oils, synthetic oils, and petroleum-based oils. The use of grinding fluids is essential to cool the wheel and workpiece, to lubricate their interface, and to remove chips. It is important that the grinding fluid be applied directly to the cutting area to insure that the fluid is not carried away by the fanlike action of the rapidly rotating grinding wheel.

Work material	Cutting fluid	Application
Aluminum	Light duty oil	Flood
Brass	Light duty oil	Flood
Cast iron	Heavy duty emulsifiable oil, light duty chemical and synthetic oil	Flood
Mild steel	Heavy duty water-soluble oil	Flood
Stainless steel	Heavy duty emulsifiable oil, heavy duty chemical and synthetic oil	Flood
Plastics	Water-soluble oil, dry, heavy duty emulsifiable oil, light duty chemical and synthetic oil	Flood

Time Calculations

Understanding the surface grinding setup is important in calculating metal removal and feed rates.

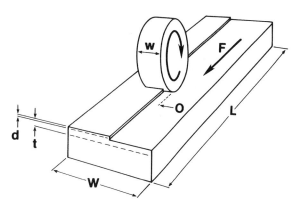

Length of workpiece (in.) = L
Width of workpiece (in.) = W
Width of grinding wheel (in.) = w
Depth of cut per pass (in.) = d
Thickness of material to be removed (in.) = t
Overlap distance (in.) = O
Feed rate (ipm) = F
Number of passes = N

$$\text{Grinding time} = \frac{N \times L}{F}$$

$$\text{Number of passes} = \frac{W}{w - O} \times \frac{t}{d}$$

Cost Elements

Cost elements include the following:

* Setup time
* Load/unload time
* Idle time
* Grinding time
* Tool change time
* Tool costs
* Direct labor rate
* Overhead rate
* Amortization of equipment and tooling

Safety Factors

The following risks should be taken into consideration:

* Personal
  - Rotating workpiece and grinding wheel
  - Eye irritation from fine chips
  - Eye and skin irritation from cutting fluids
  - Grinding wheel disintegration
* Environmental
  - Cutting fluid disposal
  - Grinding sludge disposal

Tapping

Tapping is a machining process that uses a multipoint cutting tool (tap) to produce uniform, internal, helical threads. Either the tap or the workpiece is rotated and advanced relative to the other. A hole with a diameter slightly smaller than the major diameter of the thread must already exist.

Process Characteristics

* Produces internal helical threads
* Requires an appropriate size hole in the workpiece prior to being tapped
* Uses an accurate multipoint cutting tool
* Tool lubrication is essential
* Produces small chips

Process Schematic

Tapping involves axial and rotational motions between the tap and the workpiece. Usually, the tap rotates and advances into the workpiece, but sometimes the opposite is true. With straight-flute taps, tightly coiled chips are removed by flowing along flutes in the tap where they may tend to pack. Helical or spiral taps facilitate more chip removal by creating spiral chips that flow more easily out of the flutes. Flutes also allow the passage of cutting fluid to produce a good thread finish.

TAPPING

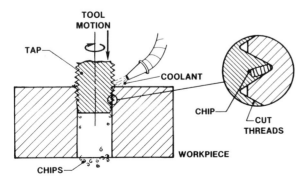

Workpiece Geometry

Shown are some typical workpieces before and after tapping. Thread cutting taps are available in a variety of styles, each one designed to perform a specific type of tapping operation efficiently. Most internal threads produced today are made with taps, although some are rolled. The smaller hole in the workpiece must provide sufficient material for the threads.

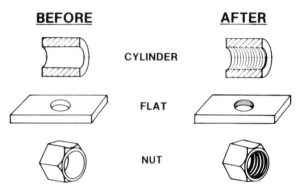

Setup and Equipment

A tapping machine typically consists of a tap head, table, column, and base. A tap head contains a drive motor, spindle, feed mechanism, and tap holder. A tap is gripped firmly in the tap holder, and the work is positioned on the table. Regularly shaped workpieces are held in a regular vise. Irregularly shaped workpieces are held in special fixtures.

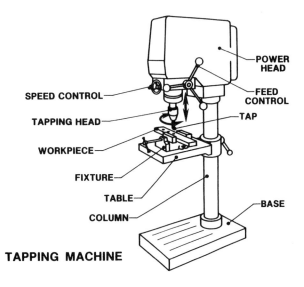

TAPPING MACHINE

Typical Tools and Geometry Produced

The wide variety of taps in use today include straight-flute taps, spiral taps, gun taps, pipe taps, adjustable taps, collapsible taps, and many other types for special purposes. For small holes, solid taps usually are used. Adjustable, collapsible, and expandable taps are for larger holes. Taps are normally made from either high-carbon steel or high speed steel.

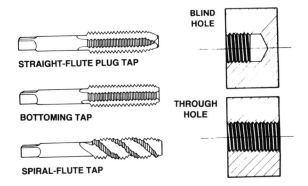

Geometrical Possibilities

Shown are some typical workpiece geometries that may be tapped. Threads as fine as 120 threads per inch in 0.01 in. diameter holes have been produced. Threads as coarse as 3 per inch in 5-in. diameter pipes are also possible; however, typically from 8 to 80 threads per inch are produced in holes ranging from 0.05 in. to 2 in. in diameter.

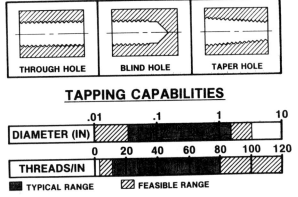

TAPPING CAPABILITIES

Workholding Methods

General purpose vises or jigs such as those shown may be used to hold the workpiece during tapping. Frequently, jigs and fixtures are designed for a specific workpiece and operation.

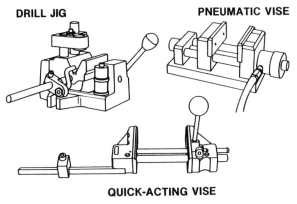

Tool Styles

Shown are two different taps: solid and collapsible. Solid taps have three types of chamfers: taper chamfers (7 to 9 threads long); plug chamfers (3 to 5 threads); and bottoming chamfers (1 to 1-½ threads). A pipe tap uses a sharp V thread; this makes possible joints that prevent fluid leakage. A collapsible tap has chasers that are set to retract radially after a thread is cut so that the tap can be withdrawn without reversing the workpiece rotation.

DESCRIPTION		STYLE	APPLICATION
SOLID TAPS	TAPER TAP	LEAD CHAMFER	STARTING TAP FOR SMALL TO MEDIUM HOLES
	PLUG TAP		INTERMEDIATE OR FINISHING TAP FOR THRU HOLES
	BOTTOMING TAP		FINISHING TAP FOR BLIND HOLES
	PIPE TAP		TAPERED TAP FOR DRY-SEAL OR PIPE THREADS
COLLAPSING TAP			SELF-CLOSING TAP FOR MEDIUM TO VERY LARGE HOLES

TOLERANCES			
TAPPING	CLASS 1	CLASS 2	CLASS 3
BASIC PITCH DIAMETER PLUS	0 to .0005	.0005 to .001	.001 to .0015

SURFACE FINISH	
PROCESS	MICROINCHES (A.A.)
TAPPING	2 4 8 16 32 63 125 250 500

■ TYPICAL RANGE   ▨ FEASIBLE RANGE

Toolholding Methods

Shown are two types of tap holders. A floating tap holder eliminates the need for bushing because the tap is automatically aligned. Also shown is a self-releasing tap holder that assures accuracy, minimum tap wear, and safety against damage to the tap.

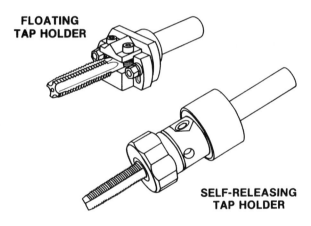

FLOATING TAP HOLDER

SELF-RELEASING TAP HOLDER

Tolerances and Surface Finish

Shown are the pitch diameter limits for class 1, 2, and 3 taps under 1 in. in diameter. Surface finishes for tapping may range from 32 to 250 microinches, with a typical range of 63 to 125 microinches. Factors that affect tolerance and surface finish are workpiece material, cutting speed, and cutting fluid.

Effects on Work Material Properties

Tapping affects the mechanical properties of the workpiece by creating residual stresses that may cause microcracking. High residual stresses make the material more susceptible to corrosion at cut surfaces.

Work material properties	Effects of tapping
Mechanical	* Residual stresses may cause microcracking
Physical	* Little effect
Chemical	* May increase susceptibility to corrosion at cut surfaces

Tool Geometry

Shown is the tool point geometry for solid high speed steel taps. The angles that are most often varied are the hook, or rake, angle and the chamfer relief angle. Threads are produced in *blind* holes with spiral-flute taps because the flutes draw the chips out of the hole. Threads are produced in *through* holes with spiral-point taps because these cut with a shearing action and throw the chips ahead of the tap.

CHAMFER RELIEF ANGLE   HOOK   RAKE

TANGENTIAL HOOK ANGLE   RAKE ANGLE

CHAMFER ANGLE 10 to 35°

MATERIAL	HOOK OR RAKE ANGLE (°)	CHAMFER RELIEF ANGLE (°)	TYPE OF TAP	
			THRU HOLE	BLIND HOLE
ALUMINUM	10 to 20	12	SPIRAL POINT HIGH HELIX	FAST SPIRAL FLUTE
BRASS	9 to 18	12	SPIRAL POINT	FAST SPIRAL FLUTE
CAST IRON	0 to 8	6	4-FLUTE	
MILD STEEL	0 to 8	6 to 8	SPIRAL POINT	FAST SPIRAL FLUTE
STNLSS STEEL	0 to 12	6 to 8	HEAVY DUTY SPIRAL POINT	
THERMO-PLASTICS	5 to 8	12	SPIRAL POINT HIGH HOOK	FAST SPIRAL FLUTE
THERMOSET	0 to 3	12	MODIFIED 4-FLUTE	

Process Conditions

Shown are the possible cutting speeds for various materials using a high speed steel tap, cutting to 75% of full thread using cutting fluids. Only cutting speed can be adjusted because optimum tapping speeds are usually based on minimum cost per hole and are a compromise between maximum tool life and maximum productivity. Feeds are governed by the pitch of the thread.

Typical speeds (sfpm)	
Workpiece material	Cutting speed
Aluminum	90 to 100
Brass	90 to 100
Cast iron	70 to 80
Mild steel	40 to 60
Stainless steel	10 to 35
Plastics	50 to 80

Typical Workpiece Materials

Aluminum, brass, and plastics have good to excellent ratings, whereas cast iron and mild steel have good machinability characteristics. Stainless steel has a poor to fair rating because it tends to work harden. As the hardness of metals to be machined increases beyond 250 HB, efficiency decreases and costs rise sharply due to tap breakage.

MATERIAL	TAPPING MACHINABILITY RATINGS			
	POOR	FAIR	GOOD	EXCEL
ALUMINUM			▨▮▨	
BRASS			▨▮▨	
CAST IRON			▨▮▨	
MILD STEEL			▨▮▨	
STAINLESS STEEL	▨▮▨			
PLASTICS			▨▮▨	

▮ TYPICAL RANGE   ▨ FEASIBLE RANGE

Tool Materials

Most machine taps are made of high speed steel (HSS). General purpose HSS, such as M1 and M2 types, are most widely used and have proved satisfactory for a majority of applications. The use of carbides is largely restricted to special applications, such as the tapping of especially abrasive grades of cast iron, or abrasive nonmetals, such as fiberglass.

Tool materials	Applications
High speed steel	* General purpose * Most widely used alloyed (HSS) * Heat-resistant alloys * Steels harder than 330 HB
Carbides	* Restricted to special applications * Abrasive grades of cast iron * Abrasive nonmetals such as fiberglass

Factors Affecting Process Results

Tolerance and surface finish depend upon the following:

* Tool geometry
* Cutting speed
* Rigidity of tool, workpiece, and machine
* Alignment of machine components and fixtures
* Cutting fluid
* Composition and hardness of the workpiece
* Class of tap used and number of passes

Lubrication and Cooling

A cutting fluid is more important in tapping than in most other machining operations because of the severity of the cutting operation. Tap teeth are highly susceptible to damage from heat, and chips are more likely to become congested in tapping than in most other cutting operations. Cutting fluids are generally used in tapping metals, except cast iron, to prevent heat buildup and to reduce the severe cutting friction.

Work material	Cutting fluid	Application
Aluminum	Soluble oil, kerosene	Flood
Brass	Soluble oil, light base oil	Flood
Cast iron	Soluble oil, none	Spray
Mild steel	Sulfur base oil	Flood
Stainless steel	Sulfur base oil	Flood
Plastics	None, cool air	Air jet

Power Requirements

Shown is the material power factor and the pitch power value used to calculate approximate tapping horsepower: hp = pitch power value (hp/sfpm) × material power factor × tapping speed × tool sharpness factor (1.5 for a sharp tap to 2 for a dull tap). For example, when tapping aluminum with a sharp tap of 16 threads per inch, 0.6 hp is needed.

Machine hp = $P \times M \times S \times T$ (approx. tapping hp)*

P = pitch power value (hp/sfpm)
M = material power factor
S = tapping speed (sfpm)
T = tool sharpness factor of 1.5 (sharp) to 2 (dull)

Threads/in.	P
32	0.002
24	0.004
20	0.006
16	0.009
13	0.012
10	0.020
8	0.030

Material	M
Aluminum	0.45
Brass	0.65
Cast iron	1.00
Mild steel	0.90
Stainless steel	1.40
Plastics	N/A

* 75% Thread

Time Calculation

Understanding the tapping setup is important in calculating the time needed for cutting, retracting, and positioning. The elements in-volved in tapping are diameter of tool, approach, depth, overtravel, threads per inch, cutting speed, retract rate, distance to next hole position, number of threads, length of cut, feed rate, positioning rate, and the number of holes to be tapped.

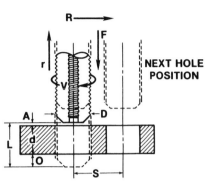

Diameter of tool (in.) = D
Approach (in.) = A
Overtravel (in.) = O
Threads per inch = T
Cutting speed (sfpm) = V
Retract rate (ipm) = r
Number of holes = H
Distance to next hole position (in.) = S
Number of threads = N
Positioning rate (ipm) = R
Length of cut (in.) = L
Depth of hole (in.) = d
Feed rate (ipr) = F

$$\text{Positioning time} = H \times \frac{S}{R}$$

$$\text{Cutting time} = \frac{L}{F}$$

$$\text{Feed rate} = N \times \frac{\text{rpm}}{T}$$

$$\text{Retract time} = \frac{L}{r}$$

$$\text{rpm} = \frac{4 \times V}{D} \text{ (approx.)}$$

Cost Elements

Cost elements include the following:

* Setup time
* Load/unload time
* Idle time
* Cutting time
* Tool change time
* Tool costs
* Direct labor rate
* Overhead rate
* Amortization of equipment and tooling

Safety Factors

The following risks should be taken into consideration:

* Personal
  - Rotating tool
  - Hot and sharp chips
  - Eye and skin irritation from cutting fluids
* Environmental
  - Disposal of tapping fluids

Thread Cutting ─────────────────────────────

Thread cutting is a machining process that uses a single-point tool to produce a uniform helical thread form on the internal or external surface of a cylinder or cone.

Process Characteristics

* Produces helical thread forms
* Requires multiple threading cuts
* Uses accurately ground single-point threading tools
* Produces stringy chips
* Requires constant positional relationship between the tool, workpiece, and the leadscrew.

Process Schematic

External threads can be cut with the workpiece either mounted between centers or held in a chuck. For internal thread cutting, the workpiece is held in a chuck. The cutting tool is usually ground to the proper shape and mounted securely in a tool holder. The tool moves longitudinally while the precise rotation of the workpiece determines the lead of the thread. Chips are produced at each pass of the tool.

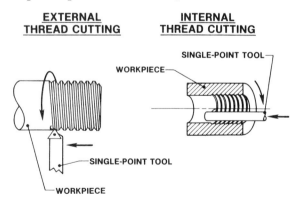

EXTERNAL THREAD CUTTING

INTERNAL THREAD CUTTING

SINGLE-POINT TOOL

WORKPIECE

SINGLE-POINT TOOL

WORKPIECE

Workpiece Geometry

Shown are examples of external and internal threading. Single-point thread cutting provides a versatile method for creating special threads in low quantities.

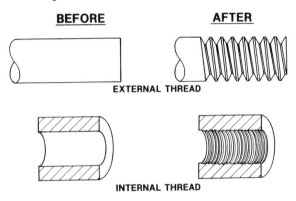

BEFORE   AFTER

EXTERNAL THREAD

INTERNAL THREAD

Setup and Equipment

Common machines used for single-point threading are engine lathes and N/C lathes. The tool is held in a toolholder; the workpiece is held in a workholding device. To cut right-hand threads, the tool is moved from right to left. For left-hand threads, the tool is moved from left to right.

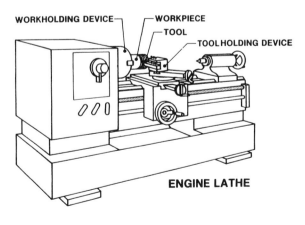

WORKHOLDING DEVICE   WORKPIECE
                         TOOL
                         TOOLHOLDING DEVICE

ENGINE LATHE

Typical Tools and Geometry Produced

Shown are some typical geometries produced in thread cutting. All threading could conceivably be done with a single-point tool but is usually carried out when only a few threads are to be cut or when special forms are desired. The thread form is usually obtained by grinding the tool to the proper shape and verifying it by means of a gage or comparator. The threads are formed by taking a series of 5 to 7 light cuts, until the desired thread depth is reached.

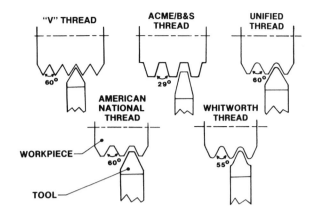

Thread Milling

Thread milling is a cutting process that uses a multipoint milling cutter to produce uniform helical threads on the internal or external surface of a cylinder or cone. The cutter, or hob, is fed radially into the workpiece while the workpiece rotates slightly over one revolution to complete the threads.

Process Characteristics

* Uses a multipoint thread milling cutter
* Feeds the cutter radially and longitudinally into the workpiece as the workpiece rotates slightly over one revolution
* Produces fine feed lines
* Can be used to make very large threaded parts

Process Schematic

In thread milling, either a single-form cutter or a multiform cutter is used. Some threads can be milled more quickly by using the multiform cutter as shown. The cutter must be slightly longer than the thread to be cut. The cutter rotates while being fed to the full thread depth. The work then rotates slowly for a little over one revolution. The rotating cutter is simultaneously moved longitudinally with respect to the workpiece according to the thread lead.

Workpiece Geometry

Shown are typical before and after workpieces. Thread milling produces a smooth, accurate thread. Thread milling is more efficient than single-point threading, when the thread has a coarse pitch, when it is near a shoulder or other interference, or when large diameter threads are required.

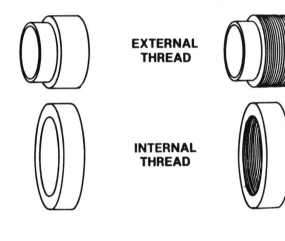

BEFORE   **AFTER**

EXTERNAL THREAD

INTERNAL THREAD

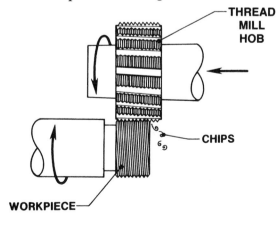

THREAD MILL HOB

CHIPS

WORKPIECE

Setup and Equipment

Shown is a universal automatic threading machine. The threading machine consists of a work spindle, cutter spindle, tailstock, bed, and base. Work is mounted either in a chuck or between centers. The milling hob is located at the rear of the machine. Universal thread mills can cut all internal or external threads, except some square threads.

Typical Tools and Geometry Produced

A variety of hobs are used for external and internal thread milling. Shown are typical multiform cutters used for cutting partial or full threads. High production rates are generally possible; however, these cutters have length limitations.

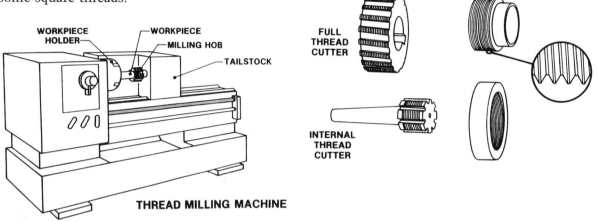

THREAD MILLING MACHINE

Turning/Facing

Turning is a material-removal process in which the major motion of the single-point cutting tool is *parallel* to the axis of rotation of the rotating workpiece. Facing is a special case of turning in which the major motion of the cutting tool is *at right angles* to the axis of rotation of the rotating workpiece. Turning is one of the oldest cutting processes known.

Process Characteristics

* Produces cylindrical external surfaces
* Produces flat surfaces during facing operation
* Uses a single-point cutting tool
* Tool motion is parallel to the axis of rotation during turning
* Tool motion is at right angles to the axis of rotation during facing
* Produces fine helical feed marks

Process Schematic

The workpiece is usually held in a workholding device such as a chuck, and the tool is mounted in a tool post. In turning, the cutting tool is advanced into the workpiece and removes material *parallel* to the axis of rotation. In facing, the cutting tool is fed *perpendicular* to the axis of rotation across the face of the workpiece. Turning/facing operations remove material in the form of chips, which poses special problems of chip removal.

TURNING/FACING

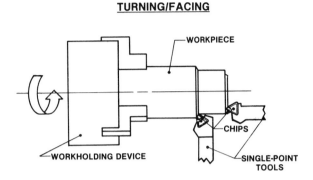

Workpiece Geometry

Shown are typical before and after workpieces. The workpiece may originally be of almost any cross-section, but the machined surface is normally straight or tapered. Fine helical feed marks are present on the machined surfaces. The surface finish obtained depends upon material and material properties, the cutting speed, feed rate, and tool nose radius.

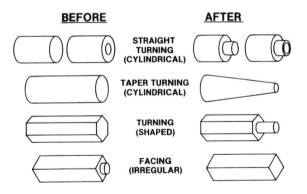

Setup and Equipment

Common machines used for turning and facing operations are the engine lathe, turret lathe, N/C lathe, and automatic lathe. Lathes are machine tools designed primarily for turning, facing, and boring. Cutting operations are performed with a cutting tool fed either parallel or at right angles to the axis of the workpiece. The tool may also be fed at an angle relative to the axis of the workpiece for machining tapers and angles.

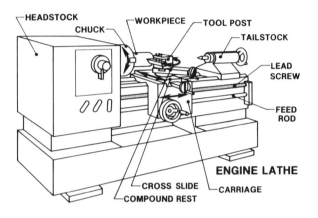

ENGINE LATHE

Typical Tools and Geometry Produced

The selection of tool depends primarily upon the type of cut desired and the workpiece material being machined. Most turning/facing operations are usually done with replaceable carbide inserts. These inserts have a number of indexable cutting edges that can be used for successive cutting and then discarded.

TURNING   **FACING**

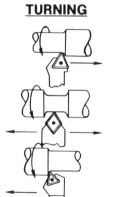 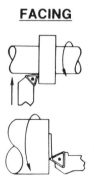

Geometrical Possibilities

Shown are some of the shapes that can be produced by turning/facing tools. Many desired shapes can be machined on a variety of engineering materials. Typical workpiece sizes range from 0.2 in. to 14 in. in diameter and from 0.1 in. to 60 in. in length. However, with appropriate equipment, it is possible to machine workpieces that are between 0.030 in. and 120 in. in diameter and 0.050 in. and 1000 in. in length.

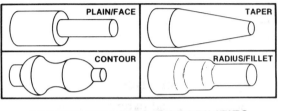

TURNING/FACING CAPABILITIES

	.01	.1	1	10	100	1000
DIAMETER (IN)						
LENGTH (IN)						

■ TYPICAL RANGE   ▨ FEASIBLE RANGE

Tolerances and Surface Finish

For most turning/facing applications, tolerances are held within ±0.001 in. for both diameter and length. For precision applications, diameter tolerances of ±0.0002 in. and length tolerances of ±0.0005 in. and less are attainable. Surface finish may range from 4 to 250 microinches, with the typical range for turned surfaces from 16 to 125 microinches.

TOLERANCES		
TURNING/FACING	TYPICAL	FEASIBLE
DIAMETER	±0.001	±0.0002
LENGTH	±0.002	±0.0005

SURFACE FINISH									
PROCESS	MICROINCHES (A.A.)								
TURNING/FACING	2	4	8	16	32	63	125	250	500

■ TYPICAL RANGE   ▨ FEASIBLE RANGE

Tool Style

Turning/facing tools are available in a wide range of styles. Shown are four common styles of carbide insert tools plus one high speed steel forming tool. Carbide inserts can be purchased in triangle, square, round, diamond, and many other shapes. Both high speed steel and carbide tools can be ground to various shapes.

DESCRIPTION	STYLE	APPLICATION
TRIANGULAR INSERT		PLAIN TURNING AND FACING
SQUARE INSERT		ROUGHING, FINISHING AND CHAMFERING
ROUND INSERT		ROUGHING, FILLETING AND CONTOURING
DIAMOND INSERT		CONTOURING
FORM TOOL		RADIUSING AND FORMING

Workholding Methods

Below are commonly used workholding devices—the three-jaw chuck and the four-jaw chuck. The latter is used for holding irregular shapes. The collet chuck is mainly used for small round workpieces. A faceplate, drive dog, and mandrel may be used to turn and face workpieces, such as gearblanks. Drive centers have hydraulic or spring-loaded teeth that "bite" into the end of workpieces and can be used when the entire length of the workpiece must be machined.

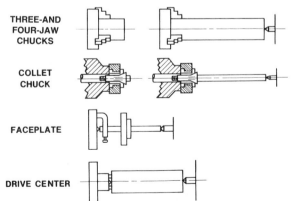

THREE-AND FOUR-JAW CHUCKS

COLLET CHUCK

FACEPLATE

DRIVE CENTER

Effects on Work Material Properties

Turning mainly affects mechanical properties of the workpiece. Some of the mechanical properties are affected by residual surface stresses, which may create microcracking. Diamond turning gives super-precision surfaces that are stress-free and resist corrosion.

Work material properties	Effects of turning/facing
Mechanical	* Creates residual surface stresses * May create microcracks * May cause surface or work-hardening on unhardened materials
Physical	* Little effect
Chemical	* Little effect

Typical Workpiece Materials

Shown are some of the typical workpiece materials and their relative machinability ratings. Aluminum, brass, and plastics have good to excellent ratings, whereas cast iron and mild steel have good machinability characteristics. Stainless steel has a poor to fair rating because of its toughness and its tendency to work harden when machined.

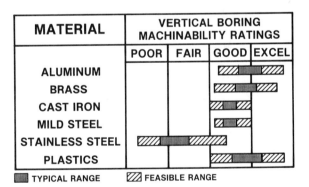

MATERIAL	VERTICAL BORING MACHINABILITY RATINGS			
	POOR	FAIR	GOOD	EXCEL
ALUMINUM			▨▨	▨▨
BRASS			▨▨	▨▨
CAST IRON			▨▨	
MILD STEEL			▨▨	
STAINLESS STEEL	▨▨	▨▨		
PLASTICS			▨▨	▨▨

▥ TYPICAL RANGE   ▨ FEASIBLE RANGE

Tool Materials

High speed steel tools have been used for some time but are being replaced by carbide, ceramic, and diamond tooling. Carbide inserts are easily replaceable and long lasting. Ceramic tools can resist high temperatures so high speed machining is possible. Ceramics are brittle, though, and often fracture if cuts are interrupted. Diamond tools produce superior surface finish and are used for nonferrous and nonmetallic materials.

Tool materials	Applications
High speed steel	* Special tool shapes * Low production * Best for interrupted cuts
Carbides (inserts)	* High production
Ceramics (inserts)	* High speed machining * Avoid interrupted cuts * Avoid positive rake angles
Diamonds (inserts)	* Abrasive machining * High surface qualities, fine tolerances * Nonferrous or nonmetallic materials

Factors Affecting Process Results

Tolerance and surface finish depend upon the following:

* Tool geometry
* Cutting speed and feed rate
* Rigidity of tool, workpiece, and machine
* Alignment of machine components and fixtures
* Cutting fluid

Tool Geometry

Turning/facing tool geometry is designated by angles on the tool. The angles that are most often varied are the back and side rake angles. The recommended starting point for back rake angles is 0° for cast iron and brass and +10° for softer materials, such as aluminum. These angles may be adjusted for each specific application. The relief angles are usually between 5° and 10° but may be adjusted according to given cuttings.

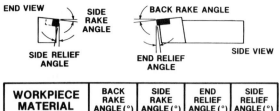

WORKPIECE MATERIAL	BACK RAKE ANGLE (°)	SIDE RAKE ANGLE (°)	END RELIEF ANGLE (°)	SIDE RELIEF ANGLE (°)
ALUMINUM	0 to 10	10 to 20	6 to 14	6 to 14
BRASS	-5 to 0	-5 to 8	6 to 10	6 to 12
CAST IRON	-7 to 5	-7 to 12	5 to 8	5 to 10
MILD STEEL	-7 to 12	-7 to 12	5 to 10	5 to 10
STAINLESS STEEL	-7 to 7	-7 to 10	5 to 10	5 to 10
PLASTICS	-5 to 5	0 to 15	20 to 30	15 to 20

Process Conditions

Shown are the suggested ranges for depth of cut, feed rate, and cutting speed using cutting fluid. Generally, cutting speeds are high for soft materials, such as aluminum and brass, and are low for hard and abrasive materials, such as cast iron and stainless steel. Depth of cut, feed rate, and cutting speed will vary, depending on the type of cut, material, and geometry of the workpiece.

Carbide tool with cutting fluid

Conditions	Rough	Finish
Depth of cut (in.)	0.094 to 0.187	0.015 to 0.094
Feed rate (ipr)	0.015 to 0.030	0.005 to 0.015

Workpiece material	Cutting speed (sfpm)	
	Rough	Finish
Aluminum	300 to 450	450 to 700
Brass	500 to 600	600 to 700
Cast iron	250 to 350	350 to 450
Mild steel	400 to 550	550 to 700
Stainless steel	250 to 300	300 to 400
Plastics	250 to 400	400 to 650

Lubrication and Cooling

The chart shows most of the typical cutting fluids used for a variety of work materials. Cutting fluids cool the tool, which reduces tool wear and makes higher cutting speeds and feed rates possible. Secondary functions include lubrication, flushing chips, and rust prevention. Cutting fluids also prevent a built-up edge on the tool and thus contribute to a better surface finish.

Work material	Cutting fluid	Application
Aluminum	None, water-soluble oil, synthetic oil, kerosene	Flood, spray
Brass	None, water-soluble oil, synthetic oil	Flood
Cast iron	None	—
Steel, all types	None, water-soluble oil, synthetic oil, sulfurized oil	Flood
Plastics	None, emulsifiable oil, synthetic oil	Flood, spray

Power Requirements

Shown is the formula used to calculate the machine horsepower, using unit power and material removal rate. For example, if mild steel is turned with a HB hardness of 85 to 200, and 10 in.^{3} of material is removed per minute, then an 11 hp engine lathe is needed (1.1 unit power × 10 in.^{3}/min = 11 hp). Unit power is based on HSS and carbide tools, a feed rate of 0.005 ipr to 0.020 ipr, and a drive with 80% efficiency.

Machine hp = unit power × removal rate (in.^{3}/min)

Material	Hardness (HB)	Unit power*
Aluminum	30 to 150	0.3
Brass	50 to 145	0.6
	145 to 240	1.0
Cast iron	110 to 190	0.7
	190 to 320	1.4
Mild steel	85 to 200	1.1
	330 to 370	1.4
	485 to 560	2.0
Stainless steel	135 to 275	1.3
	275 to 430	1.4
Plastics	N/A	0.05 est.

*Unit power based on: ● HSS and carbide tools ● feed of 0.005 ipr to 0.020 ipr ● 80% efficiency. est. = estimated.

Time Calculations

Understanding the turning/facing setup is important in calculating the time needed to turn, face, retract, and traverse. The elements involved in turning/facing are diameter of workpiece, length of turning cut, length of facing cut, approach, feed rate, cutting speed, traverse rate, traverse distance, and retract rate.

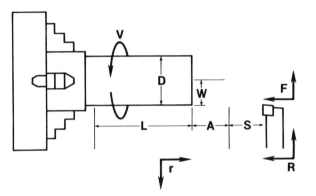

Diameter of workpiece (in.) = D
Length of turning cut (in.) = L
Length of facing cut (in.) = W
Approach (in.) = A
Feed rate (ipr) = F
Cutting speed (sfpm) = V
Retract rate (ipm) = r
Traverse distance (in.) = S
Transverse rate (ipm) = R

$$\text{Turning time} = \frac{L + A}{F}$$

$$\text{Facing time} = \frac{W}{F}$$

$$\text{Retract time} = \frac{L + A}{r}$$

$$\text{Traverse time} = \frac{S}{R}$$

$$\text{rpm} = \frac{4 \times V}{D} \text{ (approx.)}$$

Cost Elements

Cost elements include the following:

* Setup time
* Load/unload time
* Idle time
* Cutting time
* Direct labor rate
* Overhead rate
* Amortization of equipment and tooling costs

Safety Factors

The following hazards should be taken into consideration:

* Personal
  - Rotating workpiece
  - Hot and sharp chips
  - Eye and skin irritation from cutting fluids
* Environmental
  - Chip disposal
  - Cutting fluid disposal

Vertical Boring

Vertical boring is a single-point cutting operation used to produce an accurate internal cylindrical or conical surface by enlarging an existing opening. Workpieces are usually large and rotate about a vertical axis as the cutting tool is fed into the work.

Process Characteristics

* Produces internal cylindrical surfaces on large heavy parts
* Uses a single-point cutting tool
* Workpieces rotate about a vertical axis
* Is used to machine parts that have a small length-to-diameter ratio
* Produces internal helical feed marks on the workpiece

Process Schematic

A single-point cutting tool (boring bar) is advanced into a workpiece that is rotating about a vertical axis. The tool may be advanced both vertically and horizontally in order to control both the depth of cut and diameter produced. The bored hole is always concentric with the axis of rotation of the workpiece.

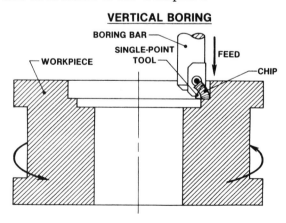

VERTICAL BORING

Workpiece Geometry

Vertical boring is customarily used to machine parts that have a small length-to-diameter ratio. These workpieces are often so large, or of such a shape, that they would be difficult to hold and rotate about a horizontal axis on a lathe.

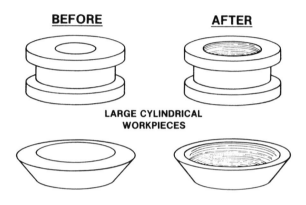

BEFORE **AFTER**

LARGE CYLINDRICAL WORKPIECES

Setup and Equipment

A vertical boring machine is similar to a lathe turned on its end. Typical machines consist of a base, a rotating horizontal work table, a column, and a crossrail. The crossrail supports one or more tool heads that are attached to a sliding ram and saddle. The machine pictured here is the single-ram type.

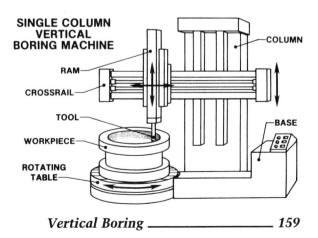

SINGLE COLUMN VERTICAL BORING MACHINE

Typical Tools and Geometry Produced

One of the most common types of boring tools used is the solid boring bar with a replaceable insert. High speed steel inserts are generally more suitable for slow speed boring of large workpieces. Carbide cutting edges are used most commonly for precision boring where speeds are high and the depth of cut is low.

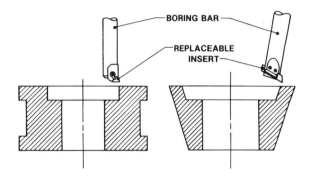

Geometrical Possibilities

Straight and tapered holes are commonly produced using vertical boring tools. However, other operations can be performed, such as turning, facing, reaming, chamfering, and counterboring. Typically, the bore diameter of the workpiece ranges from 24 in. to 200 in. and the depth ranges from 36 in. to 72 in. However, it is possible to bore holes that range from 10 in. to 1000 in. in diameter and that range from 10 in. to 100 in. in depth.

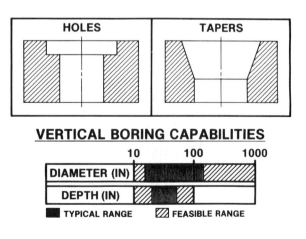

Tolerances and Surface Finish

For most vertical boring applications, tolerances are held within ±0.005 in. For precision applications, tolerances can be held within ±0.002 in. Surface finish can range from 63 to 500 microinches, with a typical range between 100 and 250 microinches. Where chatter is a problem, finishes would range between 250 and 500 microinches.

TOLERANCES		
VERT. BORING	TYPICAL	FEASIBLE
DIAMETER	±.005	±.002
DEPTH	±.005	±.002

SURFACE FINISH	
PROCESS	MICROINCHES (A.A.)
VERTICAL BORING	2 4 8 16 32 63 125 250 500

■ TYPICAL RANGE   ▨ FEASIBLE RANGE

Tool Style

Vertical boring tools are available in many sizes and shapes. Shown are three common tool styles. Adjustable insert boring bars utilize different shaped inserts, such as diamond, round, square, and triangular. Blade-type boring bars are used for high material-removal rates, and multiple heads are used for turning, facing, reaming, and chamfering.

DESCRIPTION	STYLE	APPLICATION
ADJUSTABLE INSERT		GENERAL BORING
BLADE TYPE		ROUGH BORING
MULTIPLE HEAD		BORING AND FACING

Workholding Methods

Shown is a rotating table used for vertical boring. Usually, adjustable clamps and fixtures are used to secure the workpiece to the table. A magnetic table can also be used to hold ferrous metals.

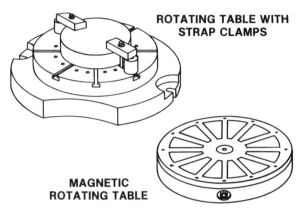

ROTATING TABLE WITH
STRAP CLAMPS

MAGNETIC
ROTATING TABLE

Effect on Work Material Properties

Residual surface stresses may create microcracking. High temperatures may also cause surface workhardening. Vertical boring has little effect on either physical or chemical properties.

Work material properties	Effects of vertical boring
Mechanical	* Creates residual surface stresses * May create microcracks * May cause workhardening
Physical	* Little effect
Chemical	* Little effect

Typical Workpiece Materials

Aluminum, brass, and plastics have good to excellent ratings, whereas cast iron and mild steel have good machinability characteristics. On the other hand, stainless steel has a poor to fair rating because of its toughness and its tendency to work harden when machined.

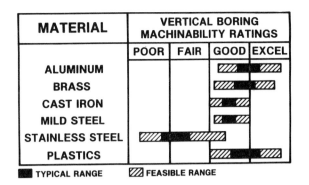

MATERIAL	VERTICAL BORING MACHINABILITY RATINGS			
	POOR	FAIR	GOOD	EXCEL
ALUMINUM			▨▨ ■	▨▨
BRASS			▨▨ ■	▨▨
CAST IRON			▨ ■ ▨	
MILD STEEL			▨ ■ ▨	
STAINLESS STEEL	▨▨ ■	▨▨▨▨		
PLASTICS			▨▨ ■	▨▨

■ TYPICAL RANGE   ▨▨ FEASIBLE RANGE

Tool Materials

High speed steel tools have been used for some time but are being replaced by carbide, ceramic, and diamond tooling. Carbide inserts are easily replaceable and are long lasting. Ceramic tools can resist high temperatures, so high speed machining is possible. Carbides and ceramics often fracture if cuts are interrupted. Diamond tools produce superior surface finish and are used for nonferrous and nonmetallic materials.

Tool materials	Applications
High speed steel	* Special tool shapes * Low production * Best for interrupted cuts
Carbides (inserts)	* Most commonly used tool * High production
Ceramics (inserts)	* High speed machining * Avoid interrupted cuts * Avoid positive rake angles
Diamonds (inserts)	* Abrasive machining * High surface qualities, fine tolerances * Nonferrous or nonmetallic materials

Factors Affecting Process Results

Tolerance and surface finish depend upon the following:

* Tool geometry
* Cutting speed and feed rate
* Rigidity of tool, workpiece, and machine
* Alignment of machine components and fixtures
* Cutting fluid
* Chip removal

Vertical Boring —————— **161**

Tool Geometry

Boring tool geometry is designated by angles on the tool. Shown are the four angles that are most often varied on the tool. Generally, the back and side rake angles are higher for soft materials. Negative rake angles are often used on hard materials.

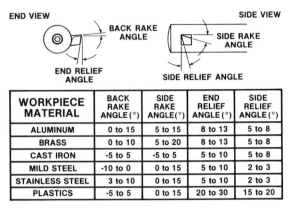

WORKPIECE MATERIAL	BACK RAKE ANGLE(°)	SIDE RAKE ANGLE(°)	END RELIEF ANGLE(°)	SIDE RELIEF ANGLE(°)
ALUMINUM	0 to 15	5 to 15	8 to 13	5 to 8
BRASS	0 to 10	5 to 20	8 to 13	5 to 8
CAST IRON	-5 to 5	-5 to 5	5 to 10	5 to 8
MILD STEEL	-10 to 0	0 to 15	5 to 10	2 to 3
STAINLESS STEEL	3 to 10	0 to 15	5 to 10	2 to 3
PLASTICS	-5 to 5	0 to 15	20 to 30	15 to 20

Process Conditions

Shown are the suggested ranges for depth of cut, feed rate, and cutting speed for vertical boring using cutting fluid. Generally, cutting speeds are high for soft materials, such as aluminum and brass and are low for the hard materials, such as cast iron and stainless steel. Depth of cut, feed rate, and cutting speed will vary, depending upon the type of cut and the geometry of the workpiece.

Carbide tool with cutting fluid

Conditions	Rough	Finish
Depth of cut (in.)	0.187 to 0.375	0.094 to 0.187
Feed rate (ipr)	0.030 to 0.050	0.015 to 0.030

Workpiece material	Cutting speed (sfpm)	
	Rough	Finish
Aluminum	200 to 300	300 to 450
Brass	400 to 500	500 to 600
Cast iron	200 to 250	250 to 350
Mild steel	300 to 400	400 to 550
Stainless steel	200 to 250	250 to 300
Plastics	150 to 250	250 to 400

Lubrication and Cooling

The chart below shows most of the typical cutting fluids. Cutting fluids cool the tool, which reduces tool wear and makes higher cutting speeds and feed rates possible. Secondary functions include lubrication, flushing chips, rust prevention, and preventing the tool from building up an edge that would mar the surface finish.

Work material	Cutting fluid	Application
Aluminum	None, water-soluble oil, synthetic oil, kerosene	Flood, spray
Brass	None, water-soluble oil, synthetic oil	Flood
Cast iron	None	—
Steel, all types	None, water-soluble oil, synthetic oil, sulfurized oil	Flood
Plastics	None, emulsifiable oil, synthetic oil	Flood, spray

Power Requirements

Shown is the formula used to calculate the machine horsepower, using unit power and material removal rate. For example, if mild steel is bored with a HB hardness of 85 to 200, and 10 in.^3 of material is removed per minute, then an 11 hp engine lathe is needed (1.1 unit power $\times$ 10 in.^{3}/min = 11 hp). Unit power is based on HSS and carbide tools, a feed of 0.005 ipr to 0.020 ipr, and 80% efficiency.

Machine hp = unit power × removal rate (in.³/min)

Material	Hardness (HB)	Unit power*
Aluminum	30 to 150	0.3
Brass	50 to 145	0.6
	145 to 240	1.0
Cast iron	110 to 190	0.7
	190 to 320	1.4
Mild steel	85 to 200	1.1
	330 to 370	1.4
	485 to 560	2.0
Stainless steel	135 to 275	1.3
	275 to 430	1.4
Plastics	N/A	0.05 est.

*Unit power based on: ● HSS and carbide tools ● feed of 0.005 ipr to 0.020 ipr ● 80% efficiency.
est. = estimate.

Cost Elements

Cost elements include the following:

* Setup time
* Load/unload time
* Idle time
* Cutting time
* Tool change time
* Tool costs
* Direct labor rate
* Overhead rate
* Amortization of equipment and tooling

Time Calculations

The following factors are used to determine process time: depth of bore, approach, feed rate, retract time, and length of cut. Retract time is a function of the length of cut divided by the retract rate.

Depth of bore (in.) = d
Approach (in.) = A
Feed rate (ipr) = F
Retract rate (ipm) = R
Length of cut (in.) = L

$$\text{Boring time} = \frac{L}{F}$$

$$\text{Retract time} = \frac{L}{R}$$

Choice of Equipment

Choice of equipment and tooling depend on workpiece size and configuration, equipment capacity (speed, feed, and horsepower range), production quantity, dimensional accuracy, and surface finish.

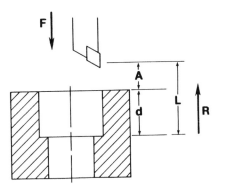

Vibratory Finishing

Vibratory finishing is a surface preparation process in which the workpiece and finish media are subjected to a controlled gyratory vibration. The interaction between the finishing media and the workpiece results in deburring, descaling, rounding of corners, blending, and polishing of the workpiece.

Process Characteristics

* Is commonly used for deburring, radiusing, cleaning, and blending
* Processes internal and external surfaces
* Lends itself to automation of work loading and unloading
* Is significantly faster than rotary barrel finishing
* Uses a wide range of abrasive and finish media

Process Schematic

In vibratory finishing, a tub containing finishing media and workpieces is vibrated. The vibratory motion scrubs the workpiece surfaces. The finishing media remove burrs and clean the workpiece. The vibration also causes the workpiece and media to roll and move in a circular motion. This motion moves the workpieces and media over a separation screen.

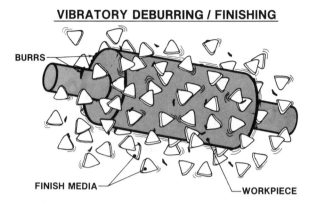

VIBRATORY DEBURRING / FINISHING

BURRS

FINISH MEDIA

WORKPIECE

Workpiece Geometry

The process is capable of removing burrs and rough location points, forming radii, cleaning, descaling, closing pores in castings, blending machine marks, changing stresses in workpieces, removing rust, and polishing.

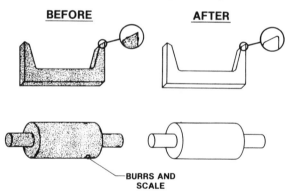

BEFORE AFTER

BURRS AND SCALE

Setup and Equipment

Vibratory finishing machines come in a wide range of sizes. The tub is the main component and can be rectangular or doughnut-shaped. Mechanical or electromagnetic methods are used to develop the vibratory motion. Most vibratory finishing machines have a screen area for media–part separation.

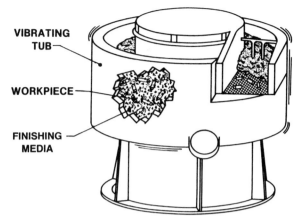

VIBRATORY FINISHING MACHINE

Labels: VIBRATING TUB, WORKPIECE, FINISHING MEDIA

Typical Tools and Geometry Produced

Finishing media are available in natural, metallic, or synthetic forms, and are either abrasive or nonabrasive. Many different shapes and sizes are available. The selection of a finish medium depends on several factors, such as workpiece size, shape, and material, along with the desired finish. Vibratory finishing can finish recesses, slots, and holes.

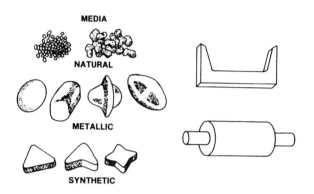

MEDIA

NATURAL

METALLIC

SYNTHETIC

Thermal Reducing

Cavity-Type Electrical Discharge Machining

Cavity-type electrical discharge machining (EDM) is a thermal mass-reducing process that uses a shaped conductive tool to remove electrically conductive material. It does this by means of rapid, controlled, repetitive spark discharges. A dielectric fluid is used to flush the removed particles, to regulate the discharge, and to keep the tool and workpiece cool.

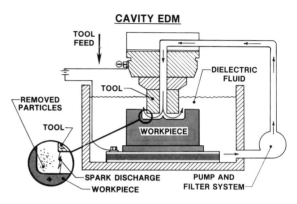

Process Characteristics

* Advances a shaped tool to within sparking (arcing) distance of the workpiece
* Tools and workpieces must be electrically conductive
* Removes material by rapid, controlled, repetitive spark discharge
* Uses a dielectric fluid to flush removed particles, control discharge, and cool tool and workpiece
* Surface finish is affected by gap voltage, discharge current, and frequency

Process Schematic

The tool and workpiece are submerged in a nonconductive dielectric fluid, usually a hydrocarbon oil. A very small gap of about 0.001 in. is maintained between the tool and the workpiece by means of a servo-system. Thousands of spark discharges occur each second, vaporizing small particles of the workpiece and slowly producing the desired cavity shape. A dielectric fluid is circulated by a pump, and workpiece particles are removed from the dielectric fluid by a filtering system.

Workpiece Geometry

In theory, any conductive material can be cut by electrical discharge machining; however, the effectiveness of the process varies widely, depending on the workpiece material and surface finish requirements. Workpiece hardness is not a concern because hard materials are machined as easily as soft ones. The process results in relatively burr-free parts.

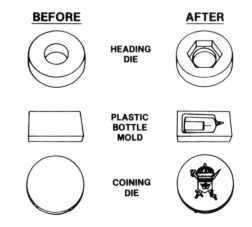

Setup and Equipment

The electrode can be held with various tool-holders. Hand wheels move the table to position the workpiece under the electrode. A servo-controlled mechanism feeds the electrode into the workpiece. Some high production machines currently in use have eight or more parallel heads, each equipped with an indexing turret electrode holder. The dielectric fluid covering the workpiece and electrode is contained in a tank attached to the X–Y table.

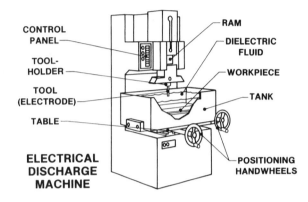

Typical Tools and Geometry Produced

Many intricate shapes can be made with EDM, as long as the material is electrically conductive. Because the electrode has no contact with the workpiece, it can be made of a soft, easily machined material, such as brass. The wear rate of the electrode depends on its material. Common electrode materials include copper–tungsten alloys, graphite, copper, brass, and zinc alloys. Tool and workpiece wear rates of 1 to 3 or more are common.

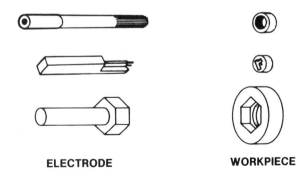

Geometrical Possibilities

The geometrical possibilities are limited by such factors as electrode shape, workpiece material, required accuracy, and amount of material to be removed. Depth of cut may typically vary from 0.1 in. to 2 in., although deeper cuts are possible. Workpiece length typically ranges from 0.25 in. to 2 in., or more, depending on the machine.

EDM CAPABILITIES

	.01	.1	1	10	100	1000
DEPTH OF CUT (IN)						
LENGTH (IN)						

■ TYPICAL RANGE   ▨ FEASIBLE RANGE

Tolerances and Surface Finish

EDM is widely used to machine dies and molds that require strict adherence to specified tolerances and surface finish. At maximum removal rate, tolerances range from ±0.005 in. to 0.002 in. At precision removal rates, tolerances range from ±0.0005 in. to 0.00015 in. Typical surface finish values range from 50 to 150 microinches, with values less than 10 microinches possible using electro-polishing techniques.

TOLERANCES (IN)

EDM	TYPICAL	FEASIBLE
MAX. REMOVAL RATE	±0.005	±0.002
PRECISION REMOVAL	±0.0005	±0.00015

SURFACE FINISH

PROCESS	MICROINCHES (A.A.)								
	2	4	8	16	32	63	125	250	500
EDM									

■ TYPICAL RANGE   ▨ FEASIBLE RANGE

Tool Style

Virtually any shape of conductive tool that can be made and held in a toolholder can serve as the electrode. Three basic shapes of tools are shown, including straight stock, multistage, and irregularly shaped tools. Common applications for each tool style are listed.

ELECTRODES

DESCRIPTION	STYLE	APPLICATION
STRAIGHT STOCK SHAPES	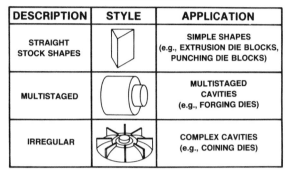	SIMPLE SHAPES (e.g., EXTRUSION DIE BLOCKS, PUNCHING DIE BLOCKS)
MULTISTAGED		MULTISTAGED CAVITIES (e.g., FORGING DIES)
IRREGULAR		COMPLEX CAVITIES (e.g., COINING DIES)

Workholding Methods

Workholding methods vary from the commonly used drill press and safety vises to special application fixturing. Workpiece shape and requirements determine workholding methods utilized in production.

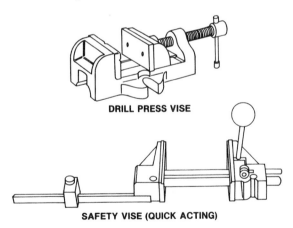

DRILL PRESS VISE

SAFETY VISE (QUICK ACTING)

Effects on Work Material Properties

One effect on mechanical properties of steel is the creation of a very thin recast carbide layer that lowers fatigue strength and creates microcracks. EDM has few other effects on material properties.

Work material properties	Effects of EDM
Mechanical	* Creates thin recast carbide layer * Lowers fatigue strength * May create microcracks
Physical	* Little effect
Chemical	* Little effect

Typical Workpiece Materials

Machinability ratings for aluminum, brass, cast iron, mild and stainless steels, and molybdenum are shown. Cast iron, brass, and mild steel rates are good, but materials with high conductivity, such as aluminum (and copper), and those with very low melting points, such as molybdenum, do not have as good ratings.

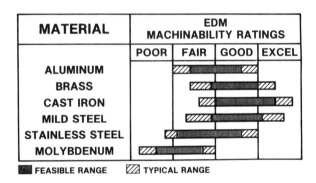

MATERIAL	EDM MACHINABILITY RATINGS			
	POOR	FAIR	GOOD	EXCEL
ALUMINUM				
BRASS				
CAST IRON				
MILD STEEL				
STAINLESS STEEL				
MOLYBDENUM				

■ FEASIBLE RANGE   ▨ TYPICAL RANGE

Tool Materials

Selection of tool or electrode materials depends on the workpiece material and required applications. Listed below are commonly used electrode materials. Brass electrodes are easy to machine but wear rapidly. Graphite electrodes are the most wear-resistant.

Electrode materials	Applications
Brass	* High accuracy for most metals
Copper	* Smooth finish * Low accuracy for holes
Zinc alloys	* Commonly used for steel, forging cavities
Copper-graphite	* General purpose work
Steel	* Used for nonferrous metals
Copper-tungsten	* High accuracy for detail work
Graphite	* Large volume/fine details

Factors Affecting Process Results

Tolerances and surface finish depend on accuracy and alignment of electrode, electrode wear, pulse frequency, spark gap, removal rate, overcut (the clearance between the electrode and the workpiece), and a properly filtered dielectric fluid.

Tool Wear

The tool end wears during both roughing and finishing cuts. The end wear ratio is a function of the amount of material removed (depth of cut) versus the amount of electrode material removed. Ratios vary from 1 to 1 for machining with brass electrodes to as much as 100 to 1 when using graphite electrodes.

$$\text{End wear ratio} = \frac{\text{depth of cut}}{\text{electrode end wear}}$$

Workpiece material	Electrode material	End wear ratio
Mild steel	Brass	1:1
	Copper	2:1
	Zinc	2:1
	Copper–graphite	4:1
	Steel	4:1
	Copper–tungsten	8:1
	Graphite	100:1

Process Conditions

Listed below are the process conditions for machining high-carbon/chrome die steel (RC 68) using a copper electrode. Removal rates and surface finish depend on the frequency of the pulse and the current. High frequency, low current processing yields the best surface finish, and high current, low frequency machining gives the highest removal rates.

Pulse frequency (kHz)	Current (amp)	Removal rate (in.^{3}/hr)	Surface roughness (microinch)
5	1 to 20	1.7	200–500
10	5 to 17	0.5	150–200
20	4 to 12	0.25	100–150
450	3 to 9	0.05	50–75
1000	0.5 to 3	0.0005–0.005	10–25

Cooling

Hydrocarbon dielectric fluids are used for most work materials, although a 90 to 10 ratio of a glycerine to water compound has proved successful. Mineral oil compounds are good fluid media for carbide workpieces. The fluid medium insulates the work area, cools the electrode and the workpiece, and flushes the particles out of the arc gap.

Dielectric fluid is used for cooling and flushing

Work material	Fluid medium	Application
Aluminum Brass Mild steel Stainless steel Tool steel Tungsten	Hydrocarbon oil or glycerine-water (90:10)	Submerge
Carbide	Mineral oil	Submerge

Power Requirements

Power requirements vary according to the electrode material, necessary current, and pulse frequency. Current ranges from less than 1 amp to 30 amps or more. Discharge frequency may vary from 16 kHz to over 130 kHz (kilohertz).

Electrode material	Current (amp)	Pulse frequency (kHz)
Copper	1–20	16–36
Brass	3–28	16–130
Graphite	3–15	65
Tungsten	0.2–1	70–100

Time Calculation

The following figures show some of the factors affecting the time involved for EDM. Calculating process time involves such variables as retract time, depth of cut, amount of overcut, number of holes, and more. See the time formulas in the figure below.

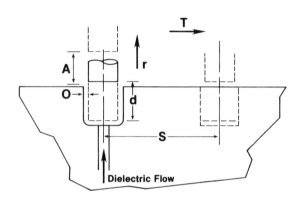

Volume to be removed (in.^{3}) = Q
Retract rate (ipm) = r
Positioning rate (ipm) = P
Removal rate (in.^{3}/min) = R
Depth of cut (in.) = d
Approach distance (in.) = A
Distance to next hole (in.) = S
Number of holes = N
Overcut (in.) = O
Traverse rate (ipm) = T

$$\text{Electrode positioning time} = N\left(\frac{S}{T}\right)$$

$$\text{Retract time} = \frac{d}{r}$$

$$\text{Cutting time} = \frac{Q}{R} + \frac{A}{P}$$

Cost Elements

Cost elements include the following:

* Workpiece preparation
* Setup time
* Idle time
* Machining time
* Tool change time
* Direct labor rate
* Overhead rate
* Amortization of equipment and tooling costs

Safety Factors

The following risks should be taken into consideration:

* Personal
  – Electrical shock
  – Vapor inhalation

Electrical Discharge Machining Grinding

Electrical discharge machining (EDM) grinding is a mass-reducing process that uses a rotating conductive wheel to remove electrically conductive material by means of controlled, repetitive spark discharges. A dielectric fluid is used to flush away the chips, regulate the discharge, and cool the wheel and the workpiece.

Process Characteristics

* Uses a slowly rotating conductive wheel that is kept within sparking (arcing) distance of the workpiece
* Wheels and workpieces must be electrically conductive
* Fragile and thin parts can be processed without distortion
* Can process very hard or difficult to machine metals
* Removes material by controlled spark erosion

Process Schematic

The basic setup for this process includes a conductive wheel mounted in a rotating spindle and a workpiece clamped to the table. The workpiece is raised to the desired position and slowly fed into the conductive wheel. Material removal rates range from 0.01 in.^{3} to 0.15 in.^{3} per hour. The result is a low stress, accurately shaped, burr-free workpiece.

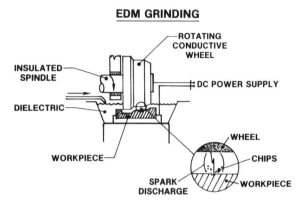

EDM GRINDING

Workpiece Geometry

Shown are typical workpieces before and after EDM grinding. This method is generally used for grinding tools composed of steel and carbide at the same time without wheel loading, grinding thin sections where pressures might cause distortion, grinding brittle and fragile parts, grinding through forms where diamond wheel costs would be excessive, and grinding circular forms in direct competition with abrasive wheel methods.

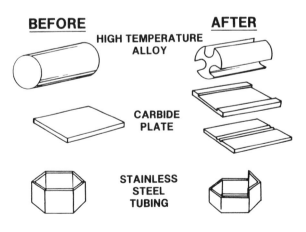

Setup and Equipment

The workpiece is fed into the rotating conductive wheel by a servo-controlled worktable. The spark gap is maintained between 0.0005 in. and 0.0030 in. by the servo-controller. The workpiece is cut by a stream of electric sparks between the negatively charged conductive wheel and a positively charged workpiece. The workpiece is immersed in a dielectric fluid to control the discharge.

Typical Tools and Geometry Produced

The conductive wheel is rotated at 100 to 600 surface feet per minute. Too high a speed causes splashing, and too low a speed may result in out-of-roundness. The power supply and the dielectric fluid are similar to cavity-type EDM, but lower amperage is used because the cutting area is usually small, and this method is used to achieve accuracy and a smooth finish.

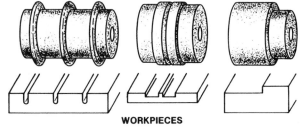

CONDUCTIVE WHEELS

WORKPIECES

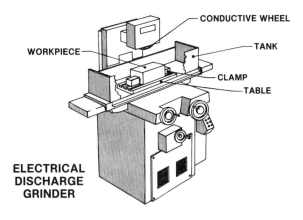

CONDUCTIVE WHEEL

WORKPIECE

TANK

CLAMP

TABLE

ELECTRICAL DISCHARGE GRINDER

Electrical Discharge Machining Wire Cutting

Electrical discharge machining wire cutting (EDM-WC) is a thermal mass-reducing process that uses a continuously moving wire to remove material by means of rapid, controlled, repetitive spark discharges. A dielectric fluid is used to flush the removed particles, regulate the discharge, and keep the wire and workpiece cool. The wire and workpiece must be electrically conductive.

Process Characteristics

* Utilizes a traveling wire that is advanced to within arcing distance of the workpiece (0.001 in.)
* Removes material by rapid, controlled, repetitive spark discharges
* Uses dielectric fluid to flush removed particles, control discharge, and cool wire and workpiece
* Is performed on electrically conductive workpieces
* Can produce complex two-dimensional shapes

Process Schematic

A moving wire electrode travels from the supply reel through the workpiece to a take-up reel. The wire is gradually advanced between the reels to compensate for the wear that occurs at the point of cutting. A dielectric fluid is supplied to the top and bottom of the workpiece. Proper gap length and desired shape are easily achieved using a numerical control system for moving the workpiece.

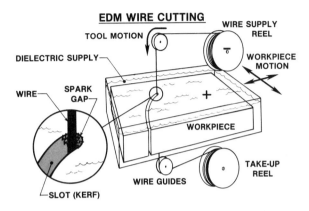

Workpiece Geometry

Punches, dies, and stripper plates can be cut in any of the hardened, conductive tool materials. Mirror-image profile work and internal contours from a starting hole are frequently produced. It is often possible to stack and cut workpieces simultaneously. The process leaves virtually no burrs on the edges of the finished workpiece.

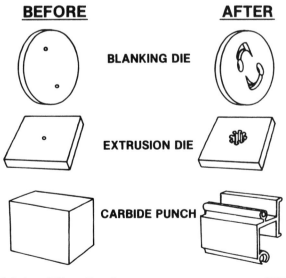

Setup and Equipment

Wire cutting is a variation of the EDM process. Initially, this type of equipment was used as a slicing machine for thin-walled structures. Now, with numerical control, complex two-dimensional shapes can be cut without using special electrodes. The narrow kerf and dimensional accuracy of the process make it possible to provide close-fitting parts.

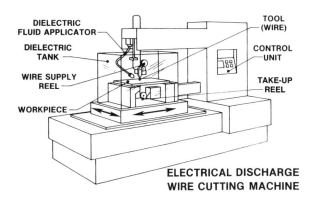

ELECTRICAL DISCHARGE WIRE CUTTING MACHINE

Typical Tools and Geometry Produced

A traveling copper or brass wire from 0.001 in. to 0.012 in. in diameter is used for the electrode. Tension in the wire and controlled positioning produce a very narrow kerf. This arrangement permits the cutting of intricate openings and tight radius contours, both internally and externally, without a specially shaped tool. Because the wire is inexpensive, it is generally used only once.

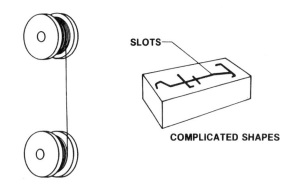

COMPLICATED SHAPES

Geometrical Possibilities

Many desired shapes can be cut on a variety of engineering materials. Workpiece thickness ranges from only a few thousandths of an inch to several inches.

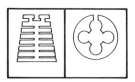

WIRE EDM CAPABILITIES

THICKNESS (IN)	0.01	0.1	1.0	10

■ TYPICAL RANGE   ▨ FEASIBLE RANGE

Tolerances and Surface Finish

For most applications, tolerances are held between ±0.002 in. and 0.005 in. For precision applications, dimensional tolerances of ±0.0001 in. to ±0.0005 in. are attainable. Surface finish can range from 8 to 200 microinches, but a typical range is between 10 and 100 microinches.

TOLERANCES (IN)		
WIRE EDM	TYPICAL	FEASIBLE
DIMENSIONS	±0.002 to ±0.005	±0.0001 to ±0.0005

SURFACE FINISH								
PROCESS	MICROINCHES (A.A.)							
WIRE EDM	4	8	16	32	63	125	250	500

■ TYPICAL RANGE   ▨ FEASIBLE RANGE

Tool Style

Electrode wire is available in many materials such as copper, steel, tungsten, brass, copper–tungsten alloys, and molybdenum alloys. The wire comes in several diameters to suit a variety of needs.

Electrode Wire

Wire diameter	Application
0.001 – 0.002	Intricate openings
0.003 – 0.004	Tight radius, slots, holes
0.005 – 0.012	Internal and external features

Workholding Methods

Two of the workholding devices most commonly used are the electromagnetic holding device for thin ferrous material and clamps for larger parts.

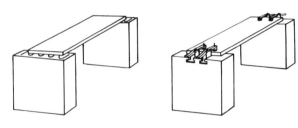

ELECTRO-MAGNETIC　　　**CLAMP**

Effects on Work Material Properties

Wire EDM mainly affects the mechanical properties of the workpiece such as surface hardness and reduced fatigue strength. The recast surface layer for steel materials is usually only 0.001 in. or 0.002 in. deep. Fatigue strength should be increased by carefully grinding or polishing away the recast layer if the part is to be used for critical applications.

Work material properties	Effects of wire EDM
Mechanical	* Thin recast surface layer * Lower fatigue strength
Physical	* Little effect
Chemical	* Little effect

Typical Workpiece Materials

Steel, super alloys, and titanium have good to excellent machinability ratings, whereas aluminum has only fair to good machinability characteristics because of its high thermal conductivity.

MATERIAL	WIRE EDM MACHINABILITY RATING			
	POOR	FAIR	GOOD	EXCEL
ALUMINUM				
STEEL				
SUPER ALLOYS				
TITANIUM				

▨ TYPICAL RANGE　　▨ FEASIBLE RANGE

Tool (Electrode) Materials

Brass, copper, and tungsten are the most common electrode wires for cutting holes and slots in nearly all metals. Copper–tungsten alloys, steel, and molybdenum alloys are also used for cutting a variety of materials, especially non-ferrous metals.

Electrode wires	Applications
Brass	* All metals, holes
Copper	* All metals, holes
Tungsten	* All metals (especially refractory metals), small slots or holes
Copper–tungsten	* All metals, carbide slots, thin slots
Steel	* Nonferrous, holes
Molybdenum	* Refractory, holes

Factors Affecting Process Results

Tolerance and surface finish depend upon the following:

* Spark gap
* Control of current
* Wire speed
* Work feed rate
* Dielectric medium
* Flush speed for dielectric medium
* Wire finish and accuracy

Tool Wear

Tool wear is minimal because the cutting action is distributed over the entire length of the wire. The wire is reusable; however, possible wire breakage should be considered as a deterrent.

Process Conditions

Sample process conditions of the specified steel are given in this table. Shown are different pulse frequencies and resulting crater size, depth, and removal rate. Removal rates range from as much as 1.7 in.^{3}/hr to as little as 0.0005 in.^{3}/hr. At low pulse frequencies, the current per discharge is high. At high pulse frequencies, the current per discharge is low, resulting in a finer surface finish.

For high-carbon, high-chromium die steel

Pulse frequency (kHz)	Crater size (mils)		Removal rate (in.^{3}/hr)
	Depth	Width	
5	1.9 to 4.0	5.3	1.7
10	1.5 to 1.9	2.4	0.5
20	1.0 to 1.5	2.0	0.25
450	0.5 to 0.6	0.65	0.1
1000	0.1	0.15	0.005 to 0.0005

Coolants and Flush Media

The chart shows most of the typical coolant and flushing media used for a variety of ferrous and nonferrous work materials. These solutions are applied either by bath or flush nozzle. Their main function is to remove particles that have been eroded from the workpiece by the spark discharge.

Work material	Dielectric media	Application
Ferrous and nonferrous metals	Deionized water, oil, plain water	Bath or flush
	Gas and air	Flush or stream

Power Requirements

The power requirements for cutting the specified die steel are given in this table. Generally, the higher the frequency, the lower the power level and the smaller the crater size. Currents range from 0.5 amp to 20 amps, with higher current levels used for higher removal rates at a given pulse frequency.

For high-carbon, high-chromium die steel

Pulse frequency (kHz)	Crater size (mils)		Current (amps)
	Depth	Width	
5	1.9 to 4.0	5.3	1 to 20
10	1.5 to 1.9	2.4	5 to 17
20	1.0 to 1.5	2.0	4 to 12
450	0.5 to 0.6	0.65	3 to 9
1000	0.1	0.15	0.5 to 3

Cost Elements

Cost elements include the following:

* Setup time
* Load/unload time
* Cutting time
* Direct labor rate
* Overhead rate
* Amortization of equipment and tooling

Time Calculation

Although removal rate per hour for wire EDM is very small because of the tiny wire size, productivity rates can be relatively high as compared with other competitive processes. Parameters involved with calculating process times are given below.

Removal rate (in.^{3}/hr) = R
Removal volume (in.^{3}) = Q
Length of cut (in.) = L
Material thickness (in.) = t
Width of cut (in.) = W
Feed rate (in./min) = F

$$\text{Cutting time} = \frac{Q}{R}$$

$$\text{Removal volume} = L \times W \times t$$

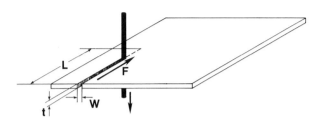

Safety Factors

The following hazards should be taken into consideration:

* Personal
  – Electrical shock
* Environmental
  – Disposition of used dielectric fluid
  – High frequency radiation

Gas Flame Cutting

Gas flame cutting is a metal cutting process that utilizes rapid oxidation of the metal by a jet of pressurized oxygen.

Process Characteristics

* Is used to cut ferrous metals by rapid oxidation of narrow heated zone
* Uses acetylene, propane, or natural gas as the flame source
* Produces a small amount of slag
* Workpiece thickness ranges from 1/8 in. to a foot or more
* Is primarily a roughing cut operation
* Is easily automated
* Produces a narrow (1/16 in. to 1/4 in.) kerf

Process Schematic

In gas flame cutting, a small area of ferrous metal is heated to 1400°F–1600°F by the preheating oxygen and fuel mixture. The elevated temperature causes the steel to oxidize quickly in the presence of the oxygen. The pressure of the gases is regulated according to workpiece thickness, which ranges from 0.125 in. to 8.0 in. Oxygen pressures range from 20 psi to 50 psi, and the fuel gas pressures range from 3 psi to 7 psi.

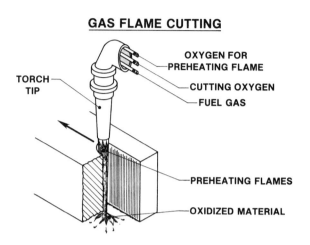

GAS FLAME CUTTING

Workpiece Geometry

Gas flame cutting is often used in the manufacturing of heavy equipment, in the construction of structural members for buildings, and in repair work. Special flame cutting machines can be used to cut structural steel and plate with considerable accuracy, speed, and economy. Flame cutters can make long straight cuts as well as circular cuts. They can also cut bevels, create intricate shapes, and prepare workpiece edges for welding.

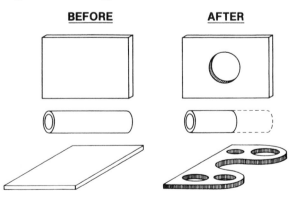

BEFORE   AFTER

Setup and Equipment

Setup and equipment include fuel, oxygen, pressure regulators, a cutting torch, an appropriate tip, and the workpiece. The fuel gas most widely used in flame cutting is acetylene. This accounts for the term oxyacetylene, which is often used to describe this process.

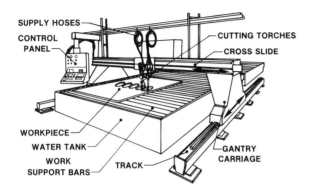

Typical Tools and Geometry Produced

The gas flame torch head is different from a welding head in that it has a channel to pass oxygen through the torch tip to perform the preheating and the cutting. Most gas flame cutting torches attach to a standard torch base (not shown). The flame cut surface is usually somewhat irregular and normally has a small slag deposit attached at the bottom edge of the kerf.

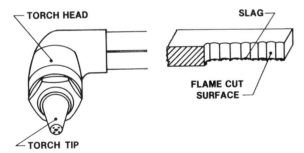

Geometrical Possibilities

The geometrical possibilities of gas flame cutting are almost unlimited; straight, circular, contour, or angular cuts are possible. Thick workpieces are generally cut using acetylene gas. This process is the most economical cutting process for ferrous materials over 3/8 in. thick.

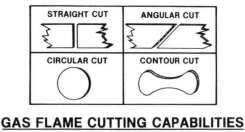

GAS FLAME CUTTING CAPABILITIES

	.01	.1	1	10	100
THICKNESS (IN.)					

■ TYPICAL RANGE   ▨ FEASIBLE RANGE

Tolerances and Surface Finish

Tolerances range from ±1/8 in. to ±1/32 in., depending on the method used. Typical methods include manual, automatic straight line, and automatic template. Surface finish depends on the method used, and ranges typically from 250 to 1000 microinches.

TOLERANCES

GAS FLAME CUTTING	TYPICAL	FEASIBLE
MANUAL	±1/8	±1/16
AUTOMATIC STRAIGHT LINE	±1/16	±1/32
AUTOMATIC TEMPLATE CONTROLLED	±1/16	±1/32

SURFACE FINISH

PROCESS	MICROINCHES (A.A.)							
	150	300	450	600	750	900	1050	1200
GAS FLAME CUTTING								

■ TYPICAL RANGE   ▨ FEASIBLE RANGE

Tool Style

There are three common tool styles. The one-piece straight tip is common for general purpose cutting with acetylene. A two-piece recessed tip is used for general purpose cutting with natural gas and propane. A third type is the one-piece divergent-bore tip, which is used for high speed cutting with acetylene.

DESCRIPTION	STYLE	APPLICATION
ONE-PIECE STRAIGHT-BORE TIP		GENERAL PURPOSE, MANUAL CUTTING WITH ACETYLENE
TWO-PIECE RECESSED TIP		GENERAL PURPOSE OR MACHINE CUTTING WITH NATURAL GAS AND PROPANE
ONE-PIECE DIVERGENT-BORE TIP		HIGH SPEED MACHINE CUTTING WITH ACETYLENE

Workholding Methods

A cutting table is provided with relief for the sparks and flame beneath the workpiece. A variety of standard clamps can be used to secure the workpiece to the cutting table.

TABLE

TOGGLE CLAMP

TWIST CLAMP

Effects on Work Material Properties

Mechanical effects are distortion, microcracks, residual stresses, and surface embrittlement. There is little effect on physical properties. An oxidation layer is present on the cut edge of the workpiece, which influences the chemical properties of the workpiece.

Work material properties	Effects of gas flame cutting
Mechanical	* Distortion * Microcracks * Residual stresses * Surface embrittlement
Physical	* Little effect
Chemical	* Oxidation layer

Typical Workpiece Materials

All carbon steels may be cut effectively by gas flame cutting. Gas flame cutting is not recommended for cutting cast iron because it resists the oxidation process, and the material melts instead of cuts. Likewise, galvanized sheet steel is difficult to cut and gives off toxic fumes. Free-machining steels containing lead do not cut well and may create toxic fumes as well.

MATERIAL	GAS FLAME CUTTING MACHINABILITY RATINGS			
	POOR	FAIR	GOOD	EXCEL
CARBON STEELS				▨▨
CAST IRON	▨▨			
GALVANIZED STEEL	▨▨			
FREE-MACHINING STEELS	▨▨			

■ TYPICAL RANGE ▨ FEASIBLE RANGE

Fuel Gases

The four most common gases used are acetylene, natural gas, propane, and methyl acetylene propadiene (MAPP). These gases are generally interchangeable, but acetylene is best when heat is to be concentrated over small areas for producing narrow kerfs. Natural gas is the least expensive but does not produce as hot a flame.

Consumable

Tool materials	Applications
Acetylene	* Heat is concentrated in primary flame over small areas
Natural gas Propane MAPP	* Large dispersion of heat with primary heat in secondary flame

Factors Affecting Process Results

Tolerance and surface finish depend upon the following:

* Thickness of material
* Material composition

* Gas pressure and feed rate
* Rigidity of tool, workpiece, and machine
* Alignment of machine components and fixtures
* Preheating variables

Tool Geometry

There are many different configurations of nozzle tips. The peripheral holes are used for the fuel gas, and the oxidizing gas comes through the center orifice.

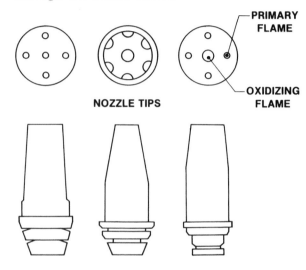

NOZZLE TIPS

Process Conditions

The different process conditions, depth of cut, and feed rates that can be expected while manually cutting low-carbon steel using standard pressure tips are listed below. The depth of cut ranges from 1/8 in. to 12 in. and the feed rate from 2.5 ipm to 30 ipm.

Gas	Depth of cut (in.)	Feed rate (ipm)
Oxyacetylene	1/8 to 12	2.6 to 20
Oxy-natural	1/8 to 12	3.0 to 20
Oxy-propane	1/8 to 4	2.5 to 3.0
Oxy-MAPP	1/8 to 4	14 to 30

Cooling

Parts that have been flame cut are hot to handle and may be quenched in water, oil, or sand, or they may be air cooled. If the workpiece is red hot when quenched, it may result in some local quench hardening. However, most workpieces have cooled enough before immersing them so that hardening does not occur.

Parts are cooled by the following methods:

* Water quench
* Oil quench
* Sand quench
* Brine water
* Air cool

Gas Consumption

Shown in this figure are the average gas consumption values of the most commonly used fuel gases. Usually, about $1.3\,ft^3$ of oxygen are required to oxidize $1\,in.^3$ of metal.

Type of gas	Consumption (cfh)		Metal removal rate (in.^{3}/hr)
	Fuel gas	Oxygen	
Acetylene	25	155	120
Natural gas	20	167	130
Propane	11	167	130
MAPP	15	120	90

cfh = cubic feet per hour = ft^3/hr.

Cost Elements

Cost elements include the following:

* Setup time
* Load/unload time
* Cutting time
* Equipment costs
* Gas costs
* Direct labor rate
* Overhead rate
* Amortization of equipment and tools

Time Calculation

Shown is a representation of a gas flame cutting operation. Major parameters for calculating process time include feed rate of the cutting nozzle, length of the cut, thickness of the workpiece being cut, and width of the kerf.

Gas Flame Cutting Setup

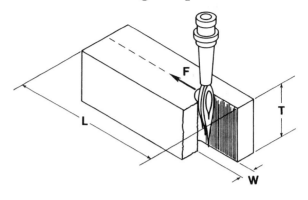

Thickness of material (in.) = T
Length of cut (in.) = L
Feed rate (ipm) = F
Width of kerf (in.) = W

$$\text{Cutting time} = \frac{L}{F}$$

$$\text{Removal volume} = L \times W \times T$$

Safety Factors

The following personal and environmental risks should be taken into consideration:

* Hot workpiece/sparks
* Inhalation of toxic fumes
* Radiation to eyes
* Flammable gases
* Smoke
* Fire
* Explosion

Laser Beam Cutting

Laser beam cutting is a high energy machining process in which the work material is melted, vaporized, and/or combusted by a narrow beam of coherent light (eventually supported by a coaxial supply of cutting gas). Laser is an acronym for light amplification by stimulated emission of radiation.

Process Characteristics

* Uses a high energy beam of coherent light that melts, vaporizes, or combusts material
* Produces small precision cuts or holes
* Usually produces rough metal surfaces; however, composite fabrics are cleanly cut
* Has good repeatability
* Leaves a narrow heat affected zone and, in some cases, heat treatment results

Process Schematic

Generation of the laser beam involves stimulating a lasing material by electrical discharges or lamps within a closed container. As the lasing material is stimulated, the beam is reflected internally by means of a partial mirror, until it achieves sufficient energy to escape as a stream of monochromatic coherent light. The coherent light then passes through a lens that focuses the light into a highly intensified beam generally less than 0.0125 in. in diameter. Depending upon material thickness, kerf widths as small as 0.004 in. are possible.

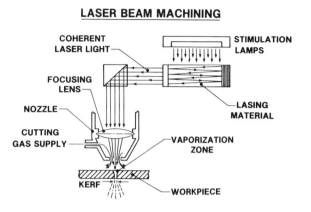

LASER BEAM MACHINING

Workpiece Geometry

Laser beams may be used to cut almost any shape. Thermal damage to adjacent regions is minimized due to very localized beam size. When cutting plastic or textile materials, the laser provides sealed edges that do not easily fray. Cutting gases, which may be oxygen, air, nitrogen, or argon, are selected depending on the specific workpiece material.

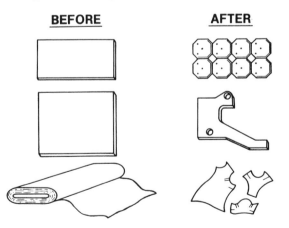

BEFORE   **AFTER**

Setup and Equipment

The laser machining system consists of a power supply for producing a laser beam, a workpiece positioning table, laser material, a method of stimulation, mirrors, and a focusing lens. In the installation shown, the workpiece is held stationary, and the focusing unit is moved to cut the desired contour.

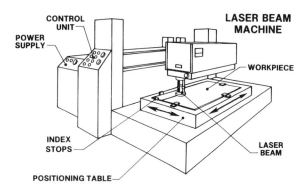

Typical Tools and Geometry Produced

This illustration shows two types of lasers. The first is a pulsed ruby type, which is used in electronic circuit fabrication. The second is a carbon dioxide (CO_2) type, which can be used in cutting steel, stainless steel, plastics, and other materials. For example, a medium powered CO_2 continuous laser can cut 0.125-in. thick low-carbon steel at a rate of 22 in. per minute.

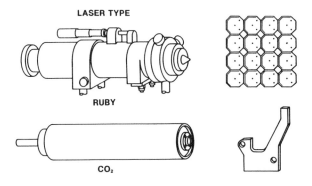

Geometrical Possibilities

Shown are a few of the intricate possibilities that this process can produce. Typical thickness capabilities range from 0.05 in. to 0.8 in.; the width of the cut typically ranges from 0.003 in. to 0.02 in.

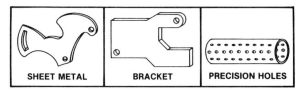

GEOMETRICAL POSSIBILITIES

SHEET METAL	BRACKET	PRECISION HOLES

LASER BEAM MACHINING CAPABILITIES

	.0001	.001	.01	.1	1	10
KERF WIDTH (IN)						
WORKPIECE THICKNESS (IN)						

■ TYPICAL RANGE   ▨ FEASIBLE RANGE

Tolerances and Surface Finish

This process is capable of holding quite close tolerances, often to within ±0.001 in. Part geometry and the mechanical soundness of the machine have much to do with tolerance capabilities. The typical surface finish resulting from laser beam cutting may range from 125 to 250 microinches.

TOLERANCES		
LASER BEAM MACHINING	TYPICAL	FEASIBLE
DRILLING (IN)	± 0.001	± 0.0005
MACHINING (IN)	± 0.001	± 0.0005

SURFACE FINISH					
PROCESS	MICROINCHES (A.A.)				
LASER BEAM MACHINING	32	63	125	250	500

■ TYPICAL RANGE   ▨ FEASIBLE RANGE

Laser Types

Shown are three types of lasers. The CO_2 laser is suited for cutting, boring, and engraving. The neodymium (Nd) and neodymium yttrium–aluminum–garnet (Nd-YAG) lasers are identical in style and differ only in application. Nd is used for boring and where high energy but low repetition are required. The Nd-YAG laser is used where very high power is needed and for boring and engraving. Both CO_2 and Nd/Nd-YAG lasers can be used for welding.

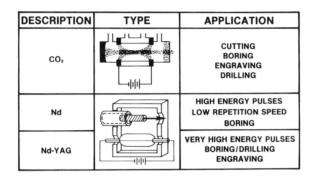

DESCRIPTION	TYPE	APPLICATION
CO₂		CUTTING BORING ENGRAVING DRILLING
Nd		HIGH ENERGY PULSES LOW REPETITION SPEED BORING
Nd-YAG		VERY HIGH ENERGY PULSES BORING/DRILLING ENGRAVING

Workholding Methods

Many types of workholding devices are used with laser beam machining. The strap clamp method works well with a wide variety of simple and more complex geometry parts.

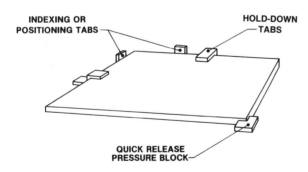

INDEXING OR POSITIONING TABS

HOLD-DOWN TABS

QUICK RELEASE PRESSURE BLOCK

Effects on Work Material Properties

The effects of laser beam machining on work material properties are quite minimal because of the small zone of metal affected by the process. The major effects of this process arise because of heat and include changes in hardness and the creation of a narrow heat-affected zone, with consequent changes in grain size.

Work material properties	Effects of laser beam cutting
Mechanical	* May affect hardness * Narrow heat-affected zone
Physical	* Grain size may change
Chemical	* No effect

Typical Workpiece Materials

Materials with sufficiently low reflectivity and thermal conductivity can be cut. Presently, mild steels, plastics, paper, and fabrics are cut. The figure shows some of the ratings.

MATERIAL	LASER BEAM MACHINABILITY RATINGS			
	POOR	FAIR	GOOD	EXCEL
ALUMINUM				
STAINLESS STEEL				
MILD STEEL				
WOOD				
PLASTICS				

▓ TYPICAL RANGE   ▨ FEASIBLE RANGE

Lasing Materials

This process uses either gas laser beams or solid state. Solid lasing is done by solid-state lasers, such as ruby, Nd-YAG, and Nd lasers. The most commonly used gas laser is the CO_2 laser. Shown are the most popular types of lasing materials with their general applications.

Lasing materials	Applications
CO₂	* Boring * Cutting/scribing * Engraving
Nd	* High energy pulses * Low repetition speed (1 kHz) * Boring
Nd-YAG	* Very high energy pulses * Boring * Engraving * Trimming

Factors Affecting Process Results

Tolerances and surface finish depend upon the following:

* Cutting speed and feed rate
* Beam geometry (mode)
* Beam intensity
* Beam focusing
* Beam/workpiece coupling
* Type of workpiece material
* Positioning accuracy

Beam Geometry

The parallel rays of coherent light from the laser source may be 1/16 in. to 1/2 in. in diameter. This beam is normally focused and intensified by a lens or a mirror to a very small spot of about 0.001 in. to create a very intense laser beam. Recent investigations reveal that the laser beam has a distinctive polarization. In order to achieve the smoothest possible finish during contour cutting, the direction of polarization must be rotated as it goes around the periphery of a contoured workpiece. For sheet metal cutting, the focal length is usually between 1.5 in. and 3 in.

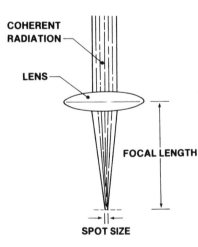

Production Rates

Cutting rates are given as functions of metal thickness and metal type. Metal thickness ranges are given from 0.020 in. to 0.50 in.

Cutting speeds for aluminum, for example, range from 800 in. per minute for 0.020 in. sheet stock to 100 in. per minute for 0.125 in. sheet stock. Laser cutting is estimated to be 30 times faster than traditional sawing.

Cutting Rates (ipm)*

Workpiece material	Material thickness (in.)					
	0.020	0.040	0.080	0.125	0.250	0.50
Stainless steel	750	550	325	10	20	—
Aluminum	800	350	150	100	40	30
Mild steel	—	177	70	40	—	—
Titanium	300	300	100	80	60	40
Plywood	—	—	—	—	180	45
Boron/epoxy	—	—	—	60	60	25

* Using a CO_2 laser.

Cooling

Cooling is required to maintain the beam generation apparatus at a reasonable operating temperature. It is necessary to cool both the lens and the medium that generates the beam. In both cases, water is the cooling medium.

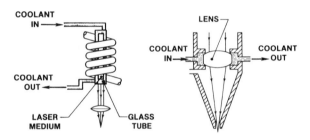

Power Requirements

Given here are several types of materials and their required heat input. The amount of kilowatts required is related to the heat input.

Watts*

Material	Material thickness (in.)				
	0.020	0.040	0.080	0.125	0.250
Stainless steel	1000	1000	1000	500	250
Aluminum	1000	1000	1000	3800	10,000
Mild steel	—	400	—	500	—
Titanium	250	210	210	—	—
Plywood	—	—	—	—	650
Boron/epoxy	—	—	—	3000	—

* Using a CO_2 laser.

Cost Elements

Cost elements include the following:

* Setup time
* Load/unload time
* Cutting time
* Direct labor rate
* Overhead rate
* Amortization of equipment and tooling

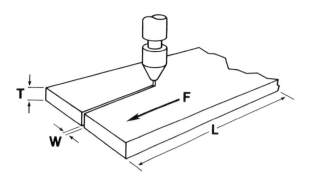

Time Calculation

Cutting time is determined by length of cut divided by the appropriate feed rate. Removal volume is the volume of material removed as a result of making the cut. Removal volume is a function of the kerf width, workpiece thickness, and length of the cut.

Length of cut (in.) = L
Material thickness (in.) = T
Feed rate (ipm) = F
Width of kerf (in.) = W

$$\text{Cutting time} = \frac{L}{F}$$

$$\text{Removal volume} = L \times W \times T$$

Safety Factors

The following risks should be taken into consideration:

* Personal
  - Contact with hot tools and workpieces
  - Eye contact with beam radiation
  - Inhalation of toxic fumes
* Environmental
  - Smoke
  - Fumes and dust particles

Plasma Arc Cutting

Plasma arc cutting is a metal cutting process that uses a high temperature, high velocity stream of ionized gas to melt and blow metal from the line of cut.

Process Characteristics

* Is used primarily for ferrous and nonferrous metals
* Produces temperatures up to 60,000°F
* Cuts at high speeds with good accuracy
* Uses an inert gas to form plasma, which also serves as a shielding gas
* Requires a cooling system for the torch and sometimes the workpiece
* Is typically automated
* Produces high noise levels

Process Schematic

In plasma arc cutting, the plasma stream is generated by directing the flow of an inert gas through an orifice in the torch tip where an electric arc ionizes the gas. The heat generated is a result of energy transfer of electrons, recombination of dissociated molecules on the workpiece, and convective heating from the high temperature plasma. Once the work material has been heated to a molten condition, the high velocity gas stream blows the material away, leaving a small kerf.

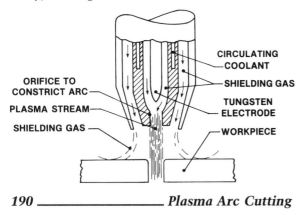

Workpiece Geometry

Because little skill is required, and no preheating of the workpiece is necessary before starting to cut, plasma arc cutting is being used widely throughout industry. The high temperature arc makes it possible to cut nonferrous metals, stainless steels, and nonmetal parts. The combination of high temperature and the jetlike action of plasma produces narrow kerfs and smooth surfaces similar to sawing. Materials that can be cut by this process range in thickness from 1/4 in. to 5 in.

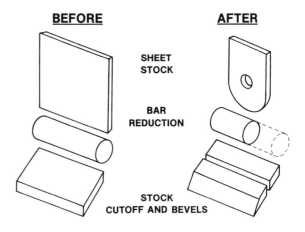

Setup and Equipment

The setup and equipment consists of the cutting torch, moving table, and control unit. The setup also includes a power supply and shielding gas storage container (not shown).

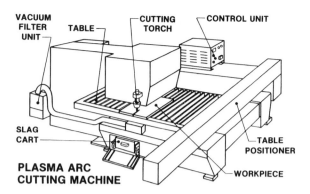

PLASMA ARC CUTTING MACHINE

Typical Tools and Geometry Produced

The nozzle and resulting stream of hot ionized gas are the main tools in the plasma arc cutting process. Gases used for plasma arc cutting include argon, helium, and nitrogen. Mixtures of 65% to 80% argon and 20% to 30% hydrogen are very common and produce temperatures of around 60,000°F.

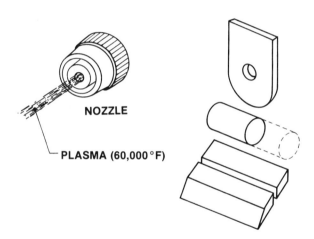

NOZZLE

PLASMA (60,000 °F)

Geometrical Possibilities

Plasma arc cutting can produce a wide variety of geometric shapes. Part size is generally limited by table or tool movement. Plasma arc cutting is often used for cutting relatively thin materials, but materials up to 5 in. thick are typically cut. A primary advantage of plasma arc cutting is its ability to cut nonferrous materials, such as aluminum and stainless steel.

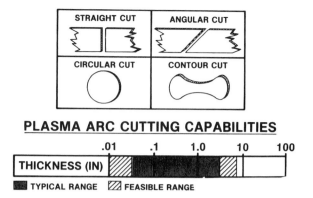

PLASMA ARC CUTTING CAPABILITIES

	.01	.1	1.0	10	100
THICKNESS (IN)					

■ TYPICAL RANGE ▨ FEASIBLE RANGE

Tolerances and Surface Finish

Typical tolerances that can be achieved with plasma arc cutting range from ±1/32 in. to ±3/16 in., depending on the type of control and the type of cut. Surface finishes typically range from about 150 to 500 microinches.

TOLERANCES (IN)		
PLASMA ARC CUTTING	**TYPICAL**	**FEASIBLE**
MANUAL	±3/16	±9/64
AUTOMATIC STRAIGHT LINE	±1/8	±5/64
AUTOMATIC TEMPLATE CONTROLLED	±1/32	±1/64

SURFACE FINISH								
PROCESS	**MICROINCHES (A.A.)**							
PLASMA ARC CUTTING	150	300	450	600	750	900	1050	1200

■ TYPICAL RANGE ▨ FEASIBLE RANGE

Torch Style

Three styles of torches used in plasma arc cutting are the air plasma torch, dual gas plasma torch, and the water injection plasma torch. These torches may be used for either ferrous or nonferrous materials. The arc can either be struck between the electrode and the workpiece (transferred arc) or between the electrode and the constricted orifice (nontransferred arc).

DESCRIPTION	STYLE	APPLICATION
AIR PLAMSA TORCH		A WIDE VARIETY OF METALS, STAINLESS STEEL, CHROME–NICKEL ALUMINUM
DUAL GAS PLASMA TORCH		FERROUS AND NONFERROUS METALS
WATER INJECTION PLASMA TORCH		THIN CROSS-SECTIONAL FERROUS AND NONFERROUS METALS

Work material properties	Effects of plasma arc cutting
Mechanical	* Distortion * Microcracks * Residual stresses * Surface embrittlement
Physical	* Carbide precipitation
Chemical	* Oxidation

Workholding Methods

Many types of supporting and workholding methods can be used in conjunction with plasma arc cutting. It is quite common to use tables that may be filled with water to catch the hot stream of molten metal. Twist clamps and fast-acting toggle clamps are frequently used for securing the work material during cutting.

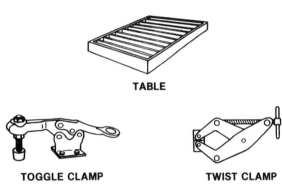

TABLE

TOGGLE CLAMP   **TWIST CLAMP**

Effects on Work Material Properties

Plasma arc cutting may have a number of significant effects on work material properties because of the high temperatures involved. Mechanical effects are distortion, warping, microcracking, and embrittlement; however microcracking and embrittlement are limited to the heat-affected zone. Carbide precipitation may occur when cutting stainless steels. The cut surface typically has a thin oxide layer that can be removed by grinding or other processes.

Typical Workpiece Materials

Plasma arc cutting can be used with most metals. Its greatest advantage is its ability to cut both ferrous and nonferrous metals very quickly.

MATERIAL	PLASMA ARC CUTTING RATINGS			
	POOR	FAIR	GOOD	EXCEL
ALUMINUM			▨▨	▨▨
BRASS			▨▨	▨▨
CAST IRON			▨▨	▨▨
MILD STEEL			▨▨	▨
STAINLESS STEEL				▨▨▨

■ TYPICAL RANGE   ▨ FEASIBLE RANGE

Shielding Gases

Common combinations of fuel and shielding gases used for plasma arc cutting are shown in this table along with typical applications. One advantage in using air as a shielding gas is cost reduction. An advantage of water as a shielding material is arc restriction and contact. However, dual gas combinations are the most common for plasma arc cutting.

Consumable

Plasma gas/ shielding gas	Applications
Argon/hydrogen	* Cuts stainless steel, aluminum, and other nonferrous metals
Nitrogen/carbon dioxide	* Cuts ferrous and nonferrous material
Nitrogen/water	* Cuts ferrous and nonferrous material
Nitrogen/air	* Cuts a variety of conductive metals: stainless steel, chrome–nickel alloy steel, aluminum, and copper

Factors Affecting Process Results

Tolerance and surface finish depend on the following:

* Thickness of material
* Gas pressure
* Feed rate
* Torch-to-work distance
* Plasma jet velocity
* Shielding medium
* Material composition
* Machine positioning accuracy

Gas Flow

The table below shows the gas flow associated with plasma cutting of steel and cast iron. Hydrogen is the plasma gas and nitrogen the shielding gas used in the dual gas plasma torch.

Workpiece thickness (in.)	Nozzle size (in.)	Gas flow (cfh)	
		Nitrogen	Hydrogen
0.25	0.125 to 0.187	120 to 150	10
0.50	0.140 to 0.250	130 to 225	15
1.0	0.161 to 0.250	130 to 225	15
2.0	0.187 to 0.250	150 to 225	15
3.0	0.218 to 0.250	170 to 225	20

Factors Affecting Tolerances and Surface Finish

Cutting speeds range from 40 ipm when cutting 0.375-in. thick mild steel to 500 ipm for cutting 0.035-in. gage mild steel. The type of fuel will affect these approximate speeds. Other factors, such as nozzles and shielding gas, will also influence cutting speeds.

Plasma Arc Cutting Speeds (ipm)

Gage (in.)	Mild steel		304 Stainless		Aluminum	
	Nom.	Max.	Nom.	Max.	Nom.	Max.
0.035	270	500	350	500	—	—
0.060	200	500	260	500	220	500
0.125	175	225	210	225	200	320
0.250	50	75	60	90	100	160
0.375	40	60	45	65	180	210

Cooling

The torch is cooled by a generous flow of water through the torch head. This is necessary to prevent the torch from overheating. The torch is also cooled by movement of surrounding air. The workpieces may be cooled by water, sand, or air.

Work materials	Torch coolant	Quench medium
Aluminum	Air	Water
Brass	Air	Water
Cast iron	Water	Sand
Mild steel	Water	Air
Stainless steel	Water	Sand
Plastics	Air	Air

Power Requirements

Power requirements range from 200 amps for cutting 1/4-in. aluminum plate at a speed of 70 inches per minute to 350 amps for cutting 1-in. aluminum plate at the same speed.

For cutting speeds of 70 ipm

Material	Thickness (in.)	Current (amps)
Aluminum	1/4	200
	1/2	250
	1	350
Steel	1/4	275
	1/2	275
	1	375
Stainless steel	1/4	270
	1/2	350
	1	400

Cost Elements

Cost elements include the following:

* Setup time
* Load/unload time
* Cutting time
* Gas costs
* Direct labor rate
* Overhead rate
* Amortization of equipment and tooling

Time Calculation

Cutting time is a function of the volume of material to be removed by cutting and the removal rate. Removal volume is a function of length of cut, material thickness, and kerf width.

Thickness of material (in.) = T
Length of cut (in.) = L
Feed rate (ipm) = F
Removal rate (in.³/hr) = R
Removal volume (in.³) = Q
Width of cut (in.) = W

$$\text{Cutting time} = \frac{Q}{R}$$

$$\text{Removal volume} = L \times W \times T$$

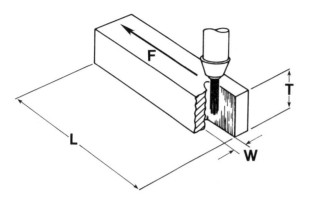

Safety Factors

The following risks should be taken into consideration:

* Personal
  - Hot workpiece and tooling
  - Inhalation of toxic fumes
  - Radiation to eyes
  - Possible hearing loss
* Environmental
  - Harmful fumes
  - Fire

Chemical Reducing

Electrochemical Grinding

Electrochemical grinding (ECG) is an electrolytic material-removal process involving a negatively charged abrasive grinding wheel, a conductive fluid (electrolyte), and a positively charged workpiece. Workpiece material deplates into the electrolyte solution. ECG is similar to electrochemical machining except that the cathode is a specially constructed grinding wheel instead of a tool shaped like the contour to be machined.

Process Characteristics

* Utilizes electrically conductive grinding wheels
* Removes material by electrochemical decomposition and abrasive action
* Deplates workpiece materials and deposits them in electrolyte
* Wheels wear extremely slowly
* Workpieces are electrically conductive

Process Schematic

ECG uses a metal disk embedded with abrasive particles. Aluminum oxide abrasives are generally used for grinding steel. Diamond is also used as an abrasive in grinding wheels for carbides and steel harder than Rockwell C65. Surface and form grinding are normally done with this process.

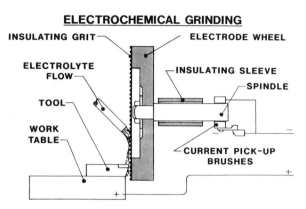

ELECTROCHEMICAL GRINDING

INSULATING GRIT — ELECTRODE WHEEL
ELECTROLYTE FLOW — INSULATING SLEEVE
TOOL — SPINDLE
WORK TABLE — CURRENT PICK-UP BRUSHES

Workpiece Geometry

An electrochemical grinder normally consists of a base, table, DC power supply, electrolyte supply, and insulated spindle. An insulated wheel is mounted on a spindle, and the workpiece is held in a vise. The spark gap is typically held at 0.005 in. to 0.0030 in. by the motion of the work table.

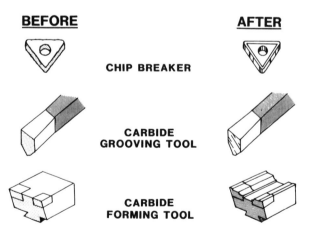

BEFORE   AFTER

CHIP BREAKER

CARBIDE GROOVING TOOL

CARBIDE FORMING TOOL

Setup and Equipment

Arrangement of the tool and workpiece is shown. The metal bond of the wheel is the cathode. The abrasive particles do three things: they serve as insulators to preserve a small gap between the cathode (tool) and the anode (work); they clean away the residue; and they cut chips that are produced should the wheel touch the workpiece. Electrolyte is flooded onto the workpiece to remove the metal from the work.

Typical Tools and Geometry Produced

When operating properly, less than 1% of the material is removed by normal chip formation. Shaping and sharpening carbide cutting tools is one of the major uses of ECG because normal grinding causes high wear rates on expensive diamond wheels. Also, it produces a smoother surface and does not increase surface stresses.

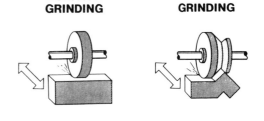

SURFACE GRINDING   FORM GRINDING

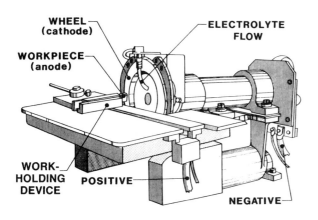

WHEEL (cathode)

ELECTROLYTE FLOW

WORKPIECE (anode)

WORK-HOLDING DEVICE   POSITIVE

NEGATIVE

Electrochemical Machining

Electrochemical machining (ECM) is an electrolytic material removal process involving a negatively charged shaped electrode (cathode), a conductive fluid (electrolyte), and a conductive workpiece (anode). ECM is characterized as "reverse electroplating." The tool must be properly shaped, and provision for waste removal is essential.

Process Characteristics

* Uses the principle of electrolysis (reverse electroplating)
* Requires no masking materials
* Requires electrically conductive workpieces
* Has shaped tools (cathode)
* Produces good tolerances and surfaces
* Has no electrode wear
* Produces complex geometries, often in one operation

Process Schematic

The negatively charged electrode is advanced into the positively charged workpiece. The pressurized electrolyte is pumped at a controlled temperature to the cutting area. The feed rate of the electrode exactly matches the rate of dissolution of the material. The gap between the workpiece and the tool ranges from 0.003 in. to 0.030 in. ECM provides faster metal-removal rates than any other nontraditional machining process.

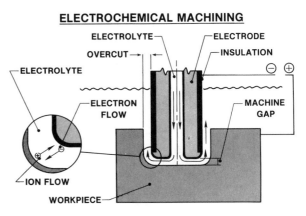

ELECTROCHEMICAL MACHINING

Workpiece Geometry

ECM is mainly used for mass production of complex shapes in difficult to machine materials. A burr-free finish is produced with no danger of metallurgical damage to the workpiece. Close tolerances can be obtained because there is very little wear on the electrode.

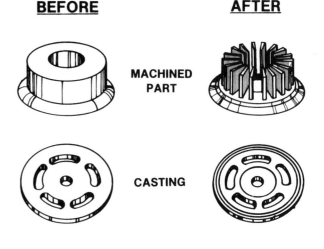

BEFORE   **AFTER**

MACHINED PART

CASTING

Setup and Equipment

Both vertical- and horizontal-type ECM machines are available in many sizes. Depending on the special work requirements, electrochemical machines are built in both standard and special sizes. Shown is a vertical electrochemical machine with a base, column, table, and spindle head. The spindle head contains a servo-mechanism that automatically advances the tool and controls the gap between the electrode and the workpiece.

Typical Tools and Geometry Produced

External or internal geometries are easily machined with ECM. Because the surface finish of the tool will be reproduced on the workpiece, electrode accuracy is very important. Copper is used quite often as an electrode material. Brass, graphite, and copper–tungsten are also used because they are easily machined, electrically conductive, and corrosion-resistant.

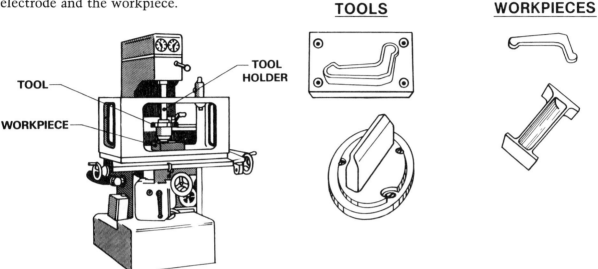

TOOL

WORKPIECE

TOOL
HOLDER

TOOLS

WORKPIECES

Immersion Chemical Milling/Blanking

Immersion chemical milling/blanking is a chemical process that dissolves material from unmasked (unprotected) areas of metallic parts immersed in a tank of heated and agitated chemical reagents. The term "blanking" denotes small, thin workpieces, and "milling" indicates relatively large workpieces.

Process Characteristics

* Parts are immersed in an acidic or basic chemical reagent
* Masks protect areas from being chemically milled
* Produces close tolerances and fine detail
* Requires no finishing of workpiece surface
* Requires no electric current to remove material

Process Schematic

During a chemical milling or blanking operation, the workpiece is first prepared for masking by a degreasing and pickling process. Then the workpiece is masked, leaving the portion to be milled exposed, or unmasked. The workpiece is then placed in a chemical etchant tank for a predetermined length of time or until the desired material has been etched away. After the workpiece is rinsed and neutralized, the mask is then removed, and the workpiece is prepared for subsequent finishing or inspection. Heating or cooling is used to maintain constant bath temperature.

IMMERSION CHEMICAL MILLING

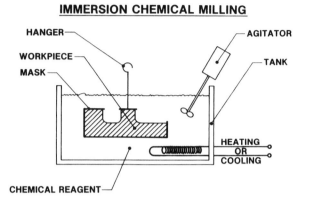

Workpiece Geometry

Shown are typical workpieces. Note the intricate geometrical possibilities and fine detail that can be milled, limited only by the quality of the original (graphic or art layout), the quality of the mask, and the depth of etch. Deep etching may cause undercutting with consequent difficulty in maintaining close dimensional tolerances.

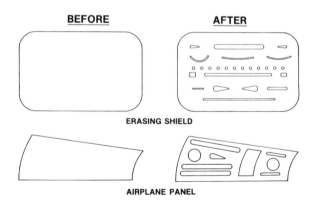

Setup and Equipment

Essential equipment for an immersion chemical milling or blanking process consists of a series of immersion tanks for degreasing, pickling, chemical etching, and rinsing; a chemical reservoir for storing the circulating etchant; and an exhaust system for ventilating the process area.

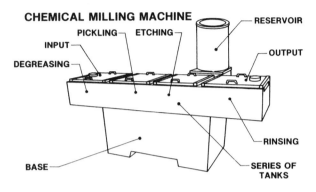

CHEMICAL MILLING MACHINE — RESERVOIR

PICKLING — ETCHING —
INPUT —
DEGREASING —
— OUTPUT
— RINSING
BASE —
— SERIES OF TANKS

Typical Tools and Geometry Produced

The principal element of this process is the chemical solution used to dissolve or etch away the unprotected areas of the workpiece. Typical shapes produced include miniature rotor parts, precision sheet metal parts, and circuit boards.

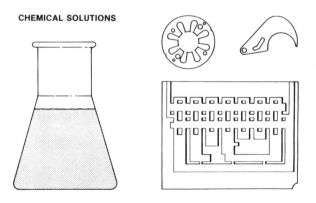

CHEMICAL SOLUTIONS

Geometrical Possibilities

Shown are some of the shapes and features that chemical milling can produce. Geometrical possibilities range from complex recesses to small, intricate designs. The depth of cut ranges from less than 0.01 in. to 0.8 in. The width of the cut is generally limited by the minimum width of the maskants or size of the tanks available.

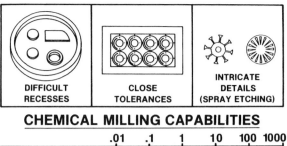

DIFFICULT RECESSES | CLOSE TOLERANCES | INTRICATE DETAILS (SPRAY ETCHING)

CHEMICAL MILLING CAPABILITIES

	.01	.1	1	10	100	1000
DEPTH OF CUT (IN)						
WIDTH OF CUT (IN)						

■ TYPICAL RANGE   ▨ FEASIBLE RANGE

Tolerances and Surface Finish

Tolerances and surface finish depend greatly on surface conditions, etch time, and purity of the work material. Depending on depth of cut, tolerances range from ±0.005 in. to ±0.001 in. The surface finish is usually from 14 to 260 microinches, depending on the original surface condition.

TOLERANCES		
IMMERSION CHEMICAL MILLING	TYPICAL	FEASIBLE
DEPTH OF CUT (IN)	±0.005	±0.001

SURFACE FINISH								
PROCESS	MICROINCHES (A.A.)							
IMMERSION CHEMICAL MILLING	2	4	8	16	32	64	125	250

■ TYPICAL RANGE   ▨ FEASIBLE RANGE

Process Materials

The type of maskants used in this process depend on two factors: the type of material that it is to be placed on and the type of etchant that it must resist. The main masking materials are poly(vinyl chloride), polyethylene, butyl rubber, neoprene rubber, and acrylonitrile. Typical etchants are listed along with the material groups for which they are intended.

Workpiece materials	Etchant	Masking materials
Aluminum/alloys Magnesium/alloys Titanium	$FeCl_3$ NaOH HNO_3* HF	Styrene butadiene, poly(vinyl chloride), polyethylene, and butyl rubber
Copper/alloys Tool steels Carbon steels Stainless steels	$FeCl_3$ HNO_3 $FeCl_3$ $FeCl_3$	Styrene butadiene, neoprene rubber, acrylonitrile, and butyl rubber

* Also used as a desmutting or cleansing agent.

Cleaning, Desmutting

Workpiece materials	Cleaning medium
Aluminum Titanium	* HNO_3 * Nonetch alkaline cleaner * Chromate deoxidizer

Workholding Methods

Shown is an overhead conveyor system used in immersion chemical milling. The workpiece is suspended in the chemical solution with specially treated racks, hooks, grips, or baskets.

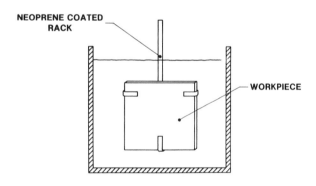

NEOPRENE COATED RACK

WORKPIECE

Effects on Work Material Properties

Immersion chemical milling has little effect on the mechanical properties of the workpiece material. It reduces surface residual stresses and fatigue strength. Certain alloys may be subject to hydrogen embrittlement (caused by hydrogen being absorbed at the grain boundaries) and must be given a postbake treatment to liberate entrapped hydrogen.

Work material properties	Effects of immersion chemical milling
Mechanical	* Surface residual stresses may be reduced * Fatigue strength may be reduced
Physical Chemical	* Little effect * Possible hydrogen embrittlement

Typical Workpiece Materials

Shown in this frame are the commonly used materials. Most materials can be chemically milled. Ratings for this process are high because a variety of etchants are available. Choice of etchant depends on the workpiece, as each material has a most suitable etchant.

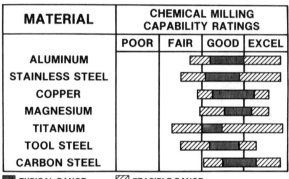

MATERIAL	CHEMICAL MILLING CAPABILITY RATINGS			
	POOR	FAIR	GOOD	EXCEL
ALUMINUM				
STAINLESS STEEL				
COPPER				
MAGNESIUM				
TITANIUM				
TOOL STEEL				
CARBON STEEL				

▓ TYPICAL RANGE   ▨ FEASIBLE RANGE

Process Conditions

Process conditions involve the etching factor, rate, temperature, and time. The etch factor is the ratio of undercut to the depth of the etch. The etching rate is the speed at which material is removed.

Etchant	Etch factor (undercut/depth of cut)	Etching rate (in./min)
$FeCl_3$	1.5 to 3.0	0.0005 to 0.002
NaOH	1.5 to 2.5	0.0010 to 0.0015
HNO_3	1.0 to 2.0	0.0005 to 0.002
HF	1.5 to 2.5	0.0007 to 0.001

Pre-/Post-Etch Treatments

Pre- and post-etch treatments include degreasing, pickling, rinsing, and mask removal. If hydrogen embrittlement occurs, a postbake treatment is applied.

Pre-etch treatments

* Degreasing
* Pickling

Post-etch treatments

* Rinsing
* Mask removal
* Postbake (optional)

Chemical Etchants

Shown are examples of etchants with their usual operating temperatures and concentrations (in volume percent). Temperature and concentration of the etchant must be closely monitored and maintained by frequent filtering and replenishing to obtain repeatable results.

Etchant	Temperature (°C)	Concentration (%)
$FeCl_3$	48.8	42
NaOH	66 to 102	10 to 20
HNO_3	49	10 to 15
HF	41	5 to 15

Factors Affecting Process Results

Tolerance and surface finish depend upon the following:

* Condition of etchant
* Type of etchant
* Type of material
* Etchant time
* Etchant temperature
* Type and quality of masking
* Workpiece surface conditions
* Workpiece impurity content

Power Requirements

Power requirements are given in calories per gallon per degree celsius. Also given are the desired bath operating temperature levels.

Etchant	Temperature (°C)	Specific heat (cal/gal per °C)
$FeCl_3$	48.8	—
NaOH	45 to 58.3	0.8
HNO_3	49	0.81
HF	37.7 to 48.8	0.79

Cost Elements

Cost elements include the following:

* Pretreatment time
* Masking/demasking time
* Etching time
* Solution analysis and maintenance
* Etchant and masking costs
* Direct labor rate
* Overhead rate
* Amortization of equipment and tooling

Time Calculation

The amount of material to be removed is the main variable affecting etching time.

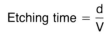

Etching rate (in./min) = V
Amount of etching (in.) = d

$$\text{Etching time} = \frac{d}{V}$$

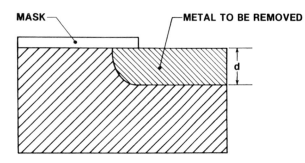

Safety Factors

The following risks should be taken into consideration:

* Personal
  - Eye and skin contact with chemicals
  - Inhalation of vapors
* Environmental
  - Chemical pollutants

Photo Etching

Photo etching is a chemical etching process that dissolves material from unmasked areas of metallic or nonmetallic parts. The maskant geometry is photographically exposed on the workpiece using ultraviolet light. Parts produced by photo etching are generally thin-gage, flat, and complex in design.

Process Characteristics

* Is a common form of chemical blanking
* Uses a chemical cutting agent
* Requires a pattern created by photoresist material
* Produces accurate, burr-free, complex, thin-gage parts
* Is typically limited to a material thickness of 0.0001 in. to 0.050 in.

Process Schematic

Here are the steps of this process: prepare the artwork or master drawing; photographically reduce the artwork to produce the image master; clean and degrease the workpiece material; apply the photoresist material; expose and develop the photoresist; etch the workpiece; rinse; remove resist material; and dry the finished workpiece.

Workpiece Geometry

Workpiece geometry is mainly influenced by the accuracy of the initial drawing and photographically reduced artwork that is used to expose the photoresist material. Other factors include material type and size, production requirements, and accuracy of photographic and processing equipment.

BEFORE   **AFTER**

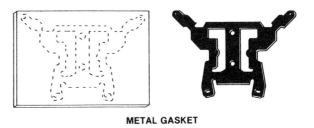

METAL GASKET

Setup and Equipment

The workpiece moves through a resist-coating area, an optional prebaking oven or drying cycle, an etching chamber, a rinse, and a final drying chamber. In the etching chamber, the workpiece is placed against a negative, exposed to ultraviolet light, and then developed in a suitable solution.

PHOTO ETCHING

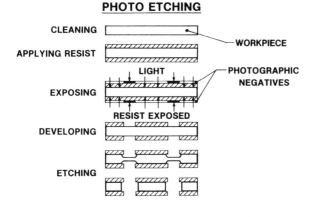

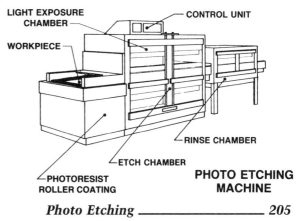

PHOTO ETCHING MACHINE

Typical Tools and Geometry Produced

Typical tools include an ultraviolet light source, photoresist material, and a developing solution, such as water or an alkaline solution. The workpiece and photoresist material will influence the choice of developing solution. Most workpieces are extremely thin, require use of brittle materials, or are complex in design. A printed circuit board is an example of one-sided etching. Gaskets represent fully or double-sided etching capabilities.

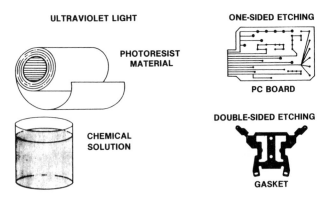

Geometrical Possibilities

Geometrical configurations are extended to virtually any shape that can be drawn and photographically reduced with quality results. Size is a process constraint with typical lengths varying from 0.001 in. to less than 2 in., with feasible lengths of up to 5 ft. Thickness values of less than 0.001 in. to 0.060 in. are typical, with feasible etching thickness to 2 inches. It is necessary to minimize the sizes of holes and slots in order to produce satisfactory results.

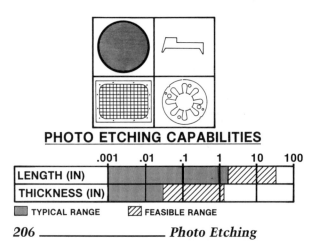

Tolerances and Surface Finish

Tolerances may vary with workpiece size and material. As material thickness increases, the tolerance value generally decreases. Typical tolerances in overall size range from ±0.005 in. to a feasible zero deviation. Material thickness values range from a typical value of ±0.002 in. to zero. Surface finish depends on the material's initial finish with typical and feasible ranges as shown.

TOLERANCES		
PHOTO ETCHING	TYPICAL	FEASIBLE
OVERALL (IN.)	±0.005	±0.000
THICKNESS (IN.)	±0.002	±0.000

SURFACE FINISH									
PROCESS	MICROINCHES (A.A.)								
PHOTO ETCHING	2	4	8	16	32	63	125	250	500

■ TYPICAL RANGE   ▨ FEASIBLE RANGE

Process Materials

Process materials include photoresist, etchant, and ultraviolet light source. Photoresist may be applied by dipping, roller coating, or spraying, depending on the workpiece requirements. Various etchants are available with material, etch depth, and resultant surface finish to be considered as variables.

Description	Style	Application
Photoresist	Dip	Simplest method
	Coat	Uniform thickness
	Spray	Most versatile
Etchant	$FeCl_3$	Aluminum, copper, steel
	HNO_3	Steel, magnesium
	HF	Titanium
Light source	Ultraviolet	All materials

Workholding Methods

A vacuum printing frame is used to register the master workpiece panel precisely. Vacuum provides intimate contact with the photoresist covered workpiece. A vacuum of 20 lbs will insure good maskant contact with the workpiece. An ultraviolet light source is able to penetrate the plastic sheeting and glass holding frame.

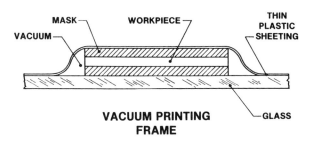

VACUUM PRINTING FRAME

Effects on Work Material Properties

Effects of photo etching on the workpiece are minimal if the process is properly controlled throughout. The mechanical effects may include possible part fatigue or surface softening.

Work material properties	Effects of photo etching
Mechanical	* Possible part fatigue * Surface softening
Physical	* Little effect
Chemical	* Little effect

Typical Workpiece Materials

Sheet, plate, or foil workpiece materials come in aluminum, brass, nickel, mild steel, molybdenum, copper, and titanium. The etchability ratings for each are shown with typical and feasible ranges.

MATERIAL	PHOTO ETCHING ETCHABILITY RATINGS			
	POOR	FAIR	GOOD	EXCEL
ALUMINUM				
BRASS				
NICKEL				
MILD STEEL				
MOLYBDENUM				
COPPER				
TITANIUM				

■ TYPICAL RANGE   ▨ FEASIBLE RANGE

Process Conditions

Process conditions include material, etchant, etch rate (in./min), and temperature. Processing time depends on the etch rate, which is controlled through temperature, and etchant concentration and composition.

Material	Etchant	Temp. (°C)	Etch rate (in./min)
Aluminum alloy	$FeCl_3$ NaOH	50 75–94	0.001 0.0005–0.001
Carbon and low Alloy steel	$FeCl_3$ CrO_3	50 50	0.001 0.001
Stainless steel	$FeCl_3$	55	0.0008
Magnesium alloy	HNO_3	30–50	0.0001–0.002
Titanium	HF	30–50	0.0005–0.001
Copper alloy	$FeCl_3$ HNO_3	50 50	0.002 0.001–0.0015

Cooling and Neutralizing Media

All work materials utilize water or a proprietary spray rinse to remove excess etchant and remaining photoresist materials.

Work material	Medium	Application
All materials	Water	* Remove excess etchant * Remove photoresist

Etchants

Part production depends on the ability of a material to be successfully removed through the etching process. Choice of an etchant depends on the material used, required quality, removal rate, and process economics. Typical etchants for various materials are shown.

Workpiece materials	Etchants
Aluminum alloy	$FeCl_3$, HCl + HNO_3, NaOH
Copper alloy	$FeCl_3$, CrO_3
Mild steel	HNO_3, $FeCl_3$
Stainless steel	$FeCl_3$
Magnesium	HNO_3
Titanium	HF, HF + HNO_3

Factors Affecting Process Results

Tolerance and surface finish depend upon the following:

* Accuracy of initial artwork and photographic processes
* Artwork undercut compensation
* Processing uniformity
* Workpiece material
* Etchant
* Type of resist

Power Requirements

A comparison chart lists the three most important features of a photoresist exposure lamp. Photoresist is not sensitive to normal lighting except fluorescent (gold fluorescent or tungsten may be used). Typical exposure times range from 25 seconds to 8 minutes, depending on the resist, coating thickness, and etchant used.

Light source	Lamp life	Replacement costs	Operating costs (kW/hr)
Mercury vapor	Excel.	Excel.	Good
Carbon arc	Fair	Fair	Fair
Pulsed xenon	Good	Good	Fair
Black light fluorescent	Excel.	Excel.	Excel.

Excel. = excellent.

Cost Elements

Cost elements include the following:

* Pretreatment time
* Masking/demasking time
* Etching time
* Direct labor rate
* Overhead rate
* Amortization of equipment and tooling

Time Calculation

Etching time is obtained by dividing the etching depth, d, by the etching rate, V.

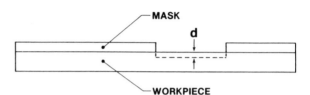

Etching rate (ipm) = V
Desired depth of etching (in.) = d

$$\text{Etching time} = \frac{d}{V}$$

Safety Factors

The following hazards should be taken into consideration:

* Personal
  - Skin irritation
  - Danger to eyes from chemicals and radiation
  - Inhalation of toxic vapor
* Environmental
  - Chemical caustic disposal
  - Pollution

Consolidation

Axial Powder Compaction _____

Axial powder compaction is a metal powder process in which fine metal particles are pressed into a desired shape in a metal die under high pressure. The "green" compact is then sintered in a furnace with a protective atmosphere to bond the powdered metal particles together metallurgically.

Process Characteristics

* Compacts powdered metal in a die
* Parts are sintered to increase strength through metallurgical bonding
* Finished parts have a smooth finish and close tolerance
* Produces very little scrap material
* Can use alloy combinations and fillers
* Cannot produce undercuts, threads, or cross holes

Process Schematic

Powdered metal is placed into a die cavity. The press forces the upper and lower punches together, which presses the powder into the shape of a tooling cavity. Pressures of 10 tons/in.^{2} to 50 tons/in.^{2} are commonly used for metal powder compaction. To attain the same compression ratio across a component with more than one level or height, it is necessary to work with multiple lower punches.

POWDER COMPACTION

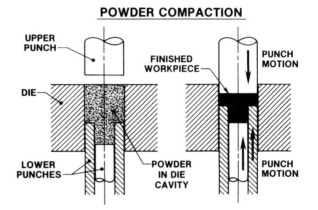

Workpiece Geometry

Both simple and complex geometries can be produced. Shown is a bushing, a blank with different diameters, and a fairly complicated gear. Production rates of 15 to 30 parts per minute are common.

BEFORE

POWDERED METAL

AFTER

Setup and Equipment

This process uses a hydraulic or mechanical press to form powdered metal into a desired shape. The press has upper and lower punches and a metal die. The die and punches are used to form the powdered metal into the desired shape.

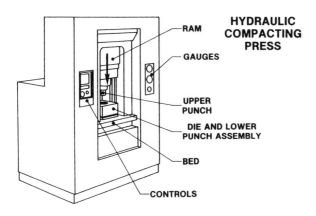

HYDRAULIC COMPACTING PRESS

Typical Tools and Geometry Produced

A cylindrical workpiece is made by single-level tooling. A more complex shape can be made by the common multiple-level tooling. Note that the multilevel part requires more tooling than the single-level part.

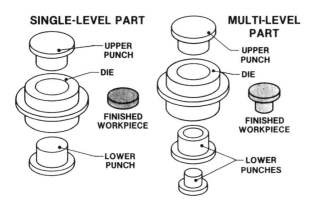

Geometrical Possibilities

One of the major advantages of this process is its ability to produce complex geometries. Parts with undercuts and threads require a secondary machining operation. Typical part sizes range from 0.1 in.^{2} to 20 in.^{2} in area and from 0.1 in. to 4 in. in length. However, it is possible to produce parts that are less than 0.1 in.^{2} and larger than 25 in.^{2} in area and from a fraction of an inch to approximately 8 in. in length.

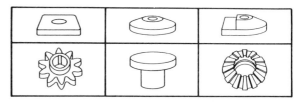

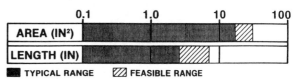

COMPACTION CAPABILITIES

Tolerances and Surface Finish

Tolerances for powdered metal components vary from ±0.0005 in. to ±0.003 in. according to the following process cycles: pressing and sintering; pressing, sintering, and sizing; pressing, presintering, coining, and sintering; and pressing, sintering, and heat treatment. Surface finish on pressed and sintered parts is usually between 2 and 8 microinches.

TOLERANCES (IN)		
PROCESS CYCLE	**AXIAL**	**RADIAL**
PRESSING, SINTERING	±.003	±.001
PRESSING, SINTERING, SIZING	±.002	±.0005
PRESSING, PRESINTERING, COINING, SINTERING	±.0015	±.0005
PRESSING, SINTERING, HEAT TREATMENT	±.003	±.0015

SURFACE FINISH								
MICROINCHES	1	2	4	8	16	32	63	125
COMPACTION								

▆ TYPICAL RANGE ▨ FEASIBLE RANGE

Tool Style

Tool style can be subdivided into four major classes: single-action compaction, which should only be used for thin, flat components; opposed double-action with two punch motions, which accommodates thicker components; double-action with floating die; and double-action withdrawal die. The latter three classes give a good density distribution.

Tooling must have a design that ensures that the pressed component will have the desired shape, size, and density. In general, double-action dies with multiple punches are required to achieve even density in complex parts.

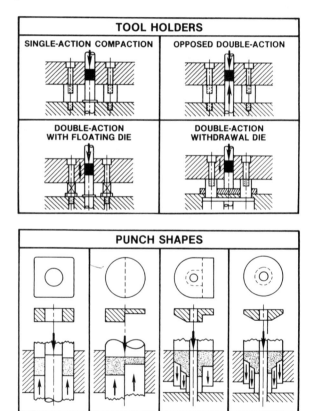

Design Considerations

* Component must be ejectable from the die
* Maximum surface area is normally below 20.0 in.^{2}
* Minimum wall thickness is 0.08 in.
* Adjacent wall thickness ratios greater than 2.5 to 1 should be avoided
* Height to diameter ratios over 7 to 1 should be avoided
* Sharp corners should be avoided
* Undercuts, reliefs, threads, and crossholes require secondary machining operations

Typical Workpiece Materials

Better properties, density, yield strength, tensile strength, and hardness may be obtained by repressing and resintering.

Obtainable properties				
Workpiece material	Density (grams/cc)	Yield strength (psi)	Tensile strength (psi)	Hardness (HB)
Iron	5.2 to 7.0	5.1×10^3 to 2.3×10^4	7.3×10^3 to 2.9×10^4	40 to 70
Low alloy steel	6.3 to 7.4	1.5×10^4 to 2.9×10^4	2.0×10^4 to 4.4×10^4	60 to 100
Alloyed steel	6.8 to 7.4	2.6×10^4 to 8.4×10^4	2.9×10^4 to 9.4×10^4	60 and up
Stainless steel	6.3 to 7.6	3.6×10^4 to 7.3×10^4	4.4×10^4 to 8.7×10^4	60 and up
Bronze	5.5 to 7.5	1.1×10^4 to 2.9×10^4	1.5×10^4 to 4.4×10^4	50 to 70
Brass	7.0 to 7.9	1.1×10^4 to 2.9×10^4	1.6×10^4 to 3.5×10^4	60

Tool/Die Materials

Dies, punches, and core rods must be polished and wear-resistant. Base plates must be dimensioned for minimum bending during compaction. Inserts are an option that may be used for improving die design.

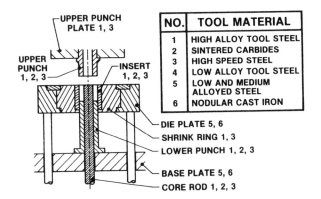

NO.	TOOL MATERIAL
1	HIGH ALLOY TOOL STEEL
2	SINTERED CARBIDES
3	HIGH SPEED STEEL
4	LOW ALLOY TOOL STEEL
5	LOW AND MEDIUM ALLOYED STEEL
6	NODULAR CAST IRON

UPPER PUNCH PLATE 1, 3
UPPER PUNCH 1, 2, 3
INSERT 1, 2, 3
DIE PLATE 5, 6
SHRINK RING 1, 3
LOWER PUNCH 1, 2, 3
BASE PLATE 5, 6
CORE ROD 1, 2, 3

Factors Affecting Process Results

Process capabilities for axial powder compaction depend on the press size, the type of material, and size of the workpiece. Given are two press ranges and the corresponding workpiece sizes. The production rate is usually inversely related to press tonnage.

Press capacity (tons)	Maximum size parts	
	Area (in.^{2})	Length (in.)
35–300	4	3
300–1000	10	4

Power Requirements

Within limits, the density of compacted powdered metal is directly proportional to the amount of pressure applied. The greater the compacting pressure, the greater the density obtained. Pressures from 1000 lbs/in.^{2} to 1,000,000 lbs/in.^{2} have been obtained, but 80 lbs/in.^{2} to 1000 lbs/in.^{2} are common.

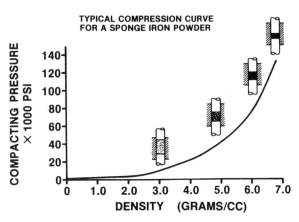

TYPICAL COMPRESSION CURVE FOR A SPONGE IRON POWDER

Process Conditions

The following parameters influence the process:

* Powder characteristics
* Powder preparation (i.e., mixing and blending)
* Type of compaction press
* Design of compacting tools and dies
* Type of sintering furnace
* Composition of sintering atmosphere
* Rate of production cycle

Subsequent Treatments

After sintering, the following treatments may be performed: machining to obtain geometries that cannot be formed, barrel tumbling to improve surface finish and remove sharp edges, coating by most conventional processes, and oil impregnation to provide lifetime self-lubrication for bushings, etc., where appropriate.

Sintering	* Phases – Heating to sintering temperature – Holding temperature in specified time – Cooling * Temperature – 70–80% of melting point * Atmosphere – Inert reducing or carburizing
Machining	* Threads * Undercuts * Crossholes * Grinding
Barrel tumbling	* Deburring * Surface finishing
Coating	* Most conventional processes
Impregnation	* Oil impregnation * Plastic (sealing) * Low alloy metal (eliminates porosity)

Cost Elements

Cost elements include the following:

* Setup time
* Load/unload time
* Idle time
* Compaction time
* Tool change time
* Tool costs
* Direct labor rate
* Overhead rate
* Amortization of equipment and tooling

Productivity Tip

Choice of equipment and procedure depends on the following:

* Projected area of the workpiece
* Complexity of part design
* Compaction principle
* Equipment capacity (power, speed)
* Production quantity
* Tolerances
* Automation requirements

Time Calculation

To calculate the cycle time for a workpiece add the fill time, the compaction time, the press opening time, and compact ejection time. Cycle time may vary from less than a fraction of a second for small tableting type machines to several seconds for high tonnage presses.

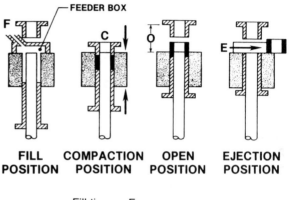

Fill time = F
Compaction time = C
Press opening time = O
Compact ejection time = E

Time per piece = F + C + O + E

Safety Factors

The following risks should be taken into consideration:

* Personal
  − Lung irritation
  − Abrasion
* Environmental
  − Explosion danger

Blow Molding

In blow molding, hollow products are formed by extruding a heated thermoplastic tube (parison) into a mold, and, under air pressure, expanding the parison to match the inner contours of the mold.

Process Characteristics

* Inflates a softened parison tube to the contour of a mold cavity
* Uses thermoplastics
* Forms thin-walled hollow products
* Parting lines are present
* Wall thickness can be increased by increasing the parison tube wall thickness
* Flash is present but is minimal

Process Schematic

This setup is typical of many blow molding processes. The parison tube is positioned in the mold cavity, and air pressure is then applied to the parison, which forces the plastic to form to the mold cavity. Parts are usually quite uniform in thickness and are formed within a relatively short cycle time.

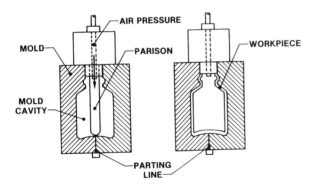

Workpiece Geometry

Raw materials for blow molding are generally in the form of either pellets or granular compounds. The workpieces produced are generally hollow containers that have a parting line and are quite uniform in thickness.

BEFORE   **AFTER**

PELLETS

HOLLOW CONTAINERS

GRANULAR COMPOUND

Setup and Equipment

In this typical setup the mold halves are in the retracted position. Granules are put in the hopper on the top and are melted and forced into the extrusion die by a rotating screw. An extruded parison is clamped between the mold halves, and pressurized air expands it to fill the mold cavity. The part is then removed from the mold by separating the mold halves.

Typical Tools and Geometry Produced

These molds form plastic bottles and other containers. A mold usually has two halves, with the cavity being in the shape of the desired container.

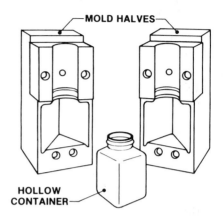

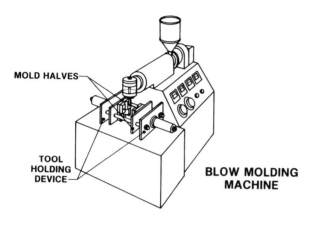

Cold Chamber Die Casting ———————

Die casting is a metal shaping process in which molten metal is forced into a reusable mold and held under pressure until solidification occurs. Of all the casting processes, die casting is considered the fastest.

Process Characteristics

* Uses pressure to force molten metal into a reusable mold
* Produces good dimensional accuracy and excellent detail
* Requires minimal machining
* May produce a small amount of flash around casting edge
* Leaves small ejector pin marks

Process Schematic

In a typical cold chamber process, a horizontal ram is used to force molten metal from the chamber into the die. Each shot of metal is fed individually into the cold chamber. Usually, water is circulated in the die to hasten cooling, to avoid air pockets in the cast part, and to preserve the life of the die.

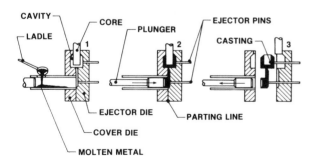

Workpiece Geometry

Die-cast parts are made from low-melting-point alloys, such as zinc/zinc alloys, magnesium/magnesium alloys, aluminum/aluminum alloys, and copper/copper alloys. Excellent details and smooth surface finishes can be achieved because of accurate metal dies and high casting pressure. Because the cost of the dies is high, die casting is only economical for large production runs.

BEFORE

MOLTEN METAL

AFTER

Setup and Equipment

This cold chamber die-casting machine is typical of the equipment currently in use. Cold chamber injection systems can be used for all metals that can be die cast. Molten metal injected under high pressure may adversely affect the die; using a cold chamber reduces this possibility and is a major advantage of this process.

Typical Tools and Geometry Produced

Dies are usually made of alloy steel and have two or more sections that meet on vertical parting lines so that they can be opened to remove the cast part. One section is called the ejector half, and the other is called the cover half. The dies must be durable to resist the high pressure and high temperature of the molten metal.

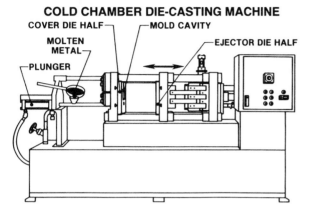

COLD CHAMBER DIE-CASTING MACHINE

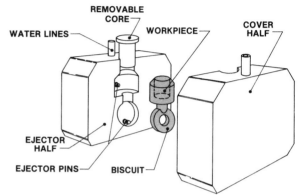

Compression Molding

Compression molding is a forming process in which a thermoset plastic material is introduced directly into a heated metal mold, is softened by the heat, and forced to conform to the shape of the mold cavity as the mold closes.

Process Characteristics

* Uses thermoset preforms or granules
* Materials are usually preheated
* Material must be accurately measured to maintain uniform size or to avoid excess flash
* Metallic inserts may be molded into the product
* Shape must not have undercuts
* Requires no sprues, gates, or runners

Process Schematic

A premeasured amount of thermosetting plastic in powder, preform, or granular form is placed in the heated female mold cavity. The mold is then closed, and the part is formed by heat and pressure. After the molded part has cured, the mold is opened, and the ejection pin pushes the part out of the mold.

Workpiece Geometry

This process is commonly used for manufacturing electrical parts, dinnerware, and gears. This process is also used to produce buttons, buckles, knobs, handles, appliance housings, radio cases, and large containers.

BEFORE   **AFTER**

GRANULAR COMPOUND

ELECTRICAL BOX

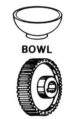

BOWL

PREFORMS

MOLDED GEAR

COMPRESSION MOLDING

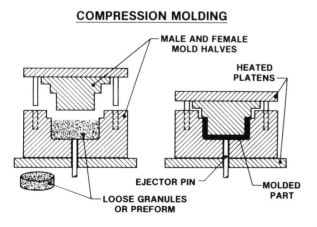

MALE AND FEMALE MOLD HALVES

HEATED PLATENS

EJECTOR PIN

MOLDED PART

LOOSE GRANULES OR PREFORM

Setup and Equipment

Compression mold presses are manufactured in a wide variety of sizes. Most presses utilize a hydraulic ram in order to produce sufficient force during the molding operation. The tools consist of a male mold plunger and a female mold.

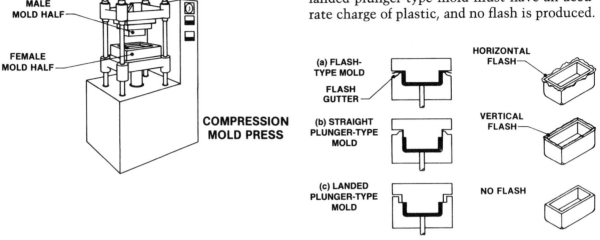

MALE MOLD HALF

FEMALE MOLD HALF

COMPRESSION MOLD PRESS

Typical Tools and Geometry Produced

Three types of molds used are the flash-type, straight plunger-type, and the "landed" plunger-type molds. The flash-type mold must have an accurate charge of plastic and produces a horizontal flash. The straight plunger-type mold allows for some inaccuracy in the charge of plastic and produces a vertical flash. The landed plunger-type mold must have an accurate charge of plastic, and no flash is produced.

(a) FLASH-TYPE MOLD

FLASH GUTTER

HORIZONTAL FLASH

(b) STRAIGHT PLUNGER-TYPE MOLD

VERTICAL FLASH

(c) LANDED PLUNGER-TYPE MOLD

NO FLASH

Cored Sand Casting

Cored sand casting is a process that involves the assembly (with glue or steel bands) of various mold sections or cores to form a casting mold. Cores made of sand and composites vary greatly in size and complexity. With cored sand casting, shapes not obtainable through other casting processes can be produced.

Process Characteristics

* Utilizes core sections to form complex molds
* Cores are made of hardened or baked sand
* Produces shapes unattainable through other casting processes
* Is a low production process
* Cores are assembled with glue or steel bands
* Can form parts requiring more than one parting line

Process Schematic

Shown in this frame is an example of an assembled mold. One advantage of this process is its ability to handle very complex shapes and yet keep fairly simple cores.

Workpiece Geometry

Creating an engine block is complex enough to require a process such as cored sand casting.

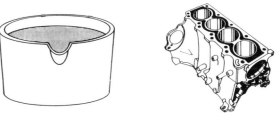

BEFORE   **AFTER**

MOLTEN METAL   ENGINE BLOCKS

Setup and Equipment

Shown here is an example of the setup and equipment used. The cores are generally assembled with glue and steel bands. Once assembled, the mold can be quite large and complex. This process is used for low production runs. Examples of such parts are special engine blocks and pump housings.

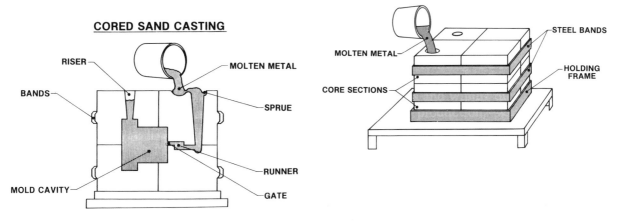

CORED SAND CASTING

RISER — MOLTEN METAL
BANDS — SPRUE
MOLD CAVITY — RUNNER
— GATE

MOLTEN METAL — STEEL BANDS
CORE SECTIONS — HOLDING FRAME

Typical Tools and Geometry Produced

Shown here are several examples of cores with the workpiece produced. There are many geometric possibilities available with cored sand casting. The cores are often designed with match points built into them to aid alignment.

CORES

WORKPIECE

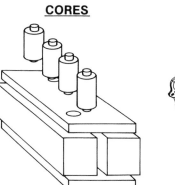

ENGINE BLOCK

Extrusion Molding

Extrusion molding is a plastic shaping process in which a continuous workpiece is produced by forcing molten thermoplastic material through a shaped die orifice. As the hot plastic workpiece is carried along a conveyor, it is cooled and cut to the desired length.

Process Characteristics

* Is a continuous, high volume process
* Accurately controls material thickness
* Products are cut to desired lengths
* Has low tooling costs
* Can produce intricate profiles

Process Schematic

Thermoplastic materials are fed from a hopper into the heated barrel of an extruder. A rotating helical screw inside the barrel pushes the plastic through the barrel toward the die located at the end of the machine as the plastic progresses along the barrel. A heating jacket carefully controls the temperature of the plastic. Molten plastic is then forced through the die opening and around the mandrel to produce a hollow workpiece.

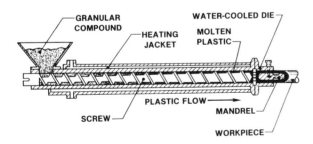

Workpiece Geometry

Granular compounds and pellets are used in extrusion molding. Scrap parts may be chopped and mixed with virgin materials. Generally, only thermoplastic materials are extruded. Typical profile extrusions are pipe, film or sheet, rain gutter components, and window components.

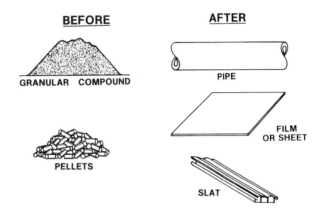

Setup and Equipment

Equipment includes an extrusion machine, a cooling trough, and a puller. As molten plastic is forced through the die opening, it assumes its desired shape. To maintain this desired shape, cooling fixtures are used, such as the water trough shown. When the plastic has cooled enough to regain some strength properties, it is pulled by the *puller* and cut to the desired length.

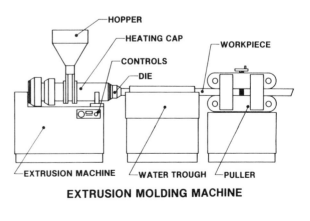

EXTRUSION MOLDING MACHINE

Typical Tools and Geometry Produced

Shown are some typical dies used for extrusion molding. Each extrusion die must be tested and modified to achieve the desired workpiece profile. Dies are usually made of tool steel, and the simplicity of the die makes it cheaper than the tooling of other plastic processes.

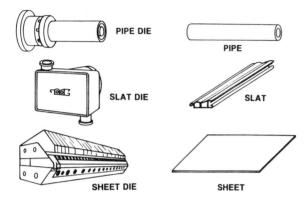

Geometrical Possibilities

There are many geometrical possibilities with extrusion molding. Simple shapes, such as round pipes and square extrusions, are common. More complex shapes, such as tracks and profiles, are also frequently produced by extrusion molding. The thickness for extrusions ranges from 0.005 in. to around 0.3 in. The length of an extrusion is typically between 10 ft and 100 ft.

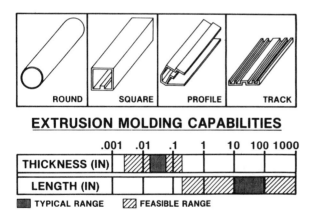

Tolerances and Surface Finish

Dimensional tolerances for extrusions typically range from ±8% to ±10% of workpiece thickness and around ±4° to ±5° on angles. The surface finish can range from 8 to 75 microinches. Extruded parts usually require very little finishing work.

TOLERANCES		
EXTRUSION MOLDING	**TYPICAL**	**FEASIBLE**
THICKNESS (%)	±8 to 10	±6 to 8
ANGLES (°)	±4 to 5	±2 to 3

SURFACE FINISH	
PROCESS	**MICROINCHES (A.A.)**
EXTRUSION MOLDING	2 4 8 16 32 63 125 250 500

■ TYPICAL RANGE   ▨ FEASIBLE RANGE

Tool Style

Shown are three different types of dies. The plate die is a low volume, low cost die used for prototype or developmental dies. The semi-streamlined die is a modified plate die. Slight radii are added to enhance material flow. The fully streamlined die is a high production die designed to distribute material flow evenly and produce uniform quality parts.

DESCRIPTION	STYLE	APPLICATION
PLATE DIE		LEAST EXPENSIVE LOW VOLUME PRODUCTION
SEMISTREAMLINED DIE		RADII ALLOWS FOR BETTER MATERIAL FLOW
FULLY STREAMLINED DIE		FULLY DISTRIBUTED MATERIAL FLOW FOR HIGH VOLUME PRODUCTION

Design Considerations

Various designs are shown. The first column illustrates examples of difficult shapes to produce because of hollow interior sections, the middle column shows modified solid designs, and the last column shows designs that not only are of the desired shape but also are easy to fabricate. As can be seen, a more uniform material thickness is preferred, and open sections are preferred to closed sections.

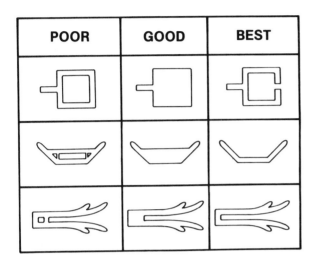

Effects on Work Material Properties

If the processing temperature becomes too high, chemical breakdowns may occur. There may also be shrinkage and warpage.

Work material properties	Effects of extrusion molding
Mechanical	* Shrinkage and warpage
Physical	* Little effect
Chemical	* Slight chemical breakdowns under extreme heat

Typical Workpiece Materials

Acetal, acrylic, nylon, and polystyrene are the best materials, whereas acrylonitrile butadiene styrene (ABS) and polycarbonate are slightly lower in their moldability ratings.

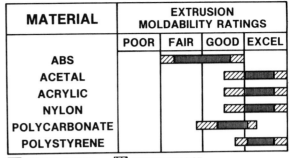

MATERIAL	EXTRUSION MOLDABILITY RATINGS			
	POOR	FAIR	GOOD	EXCEL
ABS		▨▨		
ACETAL			▨▨	▨▨
ACRYLIC			▨▨	▨▨
NYLON			▨▨	▨▨
POLYCARBONATE		▨▨		
POLYSTYRENE			▨▨	▨▨

▮▮ TYPICAL RANGE   ▨▨ FEASIBLE RANGE

Tool Materials

Common extrusion die materials consist of brass or copper for low production or prototype work, tool steel for medium to high production runs, and stainless steel for corrosive materials.

Die materials	Applications
Brass and copper	* Low production * Prototype work
Tool steel	* Most common * For medium to high production
Stainless steel	* For corrosive materials

Factors Affecting Process Results

Tolerance and surface finish depend upon the following:
* Condition and geometry of profile dies
* Type of workpiece material
* Working pressures
* Temperatures of process
* Sizing dies and cooling fixtures

Tool Geometry

An extrusion die is placed between a clamping ring and a die base, and the subassembly is then placed in an adapter ring. With this type of setup, the die can be easily changed without the need for redundant die tooling.

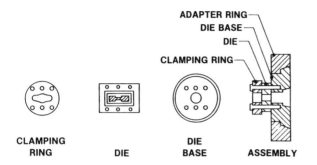

Production Rates

Production rates are a factor of screw speed and material temperatures. Different rates are given in the table below according to the material to be extruded; screw speed is 100 rpm.

Material	Exit temp (°F)	Production (lbs/hr)
High impact polystyrene	483	180
Low density polyethlene	357	189
High density polyethylene	395	192

Lubrication and Cooling

Cooling for various work materials is achieved through cold air blasts and cold water applied by either spray or immersion. All these methods are suitable for most materials.

Work material	Cooling medium	Application
ABS Acetal	Cold air	Blast
Acrylic Nylon	Cold water	Spray
Polycarbonate Polystyrene	Cold water	Immersion

Power Requirements

Power requirements are based on screw speed, material type, and operating temperature.

Given in the table below are machine horsepower requirements according to the material to be extruded; screw speed is 100 rpm.

Material	Pressure (psi)	DC drive (hp)
High impact polystyrene	1975	14.5
Low density polyethylene	1300	15.3
High density polyethlyene	700	21.2

Cost Elements

Cost elements include the following:

* Setup time
* Unload time
* Extrusion time
* Cooling time
* Die change time
* Direct labor rate
* Overhead rate
* Amortization of equipment and tooling costs

Time Calculation

Extrusion time may be calculated by multiplying the weight per unit length by the length of the extrusion and then dividing by the extrusion rate. Total process time is the sum of the heating, extrusion, and cooling times.

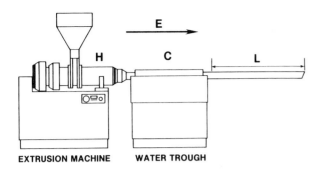

EXTRUSION MACHINE   **WATER TROUGH**

Heating time (min) = H
Extrusion time (min) = E
Cooling time (min) = C
Extrusion rate (lbs/hr) = R
Workpiece material weight (lbs/ft) = W
Workpiece length (ft) = L

$$\text{Extrusion time} = \frac{W \times L}{R} \times 60$$

$$\text{Total time} = H + E + C$$

Safety Factors

The following hazards should be taken into consideration:
* Personal
    - Hot workpieces and equipment
    - Exposure to chemicals

Filament Winding

Filament winding is a laminating process wherein precoated strands of roving woven tape, glass, graphite, or boron are wound over a mandrel and then heated and cured. The resulting composite part utilizes various winding patterns to create workpieces of exceptional strength-to-weight ratio.

Process Characteristics

* Uses a continuous length of fiber strand, roving, or tape
* Results in a shell of materials with a high strength-to-weight ratio
* Requires thermal curing of workpieces
* Patterns may be longitudinal, circumferential, or helical

Process Schematic

A precoated continuous fiber roving is wound onto a rotating mandrel following a carefully controlled path. The workpiece is oven cured and inspected, prior to removal of the mandrel. This process is both economical and flexible, producing workpieces on a large scale with uniform quality.

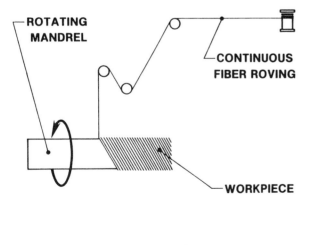

Workpiece Geometry

Fiber material may be specified as strand, roving, or woven tape. Shown are roving and single-strand filament. The pressure vessel shows the various fiber layers and the different patterns used for each layer. Longitudinal, circumferential, and helical patterns are utilized to increase workpiece strength.

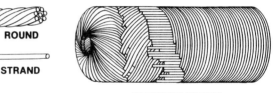

Setup and Equipment

The filament winding machine is composed of a movable filament carriage and a machine drive unit powering a rotating mandrel. The filament carriage consists of fiber spools, a filament feed unit, and a resin bath. The equipment is highly mechanized and operates under computer control.

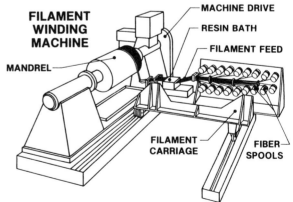

Typical Tools and Geometry Produced

Mandrels for this process are usually one of the three types shown; plaster mandrels, which are nonreusable; collapsible, reusable rubber mandrels; and nonreusable foam mandrels. These mandrels may be used in the production of parts such as spherical tanks, pressure vessels, and aircraft wing flaps.

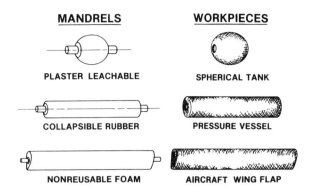

MANDRELS   WORKPIECES

PLASTER LEACHABLE   SPHERICAL TANK

COLLAPSIBLE RUBBER   PRESSURE VESSEL

NONREUSABLE FOAM   AIRCRAFT WING FLAP

Green Sand Casting

Green sand casting is a forming process in which molten metal is poured into a non-reusable, compacted sand mold and held until solidification occurs. The resulting product has a draft angle and rough textured surface. The sand mold is composed of sand mixed with small amounts of binding materials to improve its moldability and cohesive strength.

Process Characteristics

* Utilizes gravity to feed molten metal into a nonreusable mold
* Sands contain binding materials
* Requires a reusable mold pattern
* Produces a parting line on the workpiece
* Requires drafts and fillets on pattern
* Produces rough textured surfaces
* Sprues, risers, and runners must be removed

Process Schematic

A depression called a pouring basin is provided at the top of the mold. Molten metal runs into the sprue and enters the mold cavity through the runners and an opening called a gate. The *gate* (or in-gate as it is sometimes called) controls the rate of metal flow. After the mold cavity is filled, metal fills the riser. The riser provides supplemental feeding to the part during solidification.

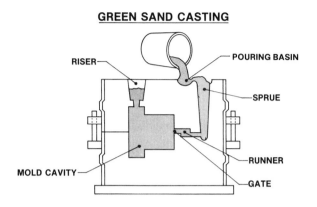

GREEN SAND CASTING

Workpiece Geometry

Because the metal shrinks after solidification, shrinkage allowances are added to the pattern size so the desired tolerances can be maintained on the part. A parting line and draft are also needed to facilitate the removal of the pattern from the sand. Almost any metal that can be melted can be cast in sand molds.

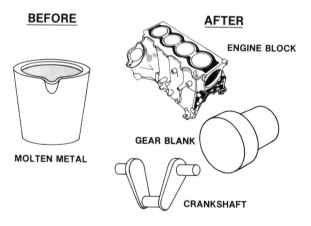

BEFORE — MOLTEN METAL

AFTER — ENGINE BLOCK, GEAR BLANK, CRANKSHAFT

Setup and Equipment

The setup for sand casting normally consists of a melting furnace containing hot, molten metal (not shown), a crucible, flask, pattern, and sand. The flask is split into two halves. The top half is called the cope, and the bottom half is called the drag. After the pattern has been removed, the cope and drag are clamped together, and molten metal is poured into the mold cavity left by the pattern.

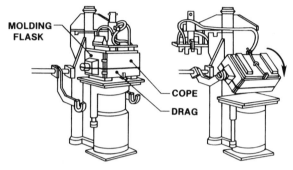

JOLT-SQUEEZE TURNOVER MACHINE

Typical Molds and Geometry Produced

The patterns used are usually made of wood or metal, depending on the number of duplicate castings required and the complexity of the part. Shown is a split pattern mold, which is used for moderate quantities of castings. There are also solid, matchplate, cope, and drag patterns.

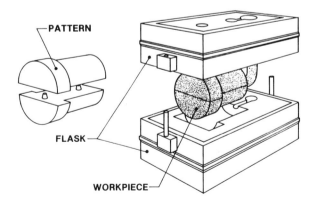

Geometrical Possibilities

One of the advantages of this process is its ability to produce a wide variety of geometries and different-sized parts economically. Typical castings range from 0.25 in. to 1 in. for wall thickness and from approximately 1 lb. to 50 lbs. in weight. However, it is possible to have castings that have wall thickness between 0.030 in. and 3 in. that weigh from a few ounces to several hundred pounds.

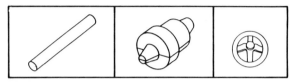

SAND CASTING CAPABILITIES

	0.01	0.1	1.0	10	100
WALL THICKNESS (IN)					
CASTING WEIGHT (LBS)					

■ TYPICAL RANGE ▨ FEASIBLE RANGE

Tolerances and Surface Finish

For most green sand casting applications, tolerances are held within ±1/8 in. For more precise applications, tolerances may be held within ±1/16 in. Surface finish may range from 200 to 1000 microinches, with the typical range between 300 and 600 microinches.

TOLERANCES (IN)		
SAND CASTING	TYPICAL	FEASIBLE
OVERALL	± 1/8	± 1/16

SURFACE FINISH					
PROCESS	MICROINCHES (A.A.)				
SAND CASTING	63	125	250	500	1000

■ TYPICAL RANGE ▨ FEASIBLE RANGE

Pattern Type

The three types of patterns generally used in green sand casting are matchplate patterns, split patterns, and loose patterns with a follow board. Matchplate patterns are used when large quantities of small castings are desired, split patterns are used for large castings, and loose patterns with follow boards are used with loose patterns that have an irregular parting line.

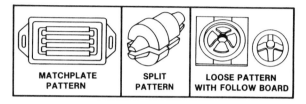

Design Considerations

When castings are designed, all sections should have a uniform thickness and be no thicker than necessary to give the castings the desired structural strengths. Fillets make corners more moldable and reduce stresses in the castings while in service. For reductions in weight and machining costs, *bosses* are often used. It is also good practice to stagger ribs in order to reduce distortion caused by thermal contraction and to minimize the possibility of hot spots and tearing.

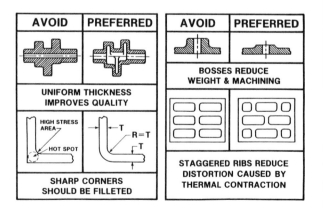

Effects on Work Material Properties

The effects of green sand casting on mechanical properties of the cast material include surface hardening and internal porosity due to entrapped gas created by moisture in the green sand. Casting has little effect on physical and chemical properties of the work material.

Work material properties	Effects of sand casting
Mechanical	* Surfaces of casting tend to be harder with imbedded sand * Internal porosity from entrapped gas
Physical	* Little effect
Chemical	* Little effect

Typical Workpiece Materials

Shown are the sand casting "castability" ratings for a few materials. Aluminum has good to excellent ratings, brass and cast steel rate fair to good, and cast iron rates a little better than aluminum. Silicon additives added to aluminum cast iron make the material very fluid so that it easily fills the molds.

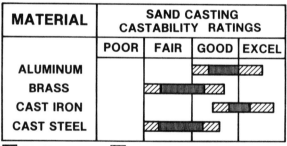

Pattern Materials

Pattern materials for green sand casting may be constructed from wood, metal, or plastic. Wood patterns are normally made from pine or mahogany and are used for both low production and master patterns. They are also inexpensive and easy to shape. Metals patterns are made from aluminum, gray iron, or steel and are used where high production and close dimensional tolerances are needed. They also have a greater resistance to abrasion and a greater stability under changing humidity. Plastic patterns can have complex surfaces and are for medium production. They have close dimensional tolerances.

Pattern materials	Application
Wood (pine) (mahogany)	* Low production and master patterns * Low precision
Metal (aluminum) (gray iron) (steel)	* High production * Close dimensional tolerances
Plastic (epoxy resins)	* Complex surfaces * Medium production * Close dimensional tolerances

Factors Affecting Process Results

Tolerance and surface finish depend upon the following:

* Accuracy of pattern including cores and inserts
* Casting shrinkage
* Dimensional stability of pattern
* Pattern smoothness
* Pattern wear
* Sand compaction
* Dimensional stability of casting alloy
* Gating and rising system

Tool Geometry

Draft is the pattern taper that must be allowed on all vertical faces to permit the pattern to be removed from the sand without tearing the mold walls. The amount of draft necessary depends on the size of the casting, the method of production, and the type of pattern and core equipment. Shown is the amount of draft in inches for draft angles of 1° to 5° for a given depth of wall. Smooth patterns need less draft than rough patterns.

DRAFT ANGLES

Process Yield

Shown are some yield percentages for a medium size casting (50 to 100 lbs) with a moderate gating and riser system. The yield is a percentage that indicates the relationship of the weight of a casting to the total weight of the casting and its gating system. For example, if the casting and gating system weigh 125 lbs and the casting weighs 100 lbs, the yield is 80%.

Workpiece material	Shrinkage (in./ft)	Yield percent*
Aluminum	1/8 to 3/16	75 to 85
Brass	5/32 to 3/16	60 to 70
Cast iron	1/10 to 1/8	80 to 85
Cast steel	3/16 to 1/4	50 to 60

* Yield is the percentage of nondefective castings per 100 pounds of melted material.

Mold Release Agents

Some of the materials used as mold release agents are mineral oils, greases, molybdenum disulfide, lard oil, oil–graphite, and paraffin. Some of the coating materials that are mixed with water and then applied to the surface of the mold to control casting cooling rate are colloidal graphite, asbestos, vermiculite, aluminum oxide, and fine silocel and clay.

Mold release agents	Coating materials
Mineral oils	Colloidal graphite
Greases	Asbestos
Molybdenum disulfide	Vermiculite
Lard oil	Aluminum oxide
Oil–graphite	Fine silocel and clay
Paraffin	

These materials are mixed with water and applied as a thin layer to control cooling rate.

Energy Requirements

The density, melting point, and specific heat are shown for typical sand casting materials. The total energy requirements are calculated by multiplying specific heat by the material density and required temperature.

Material	Density (lb/in.^{3})	Melting point (°F)	Specific heat (Btu/lb-°F)
Aluminum	0.097	1220	0.214
Brass	0.306	1980	0.450
Cast iron	0.280	1990–2300	0.130
Cast steel	0.283	2500	0.117

Btu = British thermal unit.

Cost Elements

Cost elements include the following:

* Setup time
* Load/unload time
* Casting time
* Material costs
* Direct labor rate
* Overhead rate
* Amortization of equipment and tooling costs

Time Calculation

Filling time is a function of cavity volume and metal flow rate. The total time is the sum of the fill, cooling, workpiece removal (shake-out time), and cleanup times.

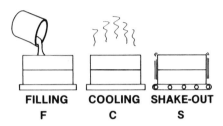

FILLING   **COOLING**   **SHAKE-OUT**
  F   **C**   **S**

Filling time (sec) = F
Cooling time (sec) = C
Shake-out time (sec) = S
Cleanup time (riser, gates, flash) (sec) = R
Metal flow velocity (in.^{3}/min) = Q
Mold cavity size (in.^{3}) = V

$$\text{Filling time} = \frac{V}{Q}$$

$$\text{Total time} = F + C + S + R$$

Safety Factors

The following risks should be taken into consideration:

* Personal
  – Burns
  – Dust and smoke inhalation
  – Noise level
* Environmental
  – Explosion danger if moisture contacts the molten metal
  – Fire risk

Hot Chamber Die Casting

Die casting is a metal shaping process in which molten metal is forced into a reusable mold from a hot chamber and then held under pressure until solidification occurs.

Process Characteristics

* Uses a reusable mold into which molten metal is forced under pressure
* Achieves dimensional accuracy and excellent detail
* Minimizes additional machining
* May produce a small amount of flash around casting edge
* Produces ejector marks
* Is used mainly for low-melting-point alloys
* Has a high production rate

Process Schematic

Molten metal is injected into a die cavity through a gooseneck channel by downward movement of a plunger. After a preset solidification time, the plunger reverses direction, the part is ejected, and the machine is ready for the next cycle.

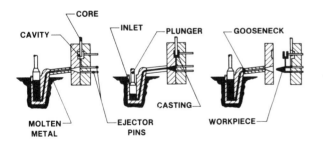

Workpiece Geometry

Complex shapes and dimensionally accurate workpieces can be produced by this process. Casting weight varies from a fraction of an ounce to 50 lbs.

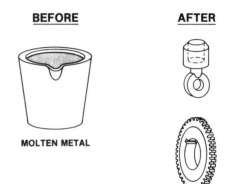

Setup and Equipment

A hot chamber machine consists of pressure and power cylinders, plunger, and mold cavity. Typically, production rates range from 50 to 500 shots per hour, although high production outputs of 2000 to 5000 shots per hour are feasible.

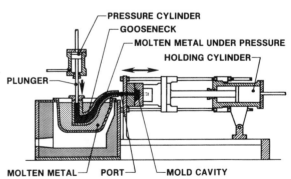

Typical Tools and Geometry Produced

Dies used for this process consist of a stationary cover portion with a sprue opening and a movable ejector portion that includes a biscuit, cores, ejector pins, and water lines. There are two types of dies available, single cavity or multiple cavity, which are used according to production and part requirements.

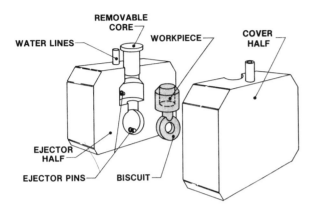

Geometrical Possibilities

Geometry of the die casting components should have, in general, a uniform section size. Shown is an engine block, a type of bushing, and a fly wheel. One advantage of die casting is its ability to produce a wide variety of complex shapes. Wall thickness ranges from 0.05 in. to around 1.5 in. Casting weights range from 0.07 lbs to several hundred pounds.

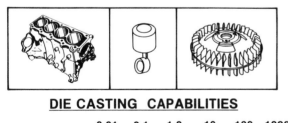

DIE CASTING CAPABILITIES

	0.01	0.1	1.0	10	100	1000
WALL THICKNESS (IN)						
WEIGHT (LBS)						

▆ TYPICAL RANGE   ▨ FEASIBLE RANGE

Tolerances and Surface Finish

Tolerances for die casting components vary according to the method and material used. In general, both hot chamber and cold chamber die casting machines achieve tolerances of ±0.015 in. to ±0.0005 in. in length. Surface finish values range from approximately 16 to 125 microinches.

TOLERANCES		
DIE CASTING	TYPICAL	FEASIBLE
HOT CHAMBER	±.015	±.0005
COLD CHAMBER	±.015	±.0005

SURFACE FINISH									
PROCESS	MICROINCHES (A.A.)								
DIE CASTING	2	4	8	16	32	63	125	250	500

▆ TYPICAL RANGE   ▨ FEASIBLE RANGE

Tool Style

Die casting molds can be almost any shape. Cores are used to form internal holes and cavities, and the mold cavity forms the exterior geometry of the castings. A plunger forces the molten metal into the die, and ejector pins remove the casting from the mold cavity.

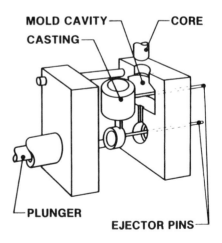

Design Considerations

Sharp corners should be avoided. Uniform wall thickness helps. Staggering or offsetting patterns, such as ribs or fins, also reduce distortions and tearing.

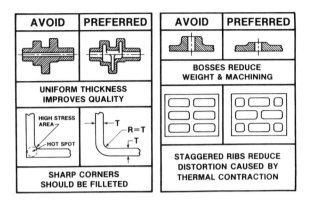

Work material properties	Effects of die casting
Mechanical	* Porosity * Stress risers
Physical	* Little effect
Chemical	* Little effect

Tool Materials

In general, tool and die materials are made of special, hot-worked tool steels developed especially for economical and efficient molding. One typical die material is high grade tool steel. Other materials such as chromium, molybdenum, and nickel alloys are used to increase die toughness. Tungsten, molybdenum, chromium, and vanadium alloys are used to resist die softening. Die steels should also have a low-carbon content to resist heat cracking.

Tool materials	Application
Hot-worked tool steel: H 13	* Aluminum and magnesium * Plated castings * 100 to 150,000 cycles
H 19	* Copper or brass * 5000 to 6000 cycles
Low-carbon mold steel P20	* Zinc * 1,000,000 cycles

Effects on Work Material Properties

The workpiece may show some porosity. This process may also cause stress risers. There is little effect on physical and chemical properties.

Typical Workpiece Materials

Die-cast materials are normally nonferrous alloys that have a low melting points. Shown are some typical workpiece materials and their castability ratings.

MATERIAL	DIE CASTING CASTABILITY RATINGS			
	POOR	FAIR	GOOD	EXCEL
ALUMINUM			▨▨	▨
COPPER		▨▨		
MAGNESIUM			▨▨	▨
ZINC				▨▨

▮ TYPICAL RANGE   ▨ FEASIBLE RANGE

Factors Affecting Process Results

Tolerances and surface finish from die casting depend upon the following:

* Accuracy of die including slides and cores
* Draft angle
* Casting shrinkage
* Thermal expansion of die
* Accuracy of die closure
* Dimensional stability of alloy
* Cavity smoothness

Tool Geometry

The die cavity must reproduce a cast part with the necessary allowances for material shrinkage and thermal expansion of the part. The die must also have draft angles that allow for easy ejection. The following charts help determine the draft angle to be used for cored holes and walls in the workpiece.

DRAFT ANGLES

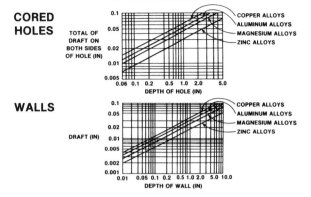

CORED HOLES

WALLS

Process Conditions

The process conditions for a reasonably complex part include material used, the operating temperature, and the necessary injection pressure. Each of these factors, along with the operating capacity of the machine, affects production rates.

Material	Temperature (°F)		Pressure (psi)
	Material	Die	
Aluminum	1200 to 1220	350 to 400	1500/2000
Zinc	800	300	1500/2000
Magnesium	1200	350 to 400	1500/2000
Brass	1980 to 2000	500	1500/2000

Lubrication and Cooling

Die lubricants prevent adherence of the molten material to the die and provide a better surface finish on the completed workpiece. Selection of the proper die lubricant depends on the work material, the die temperature, and the temperature of the molten metal.

Lubrication components	Application
Carrier fluid	* Aid metal in reaching all parts of the die
Separating film	* Reduce sticking and improve casting release
Additives	* Improve lubricant stability and ease of dilution
Solid	* Only used where conditions require a superior release

Power Requirements

The energy requirements for die casting depend on the workpiece material's density, melting point, and specific heat.

Material	Density (lb/in.^{3})	Melting point (°F)	Specific heat (Btu/lb-°F)
Aluminum	0.10	1220	0.214
Copper	0.32	1980	0.101
Tin	0.26	450	0.056
Magnesium	0.06	1200	0.217
Lead	0.42	620	0.031
Zinc	0.26	790	0.095

Cost Elements

Cost elements include the following.

* Setup time
* Load and unload times
* Casting time
* Direct labor rate
* Overhead rate
* Amortization of equipment and tooling costs

Productivity Tip

Uniform thickness improves quality. Avoid hot spots and sharp internal corners.

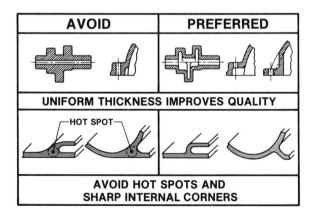

Time Calculation

Filling time for the mold cavity is a function of cavity volume and metal flow rate. The metal flow rate is generally controlled by the rate at which the plunger is advanced. The total time is the sum of mold opening and closing times, injection time, cooling time, and ejection time. The cleanup time may have to be added to the total time.

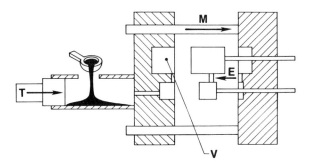

Die closing/opening time (sec) = D
Injection time (sec) = T
Cooling time (sec) = C
Ejection/removal time (sec) = E
Cleanup time (riser, gates, flash) (sec) = R
Metal flow velocity (in.^{3}/min) = Q
Mold cavity size (in.^{3}) = V

$$\text{Filling time} = \frac{V}{Q}$$

$$\text{Total time} = 2D + T + C + E + R$$

Safety Factors

The following risks should be taken into consideration:

* Personal
  - Burns
  - Die lubricant in eyes
  - Hand in die
* Environmental
  - Explosion danger
  - Fire risk
  - Heat
  - Noise

Injection Molding

Injection molding is a shaping process in which thermoplastic material is fed into a heated barrel, mixed, and forced into a mold cavity where it cools and hardens to the configuration of the mold cavity. Injection molding is used to produce more *thermoplastic* products than any other process. In some circumstances, *thermoset* plastics can also be used with injection molding.

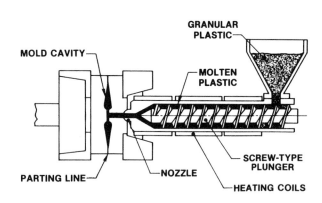

Process Characteristics

* Utilizes a ram or screw-type plunger to force molten plastic material into a mold cavity
* Produces a solid or open-ended shape conforming to the mold cavity
* Uses thermoplastic or thermoset materials
* Produces a parting line and sprue and gate marks
* Ejector pin marks are usually present

Process Schematic

With injection molding, granular plastic is fed by gravity from a hopper into a heated barrel. As the granules are slowly moved forward by a screw-type plunger, the plastic is forced into a heating chamber, where it is melted. As the plunger advances, the melted plastic is forced through a nozzle that rests against the mold, allowing it to enter the mold cavity through a gate and runner system. The mold remains cold so the plastic solidifies almost as soon as the mold is filled.

Workpiece Geometry

Shown are granular and pelletized materials before molding and resulting products after molding. Parts that leave the machine are completely finished, trimmed, and ready for immediate use. Millions of parts can be made with little or no wear on the dies.

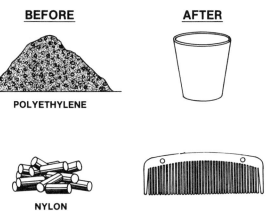

BEFORE   AFTER

POLYETHYLENE

NYLON

Setup and Equipment

An injection molding machine consists of a material hopper, an injection ram or screw-type plunger, and a heating unit. Injection machines are rated according to the maximum number of ounces of material that can be displaced by one forward stroke of the injection plunger. Almost all injection molding machines can be operated on an automatic cycle.

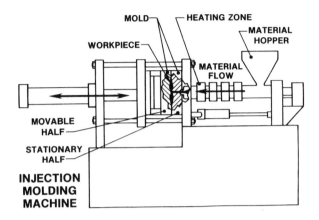

Typical Tools and Geometry Produced

Molds can be designed to produce a single part or many parts in one cycle. The mold shown produces two parts per cycle. A mold made for one plastic material usually cannot be used with another material because of the different shrinkage rates for various plastics. Heavy clamping forces are necessary to keep the mold halves together to prevent the plastic material from escaping under high injection molding pressures.

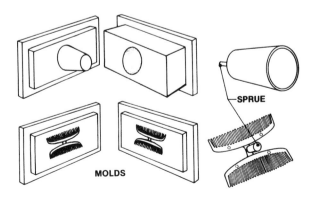

Geometrical Possibilities

One of the major advantages of injection molding is its ability to produce complex geometries. Injection molding is currently the most widely used plastic forming process. Typical size ranges for injection molding are between 2 in. and 20 in. in length or width and from 2 in. up to about 15 in. in depth. To prevent unwanted stresses and weak spots, injection molded parts should have uniform wall thickness and streamlined shape. Thick sections should be avoided. Streamlining of the part will help prevent gas pockets and blisters. This is done by rounding corners and avoiding sharp edges.

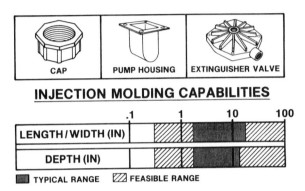

Geometrical Possibilities (Weight)

The maximum or minimum size of part that can be molded depends upon the method and type of material used. When *thermoplastics* are molded, the maximum and minimum sizes range from less than 1 oz. to 700 oz. (ounce). When *thermosets* are molded, the sizes range from less than 1 oz. to 200 oz. The initial raw material for both these methods may be in the form of granules, pellets, or powders.

Method	Raw material	Maximum size	Minimum size
Injection molding (thermo-plastic)	Granules, pellets, powders	700 oz.	Less than 1 oz.
Injection molding (thermo-setting)	Granules, pellets, powders	200 oz.	Less than 1 oz.

Tool Style

The mold cavity must have a design that insures that the molded component will have the desired shape, size, and strength. One half of a typical mold will consist of the mold cavity and guide pins, and the other half may have a core to produce hollow areas. Either half of the mold may contain the sprue, gates, and runners that guide the melted plastic to fill the mold completely. Cold shuts and improper mold fill may be avoided by providing round gates and runners. A semi-rounded gate or runner will not permit a smooth flow of material to the mold. Another factor that will insure a smooth material flow is to avoid sharp corners at runner bends.

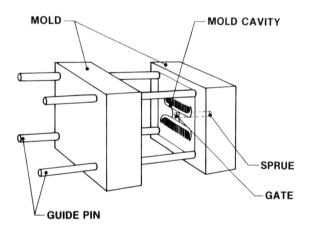

Typical Workpiece Materials

Epoxy, nylon, polyethylene, and polystyrene all typically have good to excellent ratings. Phenolic, a rapid curing thermoset, requires a special machine setup to insure that the material does not cure before it fills the mold.

MATERIAL	INJECTION MOLDING MOLDABILITY RATINGS			
	POOR	FAIR	GOOD	EXCEL
EPOXY*			▨▨	▨▨
PHENOLIC*		▨▨▨	▨	
NYLON			▨▨	▨▨
POLYETHYLENE			▨▨	▨
POLYSTYRENE			▨▨	▨▨

▨ TYPICAL RANGE   ▨ FEASIBLE RANGE
*THERMOSETTING PLASTIC

Factors Affecting Process Results

Tolerance and surface finish depend upon the following:

* Type of plastic
* Mold surface
* Proper design of the sprue runner and gate
* Mold lubricant
* Injection pressure
* Clamp pressure
* Material temperature
* Mold temperature

Lubrication and Cooling

Mold cooling is required for production molding. Water is normally channeled throughout the dies to provide a reasonable cooling time while at the same time avoiding cold-shuts. Mold temperatures vary with materials used; a cooler mold is often more desirable because it allows for a shorter cycle time. Crystalline materials require a warmer mold temperature and a longer cycle time.

Tool material	Medium	Application
Tool steel Beryllium–copper Mild steel Aluminum Nickel Epoxy	Water	Channeled flow through die

Tool Geometry

Minimum required draft angles are shown for workpieces with depths ranging from 0.2 in. to 4.0 in. Proper draft insures easy removal of the finished part from the mold.

DRAFT ANGLES

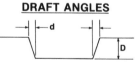

	DEPTH OF WALL (IN) D							
	0.2	0.4	0.6	0.8	1.0	2.0	3.0	4.0
REQUIRED DRAFT (IN) d	.007	.014	.022	.029	.035	.075	.125	.135

Tool Materials

Materials for injection molds include tool steels, beryllium–copper, mild steel, aluminum, nickel, and epoxy. Tool steel can be highly polished and hardened and produces the best surface finishes. Beryllium–copper is softer than steel and can be easily machined to produce intricate shapes. Mild steel, aluminum, nickel, and epoxy are suitable for prototype work or short production runs.

Tool materials	Applications
Tool steel	* Can be polished to a fine luster * Can be hardened
Beryllium–copper	* Softer than steel * Easily machined
Mild steel, aluminum, nickel, epoxy	* Suitable only for prototype or very short production runs

Tolerances and Surface Finish

Tolerances and surface finishes for injection molding depend upon the method and type of material used. In general, tolerances of ±0.008 in. to ±0.002 in. can be obtained for thermoplastics and thermosets. Surface finishes of 2 to 4 microinches or better are easily obtained as are rough textured or pebbled surfaces. A pebbled finish helps to mask scratches.

TOLERANCES		
INJECTION MOLDING	TYPICAL	FEASIBLE
THERMOPLASTIC	±.008	±.002
THERMOSET	±.008	±.002

SURFACE FINISH	
PROCESS	MICROINCHES (A.A.)
INJECTION MOLDING	1 2 4 8 16

■ TYPICAL RANGE ▨ FEASIBLE RANGE

Effects on Work Material Properties

Effects of injection molding on work material properties are usually small as far as mechanical properties go; some plastic components may develop internal stresses. Changes in physical properties include shrinkage; some plastics experience permanent shrinkage at elevated temperatures. Thermoset materials cannot be remelted after forming because of the permanent bonds created during heating.

Work material properties	Effects of injection molding
Mechanical	* Plastic components may develop internal stresses
Physical	* Permanent shrinkage at elevated temperatures
Chemical	* Thermoset materials cannot be remelted

Power Requirements

Power requirements for injection molding are determined by a material's specific gravity, melting point, thermal conductivity, part size, and molding rate. Thermal conductivity governs the cooling rate and production cycle time, which, in turn, affects the necessary horsepower requirements.

Material	Specific gravity	Melting point (°F)
Epoxy	1.12 to 1.24	248
Phenolic	1.34 to 1.95	248
Nylon	1.01 to 1.15	381 to 509
Polyethylene	0.91 to 0.965	230 to 243
Polystyrene	1.04 to 1.07	338

Cost Elements

Cost elements include the following:

* Setup time
* Load/unload time
* Molding time
* Tool change time
* Direct labor rate
* Overhead rate
* Amortization of equipment and tooling

Productivity Tip

Uniform wall thicknesses help improve quality. Streamlining helps prevent gas pockets. Gates and runners should be round, and sharp corners at runner bends should be avoided.

Additional design considerations for injection molding are as follows:

* Thin walls require less material and curing time
* Contoured parts warp less than flat parts
* Venting of molds removes trapped air
* Fillets should be used at the base of ribs or bosses

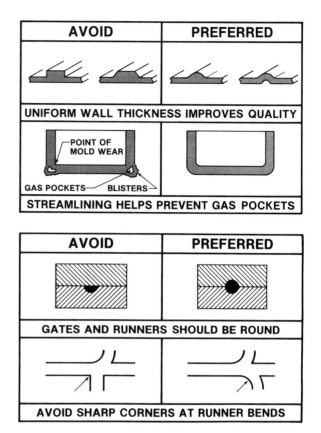

AVOID	PREFERRED
UNIFORM WALL THICKNESS IMPROVES QUALITY	
POINT OF MOLD WEAR / GAS POCKETS / BLISTERS	
STREAMLINING HELPS PREVENT GAS POCKETS	

AVOID	PREFERRED
GATES AND RUNNERS SHOULD BE ROUND	
AVOID SHARP CORNERS AT RUNNER BENDS	

Time Calculation

Filling time for the mold cavity is a function of cavity volume and material flow rate. The material flow rate can be controlled in several ways but is generally controlled by the rate at which the screw is turned or by injection plunger rate. The total time is the sum of mold opening and closing time, injection time, cooling time, and ejection time. The injection molding setup shown depicts the different times required for the injection molding process: the mold is closed, is injected with material, and is allowed to cool. Once the workpiece has sufficiently cooled, the die opens, and the workpiece is ejected from the mold.

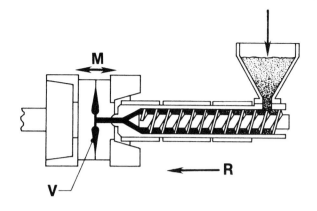

Mold closing/opening time (sec) = M
Injection time (sec) = T
Cooling time (sec) = C
Ejection/removal time (sec) = E
Material flow rate (in.^{3}/min) = R
Mold cavity size (in.^{3}) = V

$$\text{Total time} = 2M + T + C + E$$

$$T = \frac{V}{R}$$

Safety Factors
The following risks should be taken into consideration:

* Personal
  - Burns
  - Skin irritation
  - Respiratory dangers
* Environmental
  - Explosion danger
  - Fire risk
  - Heat
  - Chemical irritants

Investment Casting

Investment casting is a process in which molten metal is poured into a preheated ceramic mold producing a casting that may have intricate internal and external features with little or no draft.

Process Characteristics

* Produces intricate detail and close dimensional accuracy
* Uses wax or plastic patterns
* Requires little or no draft
* Patterns can be clustered to increase production rate
* Patterns are melted out of nonreusable ceramic patterns

Process Schematic

Investment molds for casting are preheated in a furnace prior to pouring in the molten metal. To help fill the cavity completely, pressure, vacuum, or centrifugal force may be used. After the metal has solidified, the thin refractory mold material is removed by chipping or sandblasting. The final foundry operation consists of separating the castings from the gating system by means of an abrasive cutoff machine followed by grinding.

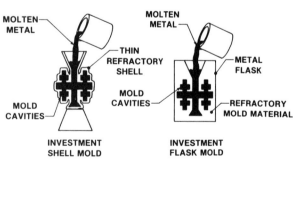

INVESTMENT CASTING

Workpiece Geometry

Almost any metal can be cast by this process. The process is especially suited for casting metals that melt at temperatures too high for plaster or metal molds. There is no parting line in the mold, because it consists of only one piece. Complex shapes can be made that cannot be made with other processes. The dimensional tolerances are excellent, and the surface finish is usually very good.

BEFORE　　　　**AFTER**

MOLTEN METAL

Setup and Equipment

The equipment required for producing an investment ceramic shell or mold is shown here. Equipment for making the pattern is also needed. Furnaces are needed for melting the wax patterns from the mold, as well as for melting the metal prior to pouring it into the heated mold.

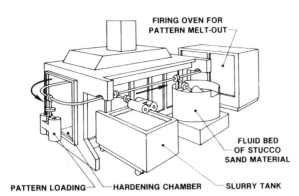

FIRING OVEN FOR PATTERN MELT-OUT

FLUID BED OF STUCCO SAND MATERIAL

PATTERN LOADING — HARDENING CHAMBER — SLURRY TANK

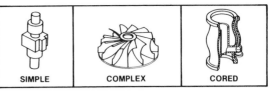

SIMPLE   COMPLEX   CORED

INVESTMENT CASTING CAPABILITIES

	0	.1	.5	1	10	100	1000
WALL THICKNESS (IN)							
WEIGHT (LBS)							

▓ TYPICAL RANGE   ▨ FEASIBLE RANGE

Typical Tools and Geometry Produced

A disposable pattern of the desired object is made by injecting wax or plastic into a metal die. These patterns are then gated to a sprue to form a tree or cluster. The cluster is repeatedly dipped in a slurry of fine refractory (ceramic) particles, stuccoed or dusted with coarse refractory particles, withdrawn, and dried until the required thickness of the shell is achieved. In the flask-type mold, the pattern is placed in a metal flask, and ceramic material is poured around it. Wax is melted from the mold in an autoclave or other furnace leaving the mold or shell ready for heating prior to casting.

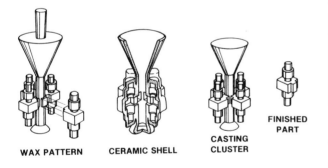

WAX PATTERN   CERAMIC SHELL   CASTING CLUSTER   FINISHED PART

Geometrical Possibilities

There is almost no limit to the shapes that can be cast by this process. Very complex workpieces can be formed having highly intricate details. Workpieces with wall thicknesses as thin as 0.05 in. can be cast. The castings typically range in weight from 0.3 lb. to 40 lbs.

Tolerances and Surface Finish

Tolerances for investment casting are typically ±0.005 in., but tolerances as low as ±0.002 in. are feasible. Surface finishes range from 32 to 300 microinches, with typical finishes ranging from 50 to 125 microinches. The quality of the mold, pouring temperature, and alloy composition determine the tolerances and the surface finish of the final workpiece.

TOLERANCES (IN)

INVESTMENT CASTING	TYPICAL	FEASIBLE
FOR EACH INCH OF MAX. DIMENSION	±.005	±.002

SURFACE FINISH

PROCESS	MICROINCHES (A.A.)						
INVESTMENT CASTING	4	16	32	63	125	250	500

▨ TYPICAL RANGE   ▓ FEASIBLE RANGE

Tool Style

Ceramic shell-type molds are the most commonly used. Solid flask molds are also used, primarily for high production. In both types of molds, the same type of wax or plastic patterns is used.

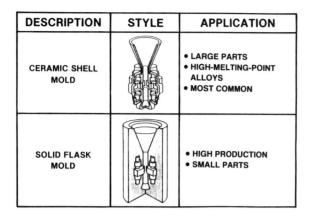

DESCRIPTION	STYLE	APPLICATION
CERAMIC SHELL MOLD		• LARGE PARTS • HIGH-MELTING-POINT ALLOYS • MOST COMMON
SOLID FLASK MOLD		• HIGH PRODUCTION • SMALL PARTS

Pattern materials	Applications
Wax	* Relatively low cost equipment * Easily formed
Plastic (polystyrene)	* High production (injection molded) * Thin sections and complex surfaces * High strength and durability * Machinable for prototype work
Frozen mercury (−70°F)	* Greater dimensional tolerances * Better surface finish * In very limited use

Design Consideration

When designing parts, it is important to maintain uniform wall thickness. Secondly, sharp corners and other stress risers should be avoided. Third, parts should be designed so that minimal machining is required to minimize thermal contraction and distortion.

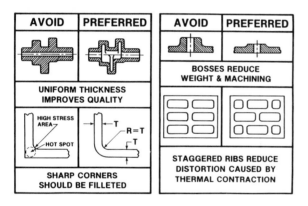

Tool Materials

Basic materials used for investment patterns are wax, plastic, and mercury. Wax is very popular because it is relatively inexpensive, easily formed, and reusable. Plastic is used in high production, for thin and complex surfaces where strength and durability are desired, and when working on prototypes. Frozen mercury is used for high dimensional tolerances and for better surface finishes. Frozen mercury is very seldom used because of the cost, difficulty in handling, and toxicity.

Effects on Work Material Properties

This process permits close control over grain size and orientation and provides directional solidification.

Work material properties	Effects of investment casting
Mechanical	* Grain size and orientation * Directional solidification
Physical	* Little effect
Chemical	* Little effect

Typical Workpiece Materials

Aluminum, cast iron, and magnesium are rated from good to excellent because of their high fluidity due to silicon additives, whereas cast steel and brass are rated from fair to good.

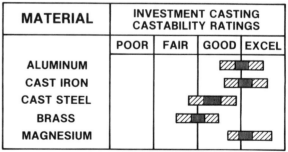

MATERIAL	INVESTMENT CASTING CASTABILITY RATINGS			
	POOR	FAIR	GOOD	EXCEL
ALUMINUM			▨▥	▥
CAST IRON			▨▥	▥
CAST STEEL		▨▥	▥	
BRASS		▨▥	▥	
MAGNESIUM				▨▥▥

▥ TYPICAL RANGE   ▨ FEASIBLE RANGE

Tool Geometry

Mold geometry and draft angles are not considerations in this process because of the methods for removing the pattern and workpiece from the mold. Both the pattern and the mold are destroyed during the casting process.

Factors Affecting Process Results

Tolerance and surface finish depend upon the following:

* Accuracy of mold and pattern
* Dimensional stability of pattern material
* Gate and riser system
* Expansion of mold during melt-out, firing, and preheating
* Stability of casting alloy
* Conditions of slurry and stucco refractory

Process Conditions

Two significant factors are part shrinkage and workpiece yield. Part shrinkage is given in inches per foot. For example, an aluminum part will shrink 1/8 in./ft during solidification and cooling. Shrinkage must be taken into consideration in making the pattern and the mold. The percent yield for various material ranges from 55% to 90%. The yield refers to the pounds of finished (nondefective) cast parts per 100 lbs of melted material.

Workpiece material	Shrinkage (in./ft)	Yield percent
Aluminum	1/8 to 3/16	80 to 90
Cast iron	1/10 to 1/8	80 to 85
Cast steel	3/16 to 1/4	55 to 65
Brass	5/32 to 3/16	65 to 75
Magnesium	1/10 to 5/32	80 to 90

Mold Release Agents

Release agents such as silicone sprays are used only in removal of the wax patterns from their molds during production of wax patterns themselves. No release agents are used in the metal casting process itself.

Work material	Release agent	Application
Wax patterns	Silicone fluid	Spray
Casting metals	None	None

Mold temperatures affect the chemical and physical characteristics of the workpiece.

Power Requirements

The energy required for investment casting is given in Btu's (British thermal units). Basically, the density and melting point of a standard material, along with its specific heat, are used to determine the number of Btu's required to raise the material to its melting point.

Material	Density (lbs/in.^{3})	Melting point (°F)	Specific heat (Btu/lb-°F)
Aluminum	0.097	1220	0.214
Cast iron	0.280	1990–2300	0.130
Cast steel	0.283	2500	0.117
Brass	0.306	1980	0.450
Magnesium	0.064	1200	0.217

Cost Elements

Cost elements include the following:

* Setup time
* Pattern making time
* Casting time
* Material costs
* Direct labor rate
* Overhead rate
* Amortization of equipment and tooling costs

Time Calculation

Fill time for the mold cavity is a function of cavity volume and metal flow rate. The total time is the sum of preheat, mold fill, and cooling times.

Mold heating time (sec) = H
Filling time (sec) = F
Cooling time (sec) = C
Cleanup time (riser, gates, flash) (sec) = R
Metal flow rate (in.^{3}/sec) = Q
Mold cavity size (in.^{3}) = V

$$\text{Filling time} = \frac{V}{Q}$$

$$\text{Total time} = H + F + C$$

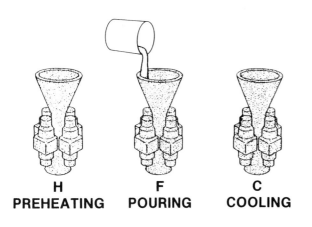

H	F	C
PREHEATING	**POURING**	**COOLING**

Safety Factors

The following risks should be taken into consideration:

* Personal
    – Burns from molten metal
    – Burns from hot molds
    – Explosions due to moisture
* Environmental
    – Fire risk

Isostatic Powder Compaction

Cold isostatic powder compaction is a mass-conserving, shaping process. Fine metal particles are introduced into the cavity of a flexible mold. The mold is then subjected to high fluid or gas pressures (from all sides) to form a workpiece. The workpiece is then sintered in a furnace to bond the powdered metal particles together metallurgically. (Sintering allows the particles to coalesce.)

Process Characteristics

* Compacts powdered metal within a flexible mold by uniformly applied, high fluid/gas pressure
* Parts are sintered to increase strength through metallurgical bonding
* Produces very little scrap material
* Can use alloy combinations and filler
* Can produce complex workpiece geometries

Process Schematic

The workpiece source material is compacted under high pressure in a flexible mold. A core rod provided for internal geometries may also be used as an ejection device.

ISOSTATIC POWDER COMPACTION

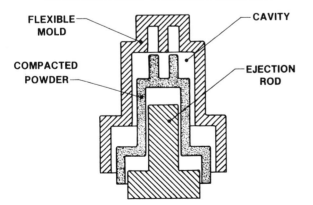

FLEXIBLE MOLD — CAVITY

COMPACTED POWDER — EJECTION ROD

Workpiece Geometry

Shown are some typical workpieces that can be produced. Many metals and alloys can be used in this process, depending on the characteristics desired in the workpiece. Iron, copper, bronze, tungsten carbide, and other materials are commonly used. Hard materials, such as tungsten carbide, may be held in a softer matrix, such as cobalt.

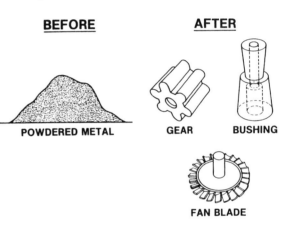

BEFORE — AFTER

POWDERED METAL — GEAR — BUSHING

FAN BLADE

Setup and Equipment

Equipment consists of a flexible mold, a pressure vessel in which the mold is placed, and equipment for supplying high pressure. It can operate with either fluid or gas. A control unit is used to regulate pressure and pressing time.

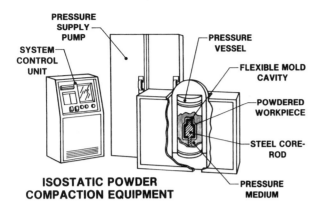

ISOSTATIC POWDER
COMPACTION EQUIPMENT

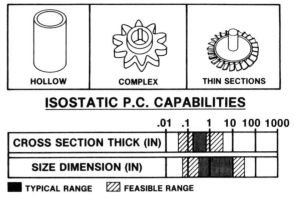

ISOSTATIC P.C. CAPABILITIES

	.01	.1	1	10	100	1000
CROSS SECTION THICK (IN)						
SIZE DIMENSION (IN)						

■ TYPICAL RANGE   ▨ FEASIBLE RANGE

Typical Tools and Geometry Produced

In the example shown, re-entrant angles are present. Form features perpendicular to the workpiece axes are also possible because of the nature of the flexible mold.

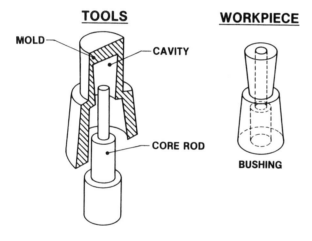

Geometrical Possibilities

Many desired shapes can be formed from a variety of engineering materials. Typical workpiece sizes range from 1/4 in. to 3/4 in. thick and 1/2 in. to 10 in. in length. However, it is possible to compact workpieces that are between 1/16 in. and 5 in. thick and 1/16 in. and 40 in. in length.

Tolerances and Surface Finish

For most isostatic powder compaction applications, tolerances are held within ±0.008 in. for axial dimensions and ±0.020 in. for radial dimensions. For precision applications, axial tolerances of ±0.005 in. and radial tolerances of ±0.025 in. are achievable. Surface finish may range from 3 to 55 microinches, with the typical range between 6 to 20 microinches.

TOLERANCES (IN)

PROCESS CYCLE	AXIAL	RADIAL
PRESSING, SINTERING	±.008	±.020
PRECISION PRESSING, SINTERING	±.005	±.015
PRESSING, SINTERING, HEAT TREATMENT	±.009	±.025

SURFACE FINISH

MICROINCHES	1	2	4	8	16	32	63	125
ISOSTATIC P.C.								

■ TYPICAL RANGE   ▨ FEASIBLE RANGE

Tool Style

Isostatic tools are available in two styles, free-mold (wet-bag) and fixed-mold (dry-bag). The free-mold style is the traditional style of isostatic compaction and is not generally used for high production work. In free-mold tooling, the mold is removed and filled outside the canister. In fixed-mold tooling, the mold is contained within the canister, which facilitates automation of the process.

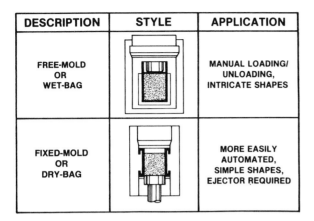

DESCRIPTION	STYLE	APPLICATION
FREE-MOLD OR WET-BAG		MANUAL LOADING/ UNLOADING, INTRICATE SHAPES
FIXED-MOLD OR DRY-BAG		MORE EASILY AUTOMATED, SIMPLE SHAPES, EJECTOR REQUIRED

Design Considerations

Listed are some of the design considerations for isostatic powder compaction. Advantages over standard powder compaction are the possibility of thinner walls and larger workpieces. Lubricants are not required (but considerable shrinkage of the workpiece occurs and must be taken into account). Height to diameter ratio has no limitation. No specific limitations exist in wall thickness variations, undercuts, reliefs, threads, and crossholes. All are possible with few restrictions.

* Minimum wall thickness 0.05 in.
* Maximum workpiece weight 40 lbs to 300 lbs
* No compacting lubricants required
* 25% to 45% shrinkage during compaction, up to 25% during sintering
* No limitations on wall thickness variations
* No limitations on height to diameter ratio
* Undercuts, reliefs, threads, crossholes are all possible with minor exceptions

Typical Workpiece Materials

Listed are some of the typical workpiece materials used in isostatic powder compaction: iron, alloyed steels, brass, stainless steel, and bronze. The obtainable densities for each are also listed, as well as the obtainable yield strengths, tensile strengths, and hardnesses. Other possible compactable materials include combination metals, composites (i.e., metals plus ceramics, lubricants, asbestos, etc.). Miscellaneous materials, such as solid fuels, can also be pressed.

Power particle size <0.006 in.

Obtainable properties (after sintering)				
Workpiece material	Density (gram/cc)	Yield strength (psi $\times 10^4$)	Tensile strength (psi $\times 10^4$)	Hardness (HB)
Iron	5.2 to 6.3	0.51 to 2.3	0.73 to 2.9	40 to 70
Alloyed steels	6.8 to 7.4	0.26 to 8.4	2.9 to 9.4	60 and up
Brass	7.0 to 7.9	1.1 to 2.9	1.6 to 3.5	60
Stainless steel	6.3 to 7.6	3.6 to 7.3	4.4 to 8.7	60 and up
Bronze	5.5 to 7.5	1.1 to 2.9	1.5 to 4.4	50 to 70

Tool/Mold Materials

Shown is an example of a fixed-mold tool used in isostatic powder compaction with the typical material for each component identified.

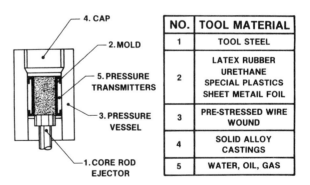

NO.	TOOL MATERIAL
1	TOOL STEEL
2	LATEX RUBBER URETHANE SPECIAL PLASTICS SHEET METAIL FOIL
3	PRE-STRESSED WIRE WOUND
4	SOLID ALLOY CASTINGS
5	WATER, OIL, GAS

Factors Affecting Process Results

Tolerance and surface finish depend upon the following:

* Type of powder material
* Size of powder particles
* Mold material and quality
* Tool style (fixed or free mold)
* Size of workpiece
* Production quantity and rate
* Subsequent treatment (i.e., sintering, barrel tumbling, coating)
* Compaction pressure

Power Requirements

This graph illustrates how much more efficient isostatic powder compaction is compared to standard powder compaction methods.

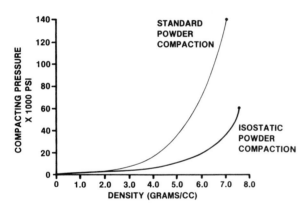

Process Conditions

Compacting pressures range from 15,000 psi to 40,000 psi for most metals and approximately 2000 psi to 10,000 psi for nonmetals. The density of isostatic compacted parts is 5% to 10% higher than other powder metallurgy processes.

Powder material	Compacting pressure (psi)	Relative density
Iron	15,000 to 30,000	99.7
Alloyed steels	20,000 to 50,000	99.6
Brass	10,000 to 30,000	99.8
Carbides	20,000 to 60,000	99.5
Ceramics	3000 to 10,000	99.9
Synthetics	2000 to 8000	99.9

Subsequent Treatments

Shown are some of the subsequent treatments for sintering and impregnation. A typical reduction of 10% to 25% in size during sintering depends on the material. Impregnating powder metallurgy parts with a lubricant provides improved lubrication qualities. A sealant can also be impregnated into the workpiece to seal the natural pores of the workpiece.

Sintering	* Phases
	  – Heating to sintering temperature
	  – Holding temperature in specified time
	  – Cooling
	* Temperature
	  – 70–80% of melting point (50–60% for hot isostatic powder compaction)
	* Atmosphere
	  – Inert gas or carburizing

Impregnation	* Oil
	  – Lifetime self-lubricating bushings
	* Plastics
	  – Sealant for fluids
	* Low alloy metal
	  – Eliminates porosity

Cost Elements

Cost elements include the following:

* Material cost
* Setup time
* Compacting costs
* Sintering costs
* Direct labor rate
* Overhead rate
* Maintenance
* Amortization of equipment and tooling cost

Time Calculation

Shown is the formula used to calculate the process time for isostatic powder compaction when using either wet- or dry-bag tooling.

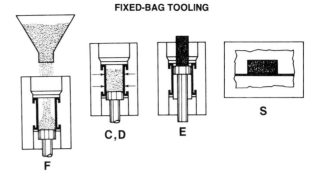

FIXED-BAG TOOLING

Fill time (sec) = F
Compaction time (sec) = C
Decompression time (sec) = D
Opening time (sec) = O
Ejection time (sec) = E
Sintering time (sec) = S

Total time = F + C + D + O + E + S

Safety Factors

The following hazards should be taken into consideration:

* Personal
  - Very high hydraulic pressures
  - Explosive hazard from powders
  - Respiratory hazards from powders
* Environmental
  - Fire
  - Explosions

No-Bake Mold Casting

In no-bake mold casting, molten metal is poured into a nonreusable mold that is made from a mixture of sand, quick-setting resin, and catalyst and is held until solidification occurs.

Process Characteristics

* Utilizes gravity to feed molten metal into a nonreusable sand mold
* Molding sand is poured into the mold instead of being rammed
* Molds are chemically cured at room temperature
* Dimensions are usually precise and repeatable
* Reduces labor costs, has high casting yields, and produces less scrap

Process Schematic

Molten metal is poured into the pouring basin, down the sprue hole, through a gate (to control the rate of metal flow), and into a mold cavity. Metal also flows into the riser, which provides additional metal to the cavity to compensate for shrinkage. After solidification has occurred, the part is removed, and the sand mixture may be discarded or reused.

NO-BAKE MOLD CASTING

MOLTEN METAL
SPRUE
SAND WITH RESIN AND CATALYST
GATE
POURING BASIN
CORE
MOLD CAVITY
RISER
COPE
DRAG

Workpiece Geometry

A wide variety of shapes and sizes can be achieved. Almost any metal that can be melted can be used with this process.

BEFORE   **AFTER**

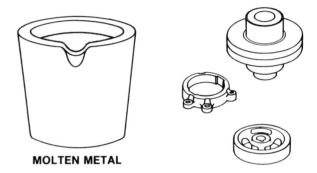

MOLTEN METAL

Setup and Equipment

A crucible is used to pour molten metal into cured mold halves, which hold mold material during the metal pouring and solidification processes.

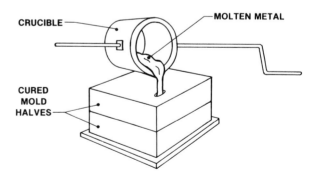

CRUCIBLE
MOLTEN METAL
CURED MOLD HALVES

Typical Tools and Geometry Produced

The patterns used are usually made of wood or metal. The material used in making patterns is determined by the complexity of the part and the number of parts that need to be made. Most patterns are similar to the one shown. The top half of the mold is referred to as the cope; the bottom half as the drag.

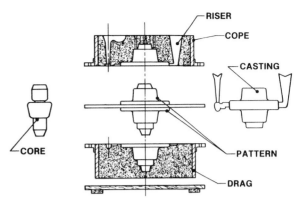

Permanent Mold Casting

Permanent mold casting is a metal shaping process in which molten metal is introduced into a permanent (reusable) mold, under gravity or low pressure, and held until solidification occurs. The molds are usually coated with a refractory wash and lampblack, which reduces the chilling effect on the metal and facilitates removal of the casting.

Process Characteristics

* Molten metal is introduced under gravity or low pressure
* Often uses close-grained cast iron as mold material
* Can produce good dimensional accuracy and smooth surface finishes
* Castings have reasonable strength with low porosity
* Complex or irregular shapes are difficult to produce

Process Schematic

The mold halves are clamped tightly or held under pressure while molten metal is poured into the cavity. Molten metal should be poured at the lowest practical temperature in order to minimize the possibility of cracks and porosity. After solidification occurs, the casting is removed, and the mold is prepared for the next cycle.

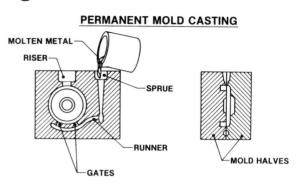

PERMANENT MOLD CASTING

MOLTEN METAL — RISER — SPRUE — RUNNER — GATES — MOLD HALVES

Workpiece Geometry

Metals that can be cast include aluminum, magnesium, zinc, copper alloys, and gray cast iron. Compared with sand casting, permanent mold casting permits uniform castings with closer dimensional tolerances, superior surface finish, and improved mechanical properties.

BEFORE

MOLTEN METAL

AFTER

Setup and Equipment

The setup typically consists of two mold halves containing a gate, a runner system, a pouring basin, alignment pins, an ejection system, and clamps to hold the mold halves together. Cores made of metal or sand are sometimes used within the mold.

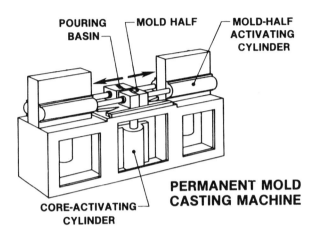

POURING BASIN — MOLD HALF — MOLD-HALF ACTIVATING CYLINDER

CORE-ACTIVATING CYLINDER

PERMANENT MOLD CASTING MACHINE

Typical Tools and Geometry Produced

Typical mold materials are fine-grain cast iron and steel. The main disadvantage of permanent molds is the relatively high initial cost. Permanent mold casting is particularly suitable for high volume production of small, simple castings that have a fairly uniform wall thickness.

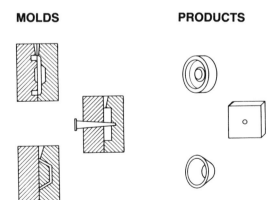

MOLDS   PRODUCTS

Geometrical Possibilities

This process can produce a wide variety of cast parts. Complex shapes require considerable expense in mold design and fabrication. Very few produced castings weigh more than 30 lbs. Aluminum and gray iron are commonly used.

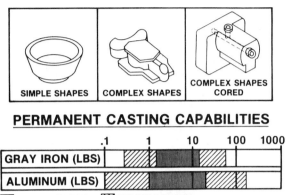

SIMPLE SHAPES   COMPLEX SHAPES   COMPLEX SHAPES CORED

PERMANENT CASTING CAPABILITIES

	.1	1	10	100	1000
GRAY IRON (LBS)					
ALUMINUM (LBS)					

▓ TYPICAL RANGE   ▨ FEASIBLE RANGE

Tolerances and Surface Finish

Dimensional accuracy is affected by mold closures and accuracy of positioning cores in the mold. Typical tolerances range from ±0.002 in. to ±0.015 in. Surface finish depends on polish on the mold wall, mold coatings used, venting, mold temperature, and gating design and size. Typical ranges for surface finish are from 125 to 300 microinches.

TOLERANCES		
PERMANENT MOLD CASTING	**TYPICAL**	**FEASIBLE**
UP TO 1 INCH EACH ADDITIONAL INCH ADDED	±.015 ±.002	±.010 ±.001
ACROSS PARTING LINES	±.015	±.010

SURFACE FINISH								
PROCESS	**MICROINCHES**							
PERMANENT MOLD CASTING	50	100	150	200	250	300	350	400

▓ TYPICAL RANGE   ▨ FEASIBLE RANGE

Tool Style

Molten metal fills the mold by gravity or low pressure. Gravity fed molds must be simple in shape and avoid thin sections and multi-branched areas. Low pressure fed molds force the molten metal into cavities by pressurized air or nitrogen, at 6 psi or 7 psi. Another method utilizes a piston to push the metal into the mold.

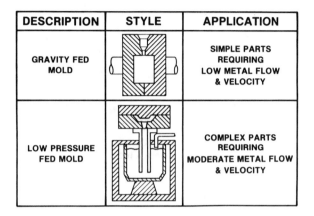

DESCRIPTION	STYLE	APPLICATION
GRAVITY FED MOLD		SIMPLE PARTS REQUIRING LOW METAL FLOW & VELOCITY
LOW PRESSURE FED MOLD		COMPLEX PARTS REQUIRING MODERATE METAL FLOW & VELOCITY

Tool Geometry

Simple mold cavities include horizontal and vertical parting lines, simple shapes, wide radii, and little variation in part thickness. Complex mold cavities may have thin sections or thick areas and often require the use of risers, chills, chaplets, or other methods to provide good thermal gradients and achieve good castings.

SIMPLE MOLD CAVITIES
- HORIZONTAL AND VERTICAL PARTING LINES
- RECTANGULAR AND CIRCULAR PARTS
- WIDE RADII AND LARGE DRAFT ANGLES
- LITTLE VARIATION IN PART THICKNESS

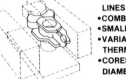

COMPLEX MOLD CAVITIES
- HORIZONTAL AND VERTICAL PARTING LINES
- COMBINATIONS OF GEOMETRIC SHAPES
- SMALLER RADII AND DRAFT ANGLES
- VARIATION IN PART THICKNESS REQUIRES THERMAL GRADIENT PLANNING
- CORES MAY BE ADDED FOR INTERNAL DIAMETERS

Design Considerations

Good designers provide uniform casting sections, do not isolate heavy sections, adhere to minimum tolerance requirements, include proper draft angles, and realize that the surface will require machining for mating parts.

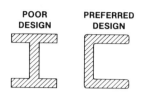

POOR DESIGN   PREFERRED DESIGN

NONUNIFORM FREEZING; SUBJECT TO HOT TEARS

* Uniform casting sections
* Heavy sections not isolated
* Minimum tolerance requirements
* Proper draft angles
* Surface finish of 125 to 300 microinches

Effects on Work Material Properties

Effects to mechanical properties include some residual stresses and may include dimensional changes due to shrinkage.

Work material properties	Effects of permanent mold casting
Mechanical	* Residual stresses * Dimensional change due to shrinkage
Physical	* Little effect
Chemical	* Little effect

Typical Workpiece Materials

Aluminum and gray iron alloys are popular casting alloys for components under 20 lbs. Gray iron alloys have excellent ratings and predictable shrinkage rates. Aluminum flows sluggishly and has low density. Magnesium has poor feeding characteristics and is prone to have hot spots. Copper and zinc parts are often die cast due to production requirements but have been successfully casted by this process.

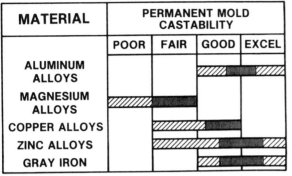

MATERIAL	PERMANENT MOLD CASTABILITY			
	POOR	FAIR	GOOD	EXCEL
ALUMINUM ALLOYS			■■■■	
MAGNESIUM ALLOYS	■■■■			
COPPER ALLOYS		■■■■		
ZINC ALLOYS		■■■■		
GRAY IRON			■■■■	

■ TYPICAL RANGE   ▨ FEASIBLE RANGE

Mold Materials

Materials used in permanent mold casting consist mainly of gray iron, cast iron alloys, sand, plaster, and graphite. Preferred applications of these materials are shown.

Mold materials	Applications
Gray iron	* Used for mold when casting most materials
Gray iron; 1020 steel	* Used for mold when casting zinc, aluminum, and magnesium
Alloy cast iron	* Used for mold when casting copper and copper alloys
Sand, plaster, and graphite	* Used as core material for most cast materials

Factors Affecting Process Results

Tolerance and surface finish depend upon the following:

* Mold material
* Mold condition and coating
* Proper gating system
* Adequate venting of gases
* Control of thermal gradients
* Proper mold and pouring temperatures

Process Conditions

One of the most critical process conditions for permanent mold casting is the material pouring temperature. This temperature varies from 700°F for zinc alloys to 2475°F for gray iron. Because of higher pouring temperatures, the mold life for gray iron parts is shorter.

Work material	Mold capacity	Pouring temperature (°F)
Aluminum alloys	1 oz. to 500 lbs	1250 to 1450
Magnesium alloys	1 oz. to 20 lbs	1300 to 1450
Copper alloys	1 oz. to 60 lbs	1800 to 2250
Zinc alloys	1 oz. to 20 lbs	700 to 1000
Gray iron	1 oz. to 30 lbs	2325 to 2475

Mold Coatings

Mold coatings are applied to the surface of the mold and allowed to dry. The coating minimizes thermal shock, prevents molten metal from directly contacting the mold, and controls the solidification rate and direction. Coating materials usually are mixed in a variety of combinations to permit spraying and brushing on the mold surface.

Coating material	Effects	Application
Sodium silicate Whiting Fireclay Metal oxide Soapstone Talc Mica	* Minimizes thermal shock * Prevents metal/mold contact * Controls solidification rate and directions	Spray or brush

Power Requirements

Melting energy requirements per unit volume depend on a metal's density, melting point, and specific heat. The energy required to melt a given metal is determined by multiplying its specific heat by its density and the pouring temperature of the metal.

Material	Density (lbs/in.^{3})	Melting point (°F)	Specific heat (Btu/lb-°F)
Aluminum	0.097	1220	0.214
Brass	0.306	1980	0.450
Zinc	0.245	790	0.095
Magnesium	0.063	1200	0.217

Cost Elements

Cost elements include the following:

* Mold fabrication and inspection
* Metal costs
* Fuel costs
* Direct labor rate
* Overhead labor rate
* Process time
* Amortization of equipment and tooling costs

Time Calculation

Filling time is a function of cavity volume and metal flow rate. Metal flow rate is controlled by the rate at which metal is poured into the cavity. The total time is the sum of mold opening and closing times, filling time, cooling time, and ejection time.

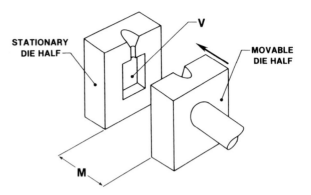

Mold closing/opening time (sec) = M
Filling time (sec) = F
Cooling time (sec) = C
Ejection/removal time (sec) = E
Cleanup time (riser, gates, flash) (sec) = R
Metal flow velocity (in.^{3}/min) = Q
Mold cavity size (in.^{3}) = V

$$\text{Filling time} = \frac{V}{Q}$$

$$\text{Total time} = 2M + F + C + E$$

Safety Factors

Molten metal can cause violent explosions and burns when in contact with moisture. Other factors are given below:
   The following risks should be taken into consideration:

* Personal
  - Explosions if the metal contains any moisture
  - Possibility of severe burns
* Environmental
  - Smoke
  - Fumes

Plaster Mold Casting

Plaster mold casting is a metal casting process in which nonferrous molten metal is poured into a nonreusable, shaped plaster mold and held until solidification occurs.

Process Characteristics

* Utilizes gravity to feed molten metal into a nonreusable plaster mold
* Is limited to nonferrous metals
* Produces smooth surface finishes and fine details compared to sand casting
* Produces minimal scrap material
* Can produce thin-walled sections and complex shapes

Process Schematic

Molten metal is poured into a pouring basin, down the sprue hole, and into the cavity, through the gate. After the cavity is full, the riser provides additional metal to compensate for shrinkage. After solidification has occurred, the plaster is broken away from the cast part.

PLASTER MOLD CASTING

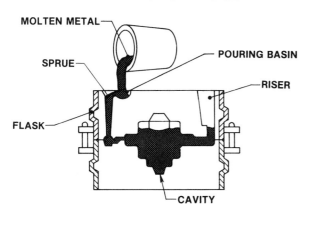

Workpiece Geometry

Plaster mold casting is most widely used when very complex parts must be made. Thin-walled sections and good surface finishes are also possible with this process.

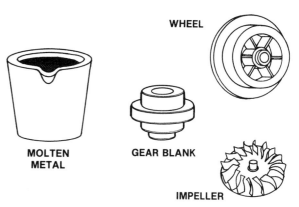

Setup and Equipment

Equipment consists of a crucible and a flask. The flask is used to hold the plaster mold securely in place until the molten metal solidifies. Metal is heated to the desired temperature inside the crucible, and then it is poured into the plaster mold.

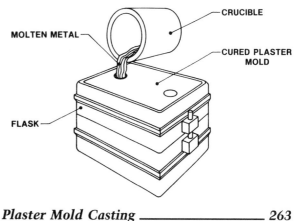

Typical Tools and Geometry Produced

Once a plaster mold has solidified, the cope and drag halves are separated, and the pattern is removed. The mold halves are then recombined to produce the desired mold ready for casting. The mold contains all essential features of the final casting, including sprues and risers.

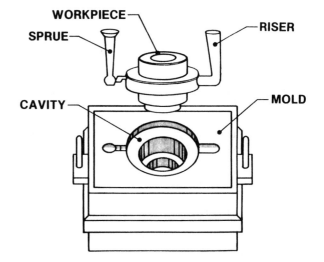

Rotational Molding

Rotational molding is a forming process in which melted plastic disperses over the inner surface of a rotating split mold, resulting in a hollow enclosed or open-ended part.

Process Characteristics

* Rotates the mold simultaneously about two axes
* Workpieces may be open-ended or hollow
* Can produce multilayered products
* May use thermoplastic or thermoset materials
* Tooling and maintenance costs are usually low

Process Schematic

In rotational molding, a premeasured amount of plastic powder is placed inside a two-piece mold. The mold is put into an oven while being rotated about two axes. The plastic, which is liquified by heat, coats the interior of the mold. The rotational motion is maintained while the mold is cooled. The mold is then opened, the workpiece removed, and the mold prepared for the next cycle.

Workpiece Geometry

Most plastics used in rotational molding are in a powder form, although some liquid plastics are used. Nearly all thermoplastics can be used for rotational molding. Workpieces are usually hollow or open-ended. Examples include commercial containers, hampers, balls, fuel tanks, and children's toys.

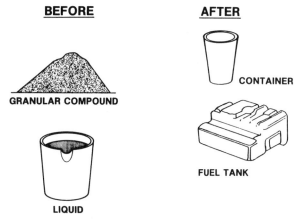

BEFORE   **AFTER**

GRANULAR COMPOUND

CONTAINER

LIQUID

FUEL TANK

Setup and Equipment

Rotational molding machines are made in a wide range of sizes. They normally consist of molds, an oven, a cooling chamber, and mold spindles. The spindles are mounted on a rotating axis, which provides a uniform coating of the plastic inside each mold.

ROTATIONAL MOLDING

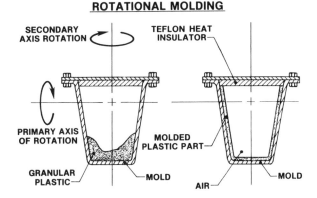

SECONDARY AXIS ROTATION

TEFLON HEAT INSULATOR

PRIMARY AXIS OF ROTATION

MOLDED PLASTIC PART

GRANULAR PLASTIC

MOLD

AIR

MOLD

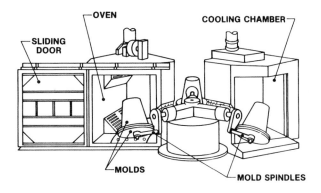

SLIDING DOOR

OVEN

COOLING CHAMBER

MOLDS

MOLD SPINDLES

Typical Tools and Geometry Produced

Molds usually consist of two pieces and are made of aluminum. The quality of the workpiece detail and finish is directly related to the quality of the mold. The workpiece may be open-ended, as shown by the container example, or may be closed, as shown by the fuel tank example.

ALUMINUM MOLDS

PRODUCTS

CONTAINER

FUEL TANK

Shell Mold Casting

Shell mold casting is a process in which molten metal is poured into a heat-cured, nonreusable shell mold made from silica sand and a resin binder and is held until solidification of the molten metal occurs.

Process Characteristics

* Is superior to other sand casting processes in the accurate duplication of intricate shapes and dimensional accuracy
* Process can be completely mechanized
* Uses a thin-walled nonreusable shell composed of a sand–resin mixture
* Requires a heated metal pattern for producing the shell molds

Process Schematic

Shell halves are clamped or bonded together, and the mold is placed in a flask and surrounded with shot, sand, or gravel. This backing material reinforces the mold during the casting process. Metals cool rapidly, producing fine-grained, strong castings that have close dimensional accuracy and a good surface finish.

SHELL MOLD CASTING

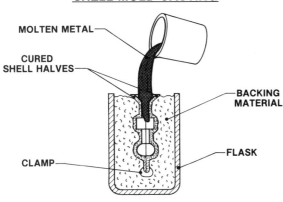

Workpiece Geometry

Workpiece shape can be basic or intricate. Cores may be used to produce holes or cavities in workpieces.

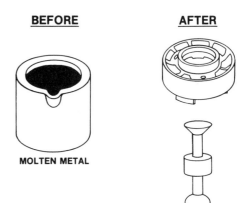

Setup and Equipment

Shell mold casting requires a shell molding machine. This machine heats a metal pattern, applies a resin and sand mixture to the pattern, and bakes the pattern until cured. A box is often needed to hold completed shell molds. Backing material is used to support molds.

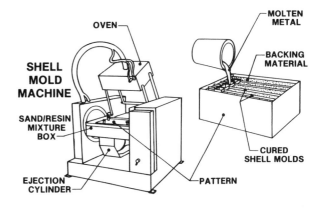

Typical Tools and Geometry Produced

The pattern consists of two halves that contain a sprue, a gate, and a runner. Patterns are made of metal and are usually tightly toleranced with a smooth surface finish. Patterns may not contain undercuts but may be very complex and detailed. Patterns are used to produce shell molds, which are joined together and used directly in the casting process.

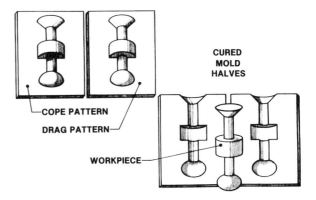

CURED MOLD HALVES

COPE PATTERN

DRAG PATTERN

WORKPIECE

Thermoform Molding

Thermoform molding is a shaping process in which a thermoplastic sheet or film, heated to its softening point, is pressed against the contours of a mold and allowed to cool until it retains the shape of the mold. Vacuum, air, or mechanical means may be used to press the heated sheet against the mold.

Process Characteristics

* Uses thermoplastic sheet or film
* Material is heated by convection or radiant heat
* Uses vacuum, air, or mechanical means to press the softened sheet against the mold
* Can produce simple shapes with large radii and no undercuts
* Produces varying thickness on the finished product

Process Schematic

A sheet of plastic that is held in place horizontally is heated to a softening point, which allows it to sag to a specified contour. The pliable sheet can be pressed into the mold by the plug assist. A vacuum is applied to pull the plastic tightly against contours of the mold, where it is held until cool and then removed.

THERMOFORM MOLDING

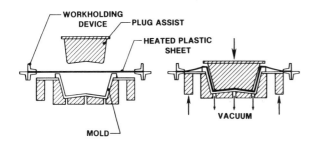

Workpiece Geometry

The workpieces are formed from plastic sheet or film and are usually simple shapes with large radii and no undercuts. Workpiece size varies and can be as large as 10 ft × 30 ft.

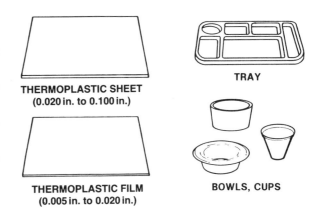

BEFORE **AFTER**

THERMOPLASTIC SHEET
(0.020 in. to 0.100 in.)

TRAY

THERMOPLASTIC FILM
(0.005 in. to 0.020 in.)

BOWLS, CUPS

Setup and Equipment

Thermoforming machines are made in a wide range of sizes. They normally consist of an oven, a mold, a plug assist, and a workpiece holding device. A workpiece is formed over or into a mold by mechanical, pneumatic, or vacuum pressure.

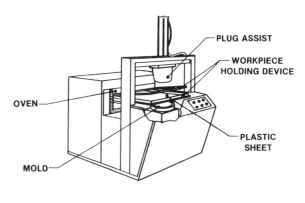

Typical Tools and Geometry Produced

Male or female molds, shaped to workpiece contours, are commonly made of wood, ceramic, or metal. An excess of material is needed for proper holding, forming, and sealing of the sheet to the mold. This excess is trimmed off in a secondary operation.

WOOD, CERAMIC OR METAL MOLDS

MOLDED SHEETS READY TO BE TRIMMED

FEMALE MOLD

MALE MOLD

Transfer Molding

Transfer molding is a shaping process in which a premeasured quantity of thermoset plastic is softened in a heated chamber and forced into a closed mold cavity where it conforms to the shape of the cavity and is cured.

Process Characteristics

* Uses a heated mold to cure and solidify plastic material
* Can mold fragile inserts in place
* Uses thermoset plastics
* Molds hot plastic under pressure
* Leaves small gate marks

Process Schematic

A premeasured amount of thermosetting plastic in powder, preform, or granular form is placed in the heating chamber. After the mold is closed, a plunger forces the plastic through the sprue and runners into the mold cavity. After curing, the mold is opened, and ejection rods lift the molded product out of the mold cavity for easy removal.

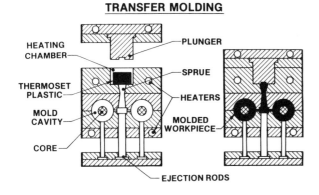

TRANSFER MOLDING

Workpiece Geometry

Transfer molding uses thermoset plastics in preform or granular form. Common products are utensil handles, electric appliance parts, electronic components, and connectors. Transfer molding is widely used to enclose or encapsulate items such as coils, integrated circuits, plugs, connectors, and other components.

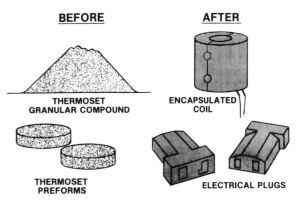

Setup and Equipment

Transfer molding machines are manufactured in a wide range of sizes and configurations. They normally consist of a heating chamber, plunger, and heated two-part mold consisting of upper and lower halves. Reusable cores are used when required.

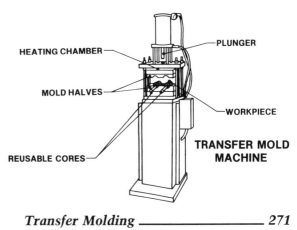

TRANSFER MOLD MACHINE

Typical Tools and Geometry Produced

Matching mold halves are precision machined and contain part cavities, gates, runners, and other configurations that may be needed to facilitate the molding process. In the example of coil encapsulation shown, reusable cores are precision machined to fit tightly within the mold, producing a specified hole or cavity within the coil. The *cull* is a small amount of material that is broken from the finished product and discarded. Its purpose is to insure an adequate reservoir of material to encapsulate the part completely.

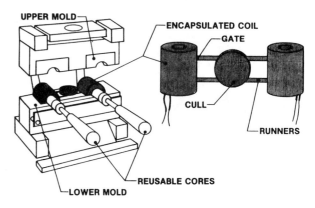

5.
Deformation

Brake Forming

Brake forming is a process in which sheet metal is formed along a straight axis by means of a U-shaped, V-shaped, or channel-shaped punch and die set.

Process Characteristics

* Forms ductile material
* Requires minimal tooling
* Produces long workpiece utilizing U, V, channel, or special punch and dies
* Is suitable for production of small parts
* Is used for low and medium production
* May use special application dies, interchangeable press brake dies, and numerically controlled machines

Process Schematic

In press brake forming, a workpiece is positioned over the die block and formed by the punch as it is forced into the die cavity. Although a secondary press motion may be required to compensate for material springback, many dies are designed to "overbend" the part and automatically compensate for material springback.

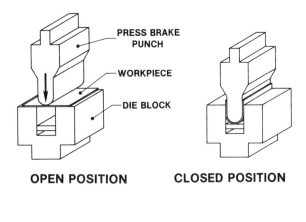

BRAKE FORMING

PRESS BRAKE PUNCH

WORKPIECE

DIE BLOCK

OPEN POSITION   **CLOSED POSITION**

Workpiece Geometry

Material thickness may range from 1/32 in. to 1/2 in. with lengths from 6 in. to 20 ft. Ductile metals, such as aluminum and mild steel, are typical material workpieces. New plastic materials may also be formed by this process.

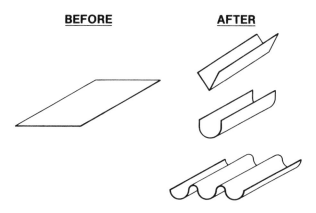

BEFORE   **AFTER**

Setup and Equipment

The punch and die used on a press brake can be changed. The punch is attached to the downward moving ram, and the die is attached to the stationary press bed. The workpiece is positioned in the press against an adjustable stop. A press brake is typically used because of its flexibility and speed. The CNC press brake shown is used for medium to high production requirements. The workpiece is fed onto the die and stopped by the preset back gage. The downward motion of the ram causes the shaped punch to form the workpiece. A secondary cutting operation may be necessary.

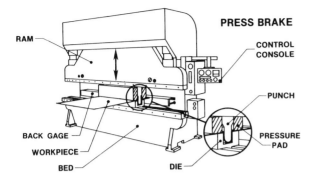

PRESS BRAKE

RAM

CONTROL CONSOLE

PUNCH

PRESSURE PAD

BACK GAGE

WORKPIECE

BED

DIE

Typical Tools and Geometry Produced

One commonly used punch and die set is used to form U-shaped workpieces. In "open-air" bending (i.e., bending without any forming done by the lower die and without any device to clamp the workpiece), die sets are relatively inexpensive and can be produced in various dimensions. Universal die blocks contain urethane rubber pads to allow a variety of forming operations with different punch shapes and one die cavity. In this case, all forming is done by the punch.

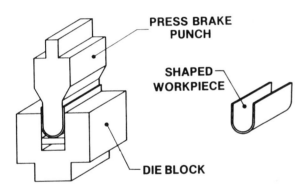

PRESS BRAKE PUNCH

SHAPED WORKPIECE

DIE BLOCK

Geometrical Possibilities

Geometrical possibilities include U-, V-, channel, and modified channel bends in a wide range of material and size options. Typical material thickness values range from 0.1 in. to 0.5 in. with feasible bending values from 0.5 in. to 1 in.

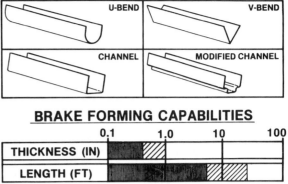

U-BEND   V-BEND

CHANNEL   MODIFIED CHANNEL

BRAKE FORMING CAPABILITIES

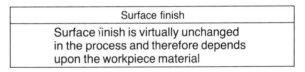

	0.1	1.0	10	100
THICKNESS (IN)				
LENGTH (FT)				

■ TYPICAL RANGE   ▨ FEASIBLE RANGE

Tolerances and Surface Finish

Locational tolerances are typically in the range of ±1/32 in. with ±1/64 in. feasible. After bending, the work material attempts to partially return to its original position; this angular springback must be allowed for to prevent inaccurate bend angles. Surface finish depends on the original condition of the workpiece material and the condition of the dies.

Tolerances		
Bending	Typical	Feasible
Locational	±1/32 in.	±1/64 in.
Angular	±3° to 5°	±0° to 2°

Surface finish
Surface finish is virtually unchanged in the process and therefore depends upon the workpiece material

Tool Style

Tool styles vary according to production requirements. The U-bend punch forms corrugations or single U-shaped parts. A V-bend punch forms 90° bends in a single stroke. The channel bend punch forms flat-bottom channel bends. Two- or four-way die blocks are available to form various sizes. A modified channel punch and changeable rubber die pad may be used to form various simple contour shapes.

DESCRIPTION	STYLE	APPLICATION
CHANNEL BEND	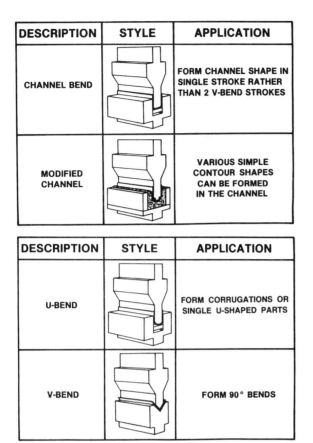	FORM CHANNEL SHAPE IN SINGLE STROKE RATHER THAN 2 V-BEND STROKES
MODIFIED CHANNEL		VARIOUS SIMPLE CONTOUR SHAPES CAN BE FORMED IN THE CHANNEL

DESCRIPTION	STYLE	APPLICATION
U-BEND		FORM CORRUGATIONS OR SINGLE U-SHAPED PARTS
V-BEND		FORM 90° BENDS

Workholding Methods

With air-bending, the workpiece is held in place by the punch, the die, and the back gage. In some instances, a pressure pad is used to restrict motion of the workpiece.

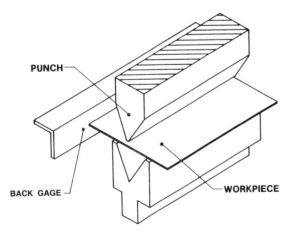

Effects on Work Material Properties

Workhardening of the metal occurs along the bend axis, with little or no effect on the physical or chemical properties of the material.

Work material properties	Effects of brake forming
Mechanical	* Workhardening along bend axis
Physical	* No effect
Chemical	* No effect

Typical Workpiece Materials

Formability of metals depends on the difference between yield stress and tensile strength. When the difference is large, forming may be difficult. Grain direction of rolled sheet metal also has a bearing on the strength of the bent part.

MATERIAL	BRAKE FORMING FORMABILITY RATINGS			
	POOR	FAIR	GOOD	EXCEL
ALUMINUM				
BRASS				
COLD ROLLED CARBON STEEL				
HOT ROLLED CARBON STEEL				
STAINLESS STEEL				

▓ TYPICAL RANGE ▨ FEASIBLE RANGE

Tool Materials

Tool materials largely depend on the application and the size of the production run. A list of materials ranging from carbide to hardwood, along with their applications, follows.

Tool materials	Applications
Low-carbon steel	* For low production * Die and punch base metal * Soft to medium materials
Tool steel	* For medium to high production * Medium to severe bending * Medium to tough materials
Carbide	* For high production * Severe bending * Very tough work materials * Usually in form of inserts
Hardwood	* For low production * Simple bends * Soft materials

Factors Affecting Process Results

Tolerance and surface finish depend upon the following:

* Tool geometry and surface conditions
* Material type
* Material thickness
* Springback
* Bottoming force

Tool Geometry

Tool geometry is influenced by punch width, punch radius, necessary clearance between the punch and die, and die opening. Recommended values for punch radius and clearance are shown for typical workpiece materials, including medium-carbon steel, aluminum, stainless steel, and brass.

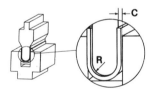

Workpiece material	Punch radius (R)	Clearance (C)
Medium-carbon steel	1 t	1.1 t
Aluminum	2 t to 4 t	1.1 t
Stainless steel	½ t to 4 t	1.1 t
Brass	2 t to 2-½ t	1.1 t

Values are in terms of material thickness (t).

Production Rates

Production rates depend on the type of press used, the size of the workpiece, and the amount of energy that can be used up in the press stroke. In this case, the amount of energy is 36%. This means that the press has to be big enough to be able to form the workpiece with only 36% of its maximum force. Production rates range from 30 bends per minute to as much as 65 bends per minute.

Workpiece material	Thickness (in.)	Required bending force (tons)	Production rate (bends/min)
Steel	1/8	4	60
	1/4	15	60
	1/2	60	30
Aluminum	1/8	3	60
	1/4	10	65
	1/2	40	40

Using a single-geared standard speed press bending a 12-in. long workpiece with a die opening width of 2 in. and a "K" value of 1.33 and a 36% energy release in the press per stroke.

Protective Media

Media such as adhesive-backed paper are used for protecting polished surfaces on work materials. After bending and other auxiliary operations, the protective sheet is stripped off the workpiece.

Work material	Protective media	Application
Aluminum Brass Steel all types	Adhesive paper	Polished surfaces

Power Requirements

A formula can be used to calculate force requirements. The definitions are shown for each parameter. These calculations are for bending 90° angles. Force requirements are calculated using a 36% energy release, which is about normal for standard presses. Bending forces range from as low as 3 tons to as high as 60 tons.

Force (tons) = F
Length of bend (in.) = L
Thickness of material (in.) = t
Material tensile strength (tons/in.^{2}) = S
Width of die opening (in.) = W
Constant for die opening distance = K;
1.2 for die opening of 16 t
to 1.33 for a die opening of 8 t

$$F = \frac{L \times t^2 \times S \times K}{W}$$

Cost Elements

Cost elements include the following:

* Setup time
* Load/unload time
* Forming time
* Direct labor rate
* Overhead rate
* Amortization of equipment and tooling costs

Time Calculations

The time calculation for the total cycle equals the punch stroke length divided by punch velocity, plus the ejector stroke length divided by ejector velocity.

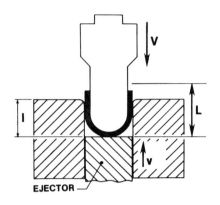

EJECTOR

Punch stroke length = L
Punch velocity (average) = V
Ejector stroke length = I
Ejector velocity (average) = v

$$\text{Forming time} = \frac{L}{V} + \frac{I}{v}$$

Safety Factors

The following personal and environmental risks should be taken into consideration:

* Pinch points
* Sharp edges on workpiece
* Workpiece motion
* Noise level

Cold Heading

Cold heading is defined as a "metal forging process used for rapidly producing enlarged (upset) sections on a piece of rod or wire held in a die." The resulting shape of the upset portion conforms to the shape of the die cavity. Workpieces are not heated prior to heading. Upsetting may be done in one or more strokes.

Process Characteristics

* Upset portion conforms to shape of the punch and/or die cavity
* Finished workpieces have improved grain structure and mechanical properties
* Finished surfaces are smooth with uniform accuracy
* May process ferrous and nonferrous metals

Process Schematic

To perform the cold heading process, a piece of rod or wire is cut to length, placed in a die-ejector assembly, and upset by one or more strokes of the heading dies. After heading, the workpiece is ejected from the die by a knockout pin. Another method of cold heading (as illustrated below) utilizes a continuous rod or wire that is fed forward, clamped, formed, further advanced, and then cut to length in a continuous cycle.

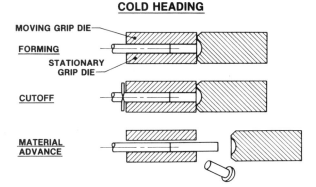

COLD HEADING

Workpiece Geometry

Typical examples of workpieces that can be formed by cold heading include bolts, rivets, valves, and similar parts. Cold heading is most commonly used to produce symmetrical parts; however, many different shapes and cross-sections can be formed. Multiple diameters may be formed by using a series of heading dies and strokes. During upsetting, the workpiece is cold worked and thus made stronger.

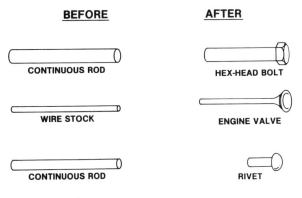

Setup and Equipment

Cold heading is a high production, automatic process. In a typical heading operation, the wire stock is fed from a continuous roll into the die, cut to the desired length, and formed. Finished parts are automatically ejected from a part drop chute.

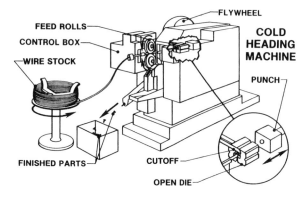

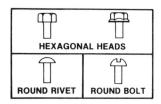

COLD HEADING CAPABILITIES

	0.1	1.0	10	100
DIAMETER (IN)				
LENGTH (IN)				

■ TYPICAL RANGE ▨ FEASIBLE RANGE

Typical Tools and Geometry Produced

Cold heading is done on two types of heading machines—solid and open die. The length of the required part usually determines the type of die used, with solid dies generally being used for relatively small wire. Upset portions may be formed in cavities in the punch, the die, or in both the punch and die. Some parts are upset between the punch and die.

Tolerances and Surface Finish

For most cold heading applications, tolerances are held within ±0.030 in. for diameter and ±0.002 in. for length. For precision applications, diameters can be held within ±0.010 in. and length within ±0.001 in. Surface finish may range from 8 to 125 microinches, with the typical range between 16 and 63 microinches.

TOLERANCES		
COLD HEADING	TYPICAL	FEASIBLE
DIAMETER	±.030	±.010
LENGTH	±.002	±.001

SURFACE FINISH									
PROCESS	MICROINCHES (A.A.)								
COLD HEADING	2	4	8	16	32	63	125	250	500

■ TYPICAL RANGE ▨ FEASIBLE RANGE

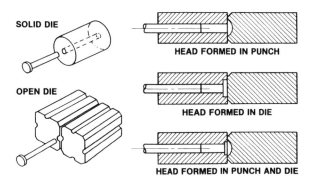

Geometrical Possibilities

Almost any shape of head can be produced. Shown are just a few of the possibilities. Some of the most common are round rivets, bolts, and nails. Others would include square or hexagonal and offset heads. Undercuts are seldom present unless produced by a secondary process. Typical and feasible upset diameters and lengths are shown.

Tool Style

Heads produced on wire or rod are usually done in one of four ways, depending on tool style. The head can be formed in the punch to produce round heads, the die to produce flat heads, in both the punch and die to produce stepped shapes, or between the punch and the die to produce parallel surfaces.

DESCRIPTION	STYLE	APPLICATION
HEAD FORMED IN PUNCH		ROUND HEAD SQUARE HEAD
HEAD FORMED IN DIE		WOOD SCREW HEAD RADIAL SLOT HEAD HEXAGONAL HEAD
HEAD FORMED IN PUNCH & DIE		CARRIAGE BOLT HEAD
HEAD FORMED BETWEEN PUNCH & DIE		COLLAR HEAD

Work material properties	Effects of cold heading
Mechanical	* Improved strength and hardness * Better surface finish * Decrease in ductility
Physical	* Slight decrease in electrical conductivity
Chemical	* Slight decrease in corrosion resistance

Workholding Methods

Several features are necessary on a cold heading machine to support the workpiece while the part is being produced. These include a wire feeding device, ejector stops, forming/heading dies, and knockout pins. Other components made include cutoff blades.

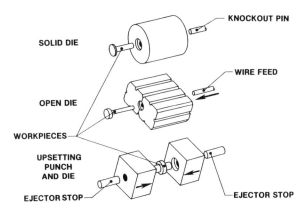

Effects on Work Material Properties

The mechanical properties affected during cold heading may include improved strength and hardness, better surface finish, and decrease in ductility. In terms of physical properties, there may be a slight decrease in electrical conductivity. Chemical properties affected may be that of slightly decreasing corrosion resistance due to the presence of high stresses in the workpiece.

Typical Workpiece Materials

The formability ratings for aluminum, brass, and copper are good to excellent because they are ductile and soft. Mild steel and stainless steel are rated fair to good because they are hard, yet ductile. Alloy steels are rated poor to fair because they often are not very ductile and have a high strain hardening tendency. Stainless steels have high ductility and high strain hardening, thus requiring higher upsetting forces.

MATERIAL	COLD HEADING FORMABILITY RATINGS			
	POOR	FAIR	GOOD	EXCEL
ALUMINUM				▨▥
BRASS				▨▥
COPPER			▨▥	
ALLOY STEEL	▨▥			
MILD STEEL		▨▥		
STAINLESS STEEL (Cr 12-17%)		▨▥		

▥ TYPICAL RANGE   ▨ FEASIBLE RANGE

Tool Materials

Typically, cold heading tool materials are used when the punch and die are made of hardened tool steel; the inserts are made of high alloy tool steel, carbides, or high speed steel; or the ejector pin is made of either high alloy tool steel or high speed steel.

Tool materials	Applications
Hardened tool steel (W2)	* Generally used for solid punch and die
High alloy tool steel	* May be used for ejector pins * May be used as a solid die insert * Typically used for open dies
Carbides	* May be used as a solid die insert
High speed steel	* May be used for ejector pins * May be used as a solid die insert

Factors Affecting Process Results

Tolerance and surface finish from cold heading depend upon workpiece material, condition of tooling, amount of deformation per stroke, and lubricant used.

Tolerance and surface finish depend upon the following:

* Workpiece material
* Condition of tooling (punch and die)
* Deformation per stroke
* Lubricant used

Tool Geometry

Shown is an example of a die block with an insert and typical workpiece materials. The tool steel die materials and subsequent die hardness values depend upon the workpiece material used.

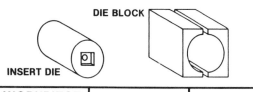

DIE BLOCK

INSERT DIE

WORKPIECE MATERIAL	TOOL STEEL DIE MATERIAL		DIE HARDNESS RC	
	SOLID	INSERT	SOLID	INSERT
ALUMINUM BRASS COPPER	W1,W2	D2,M2	58-60	60-62
ALLOY STEEL CARBON STEEL STAINLESS STEEL	D2,M2	CARBIDE	60-62	62-64

Process Conditions

Typical cold heading speeds depend primarily upon the blank (head) diameter. Blanks of up to 0.25 in. are cold headed at speeds of 400 to 600 pieces per minute. As the blank diameter increases, the speed decreases; thus, blanks of 0.5 in. to 0.25 in. are cold headed at speeds of 40 to 100 pieces per minute.

Up to 1/4 in. blank diameter

Workpiece material	Cold heading production speeds (pieces/min)
Aluminum	550 to 600
Brass	500 to 550
Copper	500 to 550
Mild steel	450 to 500
Stainless steel	400 to 450

Lubrication and Cooling

As can be seen from the chart, a variety of lubricants are used during cold heading. The lubricants are applied to the workpiece by spray, flooding, dipping, or a chemical process to secure the adherence of the lubricant. This is done so that the lubricant will stay on the component and prevent direct contact between virgin work material and the die or punch. Severe cold heading is often done with extrusion oils.

Work material	Type of fluid
Aluminum	Stearate, phosphates, lime
Copper	None, oil, drawing compounds
Carbon and alloy steels	Zinc phosphate, lime, calcium stearate, aluminum stearate, extrusion oils
Stainless steel	Extrusion oils, molybdenum disulfide

Severe cold heading should use extrusion oils. Lubricants are applied by coating the workpiece.

Power Requirements

The power requirements in a cold heading operation depend on the material used, the necessary deformation, and its complexity. The necessary work can be determined approximately (empirically) from the following parameters: true tensile strength, cross-sectional area, press velocity, and time required to produce a given shape. Although little data are available for specific die/punch loads, they are unlikely to exceed 4.5 times a material's yield stress at the maximum deformation strain. The reason for the high die load is to ensure the complete filling of the die cavity.

$$W = (\sigma_T)(A)(V)(t)$$

Materials	σ_T (ksi)	Work (ft-lbs)
Aluminum and alloys	9–50	2531 to 14,062
Brass	46–76	12,937 to 21,375
Copper and alloys	38–65	10,687 to 18,281
Alloy steels	40–55	11,250 to 15,468
Carbon steels	45–55	12,656 to 15,468
Stainless steels	90–140	25,312 to 39,375

$A = 3/16$ in.^{2}, $t = 1$ sec, $V = 1.5$ ft/sec.

Cost Elements

Cost elements include the following:

* Setup time
* Processing time
* Tool change time
* Tool costs
* Overhead rate
* Direct labor rate
* Amortization of equipment and tooling

Time Calculation

The time calculations per piece equal two times the open/close time plus cutoff and wire feed-in times.

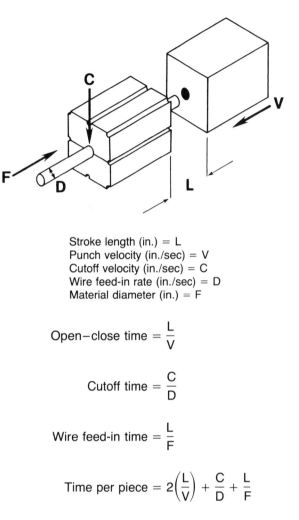

Stroke length (in.) = L
Punch velocity (in./sec) = V
Cutoff velocity (in./sec) = C
Wire feed-in rate (in./sec) = D
Material diameter (in.) = F

$$\text{Open–close time} = \frac{L}{V}$$

$$\text{Cutoff time} = \frac{C}{D}$$

$$\text{Wire feed-in time} = \frac{L}{F}$$

$$\text{Time per piece} = 2\left(\frac{L}{V}\right) + \frac{C}{D} + \frac{L}{F}$$

Productivity Tip

When cold heading, the maximum upset length should not be greater than 2.0 to 2.5 times the diameter per blow. The maximum upset diameter should not be greater than 2.5 to 3.0 times the diameter per blow.

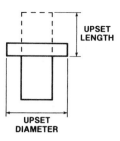

Safety Factors

In cold heading, safety risks to the operator include noise from the machine and allergies or irritation to the skin from the lubricants. Cold heading is a noisy operation and causes substantial vibrations in the floor, which may be transmitted to other machines.

The following risks should be taken into consideration:

* Personal
  - Noise
  - Possible pulmonary complications
  - Irritations from lubricants
* Environmental
  - Floor vibrations

Deep Drawing ___

Deep drawing is a cold forming process in which a flat blank of sheet metal is shaped by the action of a punch forcing the metal into a die cavity. Deep drawing differs from other drawing processes in that the depth of the drawn part can be greater than its diameter.

Process Characteristics

* Uses ductile sheet metal
* Can have low or high volume production
* Can be manually or automatically operated
* Produces various geometric shapes
* Draw depth is greater than the workpiece diameter
* Usually requires trimming of the top edge of the workpiece

Process Schematic

Deep drawing involves a descending punch that forces the workpiece into a die, and a pressure plate prevents the workpiece from wrinkling around the edges. At the end of the cycle, the punch retracts, and the workpiece is ejected out of the die by the knockout plate. A trimming operation is usually performed to remove excess material from the top edge of the drawn part.

Workpiece Geometry

Most drawn products are produced from flat, ductile sheet metal workpieces. Typical pre-drawn blanks are shown along with resulting drawn shapes.

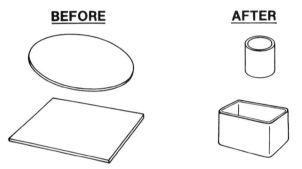

BEFORE   **AFTER**

Setup and Equipment

Shown is a typical open back inclinable (OBI) press. The punch is generally mounted on the ram of the press, and the die is mounted on the press bed. The press can be operated manually or, with special feeding equipment, it can be automated. Generally, only hydraulic presses are used due to the desired long stroke and second motion from the punch in the press bed.

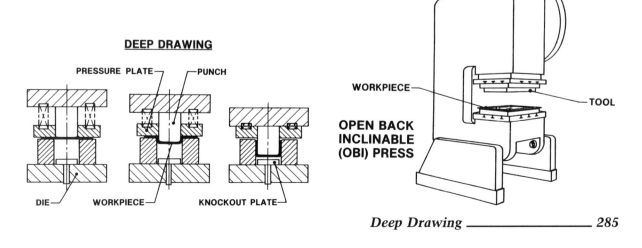

DEEP DRAWING

PRESSURE PLATE — PUNCH

DIE — WORKPIECE — KNOCKOUT PLATE —

WORKPIECE —   — TOOL

OPEN BACK INCLINABLE (OBI) PRESS

Typical Tools and Geometry Produced

A wide variety of drawn parts can be produced by deep drawing. Die sets are often of special design and construction and are thus expensive. Some simple shapes can be produced quite economically, however, using standard tooling components.

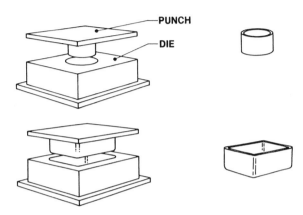

DIE SET　　　**PRODUCT**

PUNCH
DIE

Geometrical Possibilities

Representative parts may be symmetrical, asymmetrical, or bilevel. Common products are kitchen sinks, which are produced by the millions. The diameter of drawn parts may range from 2 in. to 100 in. and the height, or depth, of draw may range from approximately 1 in. to over 100 in. in special cases.

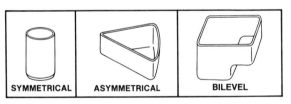

SYMMETRICAL　　ASYMMETRICAL　　BILEVEL

DEEP DRAWING CAPABILITIES

	0.1	1.0	10	100
DIAMETER (IN.)				
HEIGHT (IN.)				

▓ TYPICAL RANGE　▨ FEASIBLE RANGE

Tolerances and Surface Finish

For most deep drawing applications of relative small components, tolerances are usually held to within ±0.010 in. for the diameter and ±0.020 in. for the depth. The tolerances increase with increasing diameters. The surface finish corresponds to the original sheet metal finish, as long as proper tooling and lubrication are used. Some burnishing marks may exist along the length of the drawn portion. In some instances stretch–strain lines may be present.

Tolerances		
Deep drawing	Typical	Feasible
Diameter	±0.010	±0.05
Depth	±0.020	±0.01

Surface Finish

The surface finish corresponds to the sheet metal finish when proper tooling and lubrication are used. The process may cause stretch and strain lines.

Tool Style

Deep drawing dies are constructed in one of four basic styles: the single-action die (simple type) that uses neither a blankholder nor pressure pads and shallow draws; the single-action die, with pressure pad (blankholder), used for shallow to moderately deep draws; the double-action die, requiring two independently actuated rams on the hydraulic press; and the double-action inverted type, utilized for redrawing parts and for large, awkward shapes where gravity can aid in ejecting the workpiece.

DESCRIPTION	STYLE	APPLICATION
SINGLE-ACTION DIE SIMPLE TYPE		•SHALLOW CUPS
SINGLE-ACTION DIE WITH PRESSURE PAD		•SHALLOW AND DEEP DRAWS •ENGINE OIL PANS
DOUBLE-ACTION DIE		•REDRAWING PARTS •DRAWING PARTS THAT HAVE TWO DIAMETERS
DOUBLE-ACTION DIE INVERTED TYPE		•REDRAWING PARTS •BOX SHAPED OBJECTS

Workholding Methods

For simple, single-action dies, no blankholder is used. Some of the most commonly used methods of holding the workpiece during the drawing operation are shown in the other three figures. For single-action dies, a pressure pad with a coil spring, rubber spring, or air pressure provides the hold-down force. In the double-action method, the hydraulic press is equipped with two independent rams, one for activating the punch and the other for activating the blankholder.

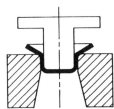

SINGLE-ACTION

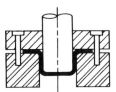

SINGLE-ACTION WITH CONSTANT GAP

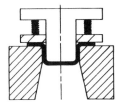

SINGLE-ACTION WITH PRESSURE PAD

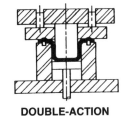

DOUBLE-ACTION

Effects on Work Material Properties

Deep drawing may produce residual stresses, microcracks, and workhardening (the latter increases the strength of the part).

Work material properties	Effects of drawing
Mechanical	* Residual stresses and microcracks * Workhardening * Increases strength
Physical	* Little effect
Chemical	* Little effect

Typical Workpiece Materials

Aluminum, mild steel, and stainless steel have fair to excellent formability ratings, depending upon their alloying composition. Brass has a fair to good rating, depending on composition and grain size. Coarse grained materials may undergo more severe deformation, and rough "orange-peel" surfaces may result.

MATERIAL	DEEP DRAWING FORMABILITY RATINGS			
	POOR	FAIR	GOOD	EXCEL
ALUMINUM			▨▨	▨▨
BRASS		▨	■■■	▨
MILD STEEL		▨	■■	▨
STAINLESS STEEL		▨	■	▨

■ TYPICAL RANGE ▨ FEASIBLE RANGE

Tool Materials

Deep drawing dies are commonly made of tool steel. Carbon steel is not as hard but is tougher and less expensive, therefore, it is used for less severe punch and die applications and for die elements such as the blankholder. Alloy steel is more durable and heat resistant, and cemented carbides are used for applications requiring high wear and abrasive resistance.

Tool materials	Applications
Tool steel	* The most widely used punch and die material
Carbon steel	* Strong and weldable punch and dies * Used for blankholder
Alloy steel	* Used for ejector * Durable and heat resistant blankholder
Cemented carbides	* Wear resistant punch and dies

Factors Affecting Process Results

Normally, drawn surfaces are quite smooth but may develop stretches, strains, orange-peel, or other defects if the conditions mentioned below are not closely controlled.

Tolerance and surface finish depend upon the following:

* Radii of punch and die
* Size of clearance
* Blankholder pressure
* Lubricant
* Draw depth
* Material condition and grain size

Tool Geometry

Tool geometry for deep drawing tools is based on the original blank diameter, the blank or workpiece thickness, the die radius, and the cup diameter. The required die radius can be calculated using the empirical formula: $r = 0.08\sqrt{(D - d)t}$, where r = die radius, D = blank diameter, d = cup diameter, and t = blank thickness.

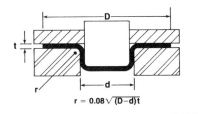

$$r = 0.08\sqrt{(D-d)t}$$

WORKPIECE MATERIAL	THICKNESS RANGE (IN)	DIE RADIUS RANGE (IN)	DIAMETER DIFFERENCE (IN)
ALL	0.012-0.160	0.040-1.600	0.400-16.00

Process Conditions

Shown are the recommended punch forces and blankholder forces for a drawing operation with a punch diameter of 5 in., a material thickness of 0.040 in., a drawing ratio or reduction (D − d/D) of 39%, and a correction factor of 0.8. The maximum punch force (before tearing occurs in the workpiece) is calculated using the given formula (see Tool Geometry), as well as the blankholder force. The maximum blankholder force must be kept to less than one-third of the maximum allowed punch force.

d = 5 in.	t = 0.040 in.	R = 39%	K = 0.8

Workpiece material	Workpiece tensile strength (psi)	Punch force (lbs)
Aluminum	45,000	22,600
Brass	60,000	30,200
Cold-rolled steel	65,000	32,700
Stainless steel	85,000	42,700

Punch force $P_{max} = \sigma_T \times \pi \times d \times t \times K$						
Blankholder force $p = 1/2\ P_{max}$						
R	50%	45%	43%	39%	36%	34%	31%
K	1	0.9	0.8	0.7	0.6	0.5	0.4

R = drawing ratio   d = punch diameter
t = blank thickness
σ_T = tensile strength
K = correction factor

Lubrication and Cooling

Lubricants used for deep drawing help reduce friction resulting from metal movement between the punch and die. They also permit greater drawing ratios and reduce problems in removing the part from the punch. Typical lubricants include heavy duty emulsions, phosphates, molybdenum disulfide, white lead, and wax films. Some considerations for choosing a lubricant are its effective temperature range, corrosion characteristics, and methods of applying and removing the lubricant. Plastic films covering both sides of the blank when combined with a suitable oil represent an excellent solution and give a part with a fine surface.

Work material	Lubrication	Application
All materials can use these lubricants	* Heavy duty emulsions * Phosphates * Metal coatings with: – Molybdenum disulfide – White lead – Wax film	* Reduction of friction * Greater drawing ratio * Easy separation of part from punch

Power Requirements

Note that soft materials, such as aluminum, require a much lower punch force than a less ductile material, such as stainless steel. In this instance (stainless steel versus aluminum), the difference is about 2 to 1, respectively.

Drawing force (lbs)

Material	Percent reduction			
	39%	43%	47%	50%
Aluminum	19,800	22,600	25,400	28,300
Brass	26,400	30,200	34,000	37,700
Cold-rolled steel	28,600	32,700	36,800	40,800
Stainless steel	37,400	42,700	48,100	53,400

Cost Elements

Cost elements include the following:

* Setup time
* Load/unload time
* Drawing time
* Tool change time
* Direct labor rate
* Overhead rate
* Amortization of equipment and tooling costs

Time Calculation

Shown are the formulas used to calculate the time for deep drawing, ejection, and positioning. Drawing time is a function of draw depth and punch velocity; ejection time is a function of draw depth and ejection velocity; and positioning time is a function of distance between blank stock and die, and feed rate.

DEEP DRAWING SETUP

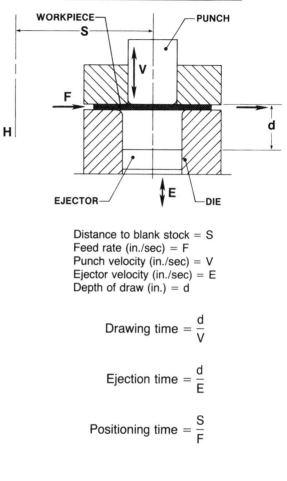

Distance to blank stock = S
Feed rate (in./sec) = F
Punch velocity (in./sec) = V
Ejector velocity (in./sec) = E
Depth of draw (in.) = d

$$\text{Drawing time} = \frac{d}{V}$$

$$\text{Ejection time} = \frac{d}{E}$$

$$\text{Positioning time} = \frac{S}{F}$$

Safety Factors

The following risks should be taken into consideration:

* Personal
  - Rapid die closing
  - Elevated noise level
  - Sharp scrap disposal
* Environmental
  - High machining noise levels

Deep Drawing ——————— 289

Drop Forging

Drop forging is a metal shaping process in which a heated workpiece is formed by rapid closing of a punch and die forcing the workpiece to conform to a die cavity. A workpiece may be forged by a series of punch and die operations (or by several cavities in the same die) to gradually change its shape. Drop forging is also called impression die or closed die forging, or rot forging.

Process Characteristics

* Gradually forms heated metal by singular or repeated blows in a sequence of individual or multistage die cavities
* Produces a parting line and flash on the workpiece; flash must be removed
* Typically requires machining to obtain dimensional tolerances and good surface finish

Process Schematic

The hot metal to be shaped is placed on a die. Impact of a ram on the workpiece compels the hot and malleable metal to conform to the shape of a punch and die cavity. The flow of metal is drastic. The flash acts as a relief valve for the extreme pressure produced by the closing die halves but must eventually be trimmed off. Typically, one blow is given in each die cavity.

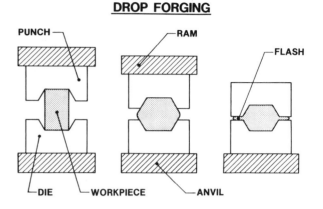

DROP FORGING

PUNCH — RAM — FLASH — DIE — WORKPIECE — ANVIL

Workpiece Geometry

Shown are successive steps for forging a connecting rod using the four-stage die process. Stage 1 is the pinch-off stage to get a correct size workpiece. Stages 2 and 3 include intermediate impressions that block the metal to approximately its final shape. Stage 4, or the final impression, gives the part its final size and shape. A trimming operation, stage 5, then removes the unwanted surplus material, or "flash."

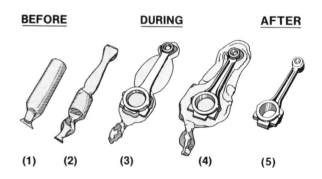

BEFORE   DURING   AFTER

(1)   (2)   (3)   (4)   (5)

Setup and Equipment

The forging press, known as a power drop hammer, may be air, hydraulic, or mechanically powered. The drop hammer shown is air actuated. Air pressure raises the weighted ram. Striking force capacities range from 11,000 to 425,000 pounds, depending on the mass of the ram and upper die and on the drop height.

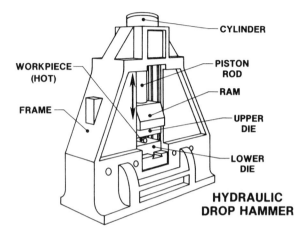

HYDRAULIC DROP HAMMER

Typical Tools and Geometry Produced

Proper metal flow is important. Therefore, the operation is divided into steps. The punch attaches to the ram and the die to the anvil. The four stages shown are as follows: first, heated metal is cut from bar stock and properly distributed; second, rough forming; third, final forming; and fourth, the product is manually removed.

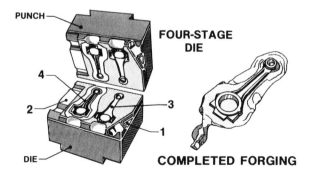

FOUR-STAGE DIE

COMPLETED FORGING

Geometrical Possibilities

Geometrical possibilities are wide ranging. Connecting rod and crankshafts are commonly forged parts. Typical parts range in weight from less than 3 pounds to 750 pounds. Overall dimensions vary from 3 in. to 50 in., typically ranging between 7 in. and 20 in.

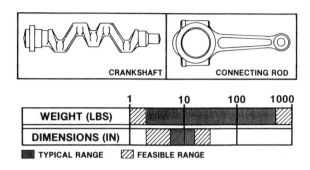

	1	10	100	1000
WEIGHT (LBS)				
DIMENSIONS (IN)				

■ TYPICAL RANGE ▨ FEASIBLE RANGE

Tolerances and Surface Finish

Typical dimensional tolerances range from ±0.02 in. to ±0.03 in. Surface finish values range from 80 to 300 microinches. As forging force increases, dimensional tolerances improve.

TOLERANCES		
DROP FORGING	**TYPICAL**	**FEASIBLE**
OVERALL DIMENSION (IN)	±0.03	±0.02

SURFACE FINISH									
PROCESS	**MICROINCHES (A.A.)**								
DROP FORGING	2	4	8	16	32	63	125	250	500

■ TYPICAL RANGE ▨ FEASIBLE RANGE

Tool Style

Tool styles vary in complexity and shape. The flat and V-dies are used for producing flat or round shapes and for improving mechanical properties. Single- and multi-impression dies are used for complex shapes using closed-die forging techniques. Trim dies for removing flash must be provided when doing closed-die forging.

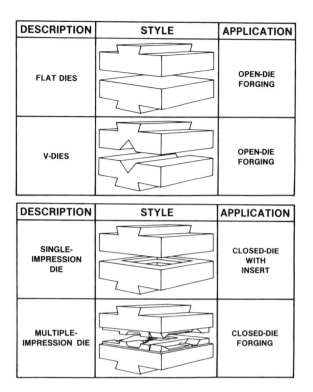

DESCRIPTION	STYLE	APPLICATION
FLAT DIES		OPEN-DIE FORGING
V-DIES		OPEN-DIE FORGING

DESCRIPTION	STYLE	APPLICATION
SINGLE-IMPRESSION DIE		CLOSED-DIE WITH INSERT
MULTIPLE-IMPRESSION DIE		CLOSED-DIE FORGING

Workholding Methods

Size and weight of the material influences the use of either manual tongs or a mechanical sling. Large parts are balanced in a sling and mechanically rotated by an overhead crane. In closed-die forging, manual tongs are normally used to hold a workpiece and move it from one stage to the next.

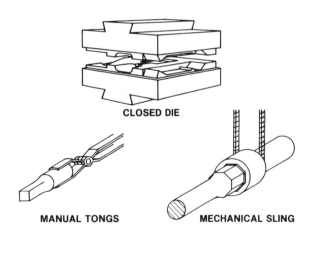

CLOSED DIE

MANUAL TONGS   MECHANICAL SLING

Effects on Work Material Properties

The mechanical properties of a workpiece are generally enhanced. Microcracks may be introduced, which are normally machined away.

Work material properties	Effects of drop forging
Mechanical	* Creates good to excellent mechanical properties including fatigue and impact resistance * May create microcracks
Physical	* Little effect
Chemical	* Little effect

Typical Workpiece Materials

Magnesium has poor to fair forgeability ratings. The remaining materials are all in the fair to excellent range. Stock workpiece materials are normally round, square, or flat with medium to high ductility. They are selected and cut from a parent workpiece to provide a favorable grain flow orientation.

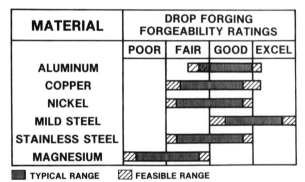

MATERIAL	DROP FORGING FORGEABILITY RATINGS			
	POOR	FAIR	GOOD	EXCEL
ALUMINUM				
COPPER				
NICKEL				
MILD STEEL				
STAINLESS STEEL				
MAGNESIUM				

TYPICAL RANGE   FEASIBLE RANGE

Tool Materials

Tool materials are selected on the basis of cost, ability to harden uniformly, ability to withstand high pressures, resistance ability to resist hot abrasion, heat cracking resistance, and ability to resist checking. Carbon steel is used for large open-die or closed-die applications because of low cost and good service life. Stainless steels are used where closer tolerances must be held on high production runs. Tool steel dies are the most expensive but have superior properties under extreme pressure and heat.

Tool materials	Applications
Carbon steel	* Used for open and closed dies * Heat intensive production * Low cost dies
Stainless steel	* Heat intensive production * Close tolerance * High production work
Tool steels	* Plugs or full inserts * Small die blocks * Trimming tools * High strength under extreme heat; pressure * Highest cost

Factors Affecting Process Results

Tolerance and surface finish depend upon the following:

* Accuracy and condition of forging dies
* Proper draft angles
* Allowance for flash
* Type and amount of lubricant
* Material grain flow orientation
* Cavity geometry
* Die alignment/guide tolerances
* Die wear

Tool Geometry

Tool design considerations include proper alignment of both halves of each die, designing die impressions to maximize material grain flow, selection of parting line location, provision for external and internal draft, consideration for supporting webs and ribs, and specification of generous fillet and corner radii.

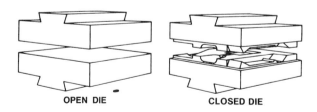

OPEN DIE   **CLOSED DIE**

•PROPER ALIGNMENT OF UPPER/LOWER DIES
•DIE IMPRESSIONS MAXIMIZE MATERIAL GRAIN FLOW
•PARTING LINE LOCATION
•EXTERNAL DRAFT: 6°
•INTERNAL DRAFT: 8-10°
•SUPPORT WEBS AND RIBS
•GENEROUS FILLET AND CORNER RADII

Process Conditions

The following parameters influence the process:

* Workpiece material and size
* Forging temperature
* Forging equipment type and capacity
* Die material and condition
* Lubricant
* Required tolerances
* Production rate

Lubrication and Cooling

Forging lubricants minimize friction between work material and die, facilitate separation after forging, serve as heat-transfer barriers, and protect against further oxidation. Common lubricants are kerosene, various oils, graphite suspensions, salts, and sawdust. Lubricants are generally applied by spraying.

Work material	Die lubricant	Application
Aluminum	Kerosene, oils, graphite suspension	Spray
Mild steel	Salts, oil, sawdust, graphite suspension	Spray
Stainless steel	None	—
Magnesium	Vegetable oils, graphite suspension	Spray
Titanium	Extreme-pressure oil	Spray

Power Requirements

Drop hammer weights and their corresponding production rates are given in the following diagram. Complexity of the part is also an important variable.

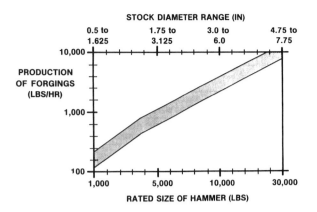

STOCK DIAMETER RANGE (IN)

Cost Elements

Cost elements include the following:

* Setup time
* Load/unload time
* Preheat time
* Forging time
* Die change and maintenance time
* Direct labor rate
* Overhead rate
* Amortization of equipment and tooling costs

Time Calculation

Forging time can be calculated by adding together the travel distance divided by the travel velocity, for both opening and closing of the die. Total forging time is the individual forging time per blow multiplied by the number of blows required.

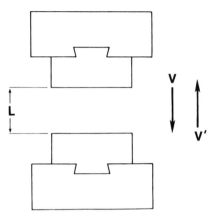

Number of blows = N
Ram velocity, closing (in./sec) = V
Ram velocity, opening (in./sec) = V'
Ram travel length (in.) = L

$$\text{Individual forging time} = \frac{L}{V} + \frac{L}{V'}$$

$$\text{Total forging time} = N\left(\frac{L}{V} + \frac{L}{V'}\right)$$

Safety Factors

The following risks should be taken into consideration:

* Personal
  - Heat, burns
  - Noise
  - Sharp edges
  - Pinch points
* Environmental
  - Vibration, noise
  - Smoke or fumes from lubricant
  - Fire hazard

Impact Extrusion

Impact extrusion is a forming process that produces workpieces by striking a cold slug of metal contained in a die cavity. A metal slug is forced to flow around a punch by a single high speed blow.

Process Characteristics

* Converts a metal slug to a final shape by a single blow
* Forms a workpiece by forcing metal to flow around the punch
* Wall thickness is controlled by clearance between punch and die
* Finished workpieces have excellent surface finishes

Process Schematic

A metal slug of predetermined size is placed in a die cavity. It remains stationary until struck with a single blow from a high speed punch. Metal flow takes place either downward through the die (forward extrusion), upward around the punch (backward extrusion), or with a combination of both. Wall thickness of the extruded part is controlled by clearance between the punch and die.

IMPACT EXTRUSION

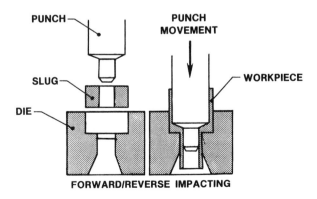

FORWARD/REVERSE IMPACTING

Workpiece Geometry

A short cylindrical workpiece or "slug" is used. Shown are some typical examples of impact-extruded parts. Although round or symmetrically cross-sectioned parts are most common, rectangular or odd cross-section parts can be extruded. The length that can be extruded is limited, in part, by the column strength of the machine. Finished part length is limited to approximately six times the inside diameter of the part.

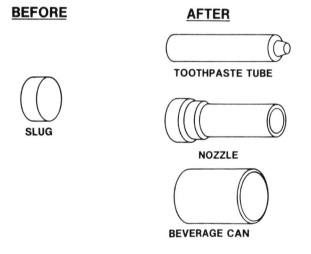

Setup and Equipment

Impact extrusion is usually performed on a high speed mechanical or hydraulic press. Press speeds may range from 20 to 60 strokes per minute. Duration of the actual working stroke is usually 1/10 to 1/15 of a second. In some cases, cooling the punch and die by compressed air is necessary to maintain continuous production.

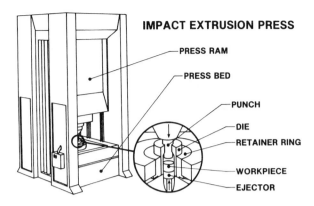

IMPACT EXTRUSION PRESS

- PRESS RAM
- PRESS BED
- PUNCH
- DIE
- RETAINER RING
- WORKPIECE
- EJECTOR

Typical Tools and Geometry Produced

The forming punch controls the inside shape of the workpiece. A die cavity controls the outside shape and may have more than one diameter. The workpiece is ejected by a counterpunch or ejector.

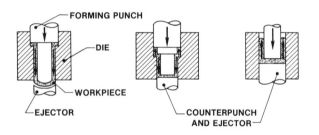

- FORMING PUNCH
- DIE
- WORKPIECE
- EJECTOR
- COUNTERPUNCH AND EJECTOR

Geometrical Possibilities

Shown are different geometrical shapes that can be efficiently produced. They are divided into three groups: combined shapes, stepped shapes, and cupped shapes. Typical and feasible dimension ranges for diameters and lengths are also shown.

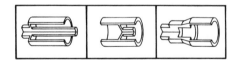

IMPACT EXTRUSION CAPABILITIES

	.01	.1	1	10	100	1000
DIAMETER (IN)						
LENGTH (IN)						

▓ TYPICAL RANGE   ▨ FEASIBLE RANGE

Tolerances and Surface Finish

This frame shows typical tolerances obtainable in diameter (±0.010 in.) and lengths (±0.015 in.). Tolerances largely depend on the size and shape of the section to be produced. Surface finishes typically range from 20 to 125 microinches.

TOLERANCES		
IMPACT EXTRUSION	**TYPICAL**	**FEASIBLE**
DIAMETER (IN)	±0.010	±0.005
LENGTH (IN)	±0.015	±0.005

SURFACE FINISH	
PROCESS	**MICROINCHES (A.A.)**
IMPACT EXTRUSION	2 4 8 16 32 63 125 250 500

▓ TYPICAL RANGE   ▨ FEASIBLE RANGE

Tool Style

There are three major types of dies (tools). They include forward extrusion, backward extrusion, and combined extrusion dies. A combined extrusion die causes both forward and backward flow of material. A hydrostatic extrusion process uses pressure, rather than a punch, to force material through the die and is usually a forward extrusion process.

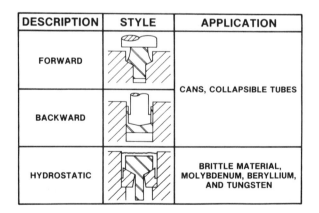

DESCRIPTION	STYLE	APPLICATION
FORWARD		CANS, COLLAPSIBLE TUBES
BACKWARD		
HYDROSTATIC		BRITTLE MATERIAL, MOLYBDENUM, BERYLLIUM, AND TUNGSTEN

Workholding Methods

Shown is the typical workholding method. An undeformed slug is placed in the die and then formed by the punch. A severe plastic flow of material occurs.

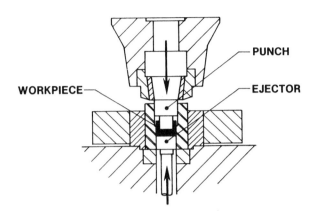

Effects on Work Material Properties

Hardness and yield strength increase, whereas cross-sectional area decreases. Some residual surface stresses are introduced, and microcracks may appear. Physical and chemical properties are only slightly influenced.

Work material properties	Effects of impact extrusion
Mechanical	* Increases hardness, and yield strength * Creates residual surface stresses, microcracks
Physical	* Little effect
Chemical	* Little effect

Typical Workpiece Materials

Listed are some typical workpiece materials and machinability ratings. There are certain alloys that will not match these ratings.

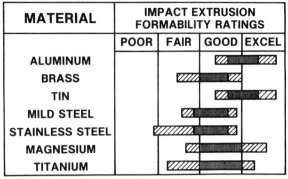

MATERIAL	IMPACT EXTRUSION FORMABILITY RATINGS			
	POOR	FAIR	GOOD	EXCEL
ALUMINUM			▨▰▨	
BRASS		▨▰▨		
TIN			▨▰▨	
MILD STEEL		▨▰▨		
STAINLESS STEEL	▨▰▨			
MAGNESIUM		▨▰▰▨		
TITANIUM		▨▰▰▨		

▰ TYPICAL RANGE   ▨ FEASIBLE RANGE

Tool Materials

The materials used and their applications are listed in order of processing difficulty.

Material AISI steel	Rockwell C hardness	Applications
W1	65 to 67	Solid die
D2	55 to 57 58 to 60 60 to 62	Ejector Punch Die sleeve
L6	56 to 58	Stripper
S1	52 to 54 54 to 56	Ejector mandrel Punch

For extruding aluminum.

Factors Affecting Process Results

Tolerance and surface finish depend upon the following:

* Press equipment condition
* Ram pressure
* Tool geometry
* Material size, shape, and alloy
* Material length-to-diameter ratio
* Material lubrication

Tool Geometry

Shown is a specific type of tool used for backward impact extrusion. The angle on the punch helps relieve pressure on the punch and prevents a dead zone. A dead zone is an area of no pressure. In forward extrusion, a radius on the punch is used to aid the workpiece material's flow.

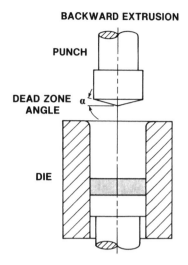

BACKWARD EXTRUSION

PUNCH

DEAD ZONE α
ANGLE

DIE

Process Conditions

This table illustrates typical press sizes and production rates as a function of part size and material for aluminum workpieces.

Workpiece	Approx. part size (in.³)	Press size (tons)	Production rate (pieces/hr)
Double-wall container	3.7	350	300
Step-walled cup	3.5	90	750
Splined housing	6.0	1000	1500
Striker	0.2	150	3900

Lubrication and Cooling

Given are the lubricants used for each type of workpiece material. Methods for applying the lubricants are also given.

Work material	Lubricant	Applications
Aluminum	Lanolin	Tumbling
Tin	Mineral or lard oil	Tumbling
Steel	Palm oil, fatty acid soap	Phosphate coating
Zinc	Lanolin	Tumbling
Magnesium	Graphite dispersions	Tumbling or spraying

Power Requirements

Listed are the working pressures for various alloys and different temperatures.

Extrusion Pressure

Alloy*	Pressure in tons/in.²						
	450°F	500°F	550°F	600°F	650°F	700°F	750°F
AZ31 B	33	33	30	27	26	25	23
AZ61 A	35	34	33	32	31	30	29
AZ80 A	36	35	34	33	32	31	30
ZK60 A	34	33	32	31	29	27	26

* Magnesium alloys reduce in area 85%.

Cost Elements

Cost elements include the following:

* Setup time
* Slug preparation time
* Lubrication time
* Load/unload time
* Impacting time
* Direct labor rate
* Overhead rate
* Amortization of equipment and tooling costs

Productivity Tip

Listed are some variables that influence productivity. Choice of equipment and tooling depends largely on the following:

* Part shape/size
* Slug preparation
* Work material
* Required surface finish
* Production rate

Time Calculation

Shown are the formulas used to calculate forming and ejection times. Total time must also include load time (for slugs or preforms), lubrication time, and unload time.

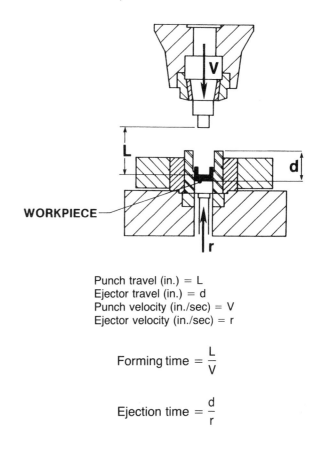

Punch travel (in.) = L
Ejector travel (in.) = d
Punch velocity (in./sec) = V
Ejector velocity (in./sec) = r

$$\text{Forming time} = \frac{L}{V}$$

$$\text{Ejection time} = \frac{d}{r}$$

Safety Factors

The following hazards should be taken into consideration:

* Personal
  – Rapid opening and closing of die
  – Noise

Plate Roll Bending

Plate roll bending is a cold forming process. Plate or steel metal is formed into cylindrical shapes by a combination of three rolls arranged in a pyramid formation. Two of the rolls are power driven, in a fixed position, and the third is adjustable to suit the desired bend radius and workpiece thickness.

Process Characteristics

* Is used for production of cylindrical-type workpieces from heavy sheets and plate metal (up to 6 in. thick and 20 ft wide)
* Typically uses material greater than 1/15 in. thick and 15 in. wide
* Utilizes any material capable of cold working
* Is primarily used in forming large cylindrical sections requiring the seam to be welded

Process Schematic

Three rolls form the material through a rolling and bending action. The top and front roll push the stock into the third roll, which bends and forms the stock. The rolling action of the rollers lowers the friction and helps maintain uniformity.

PLATE ROLL BENDING

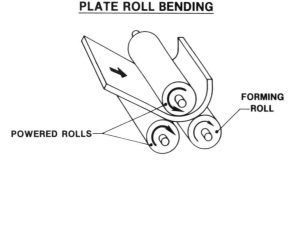

Workpiece Geometry

Shapes produced range from simple cylindrically shaped parts to more complex parts, such as conical and flattened cylinders. Materials ranging in thickness from 1/16 in. to 6 in., and lengths of up to 20 in., or more, can be formed by this method.

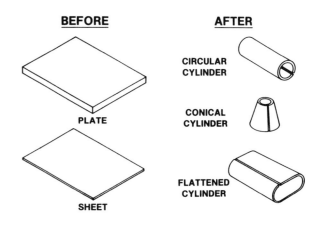

Setup and Equipment

A typical machine consists of an upper roll and two lower rolls. Machines vary in size and power, according to the workpiece material, with optional stock feeders and cutoff equipment. Roll size depends upon material formability, desired workpiece shape, and material thickness.

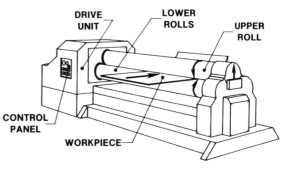

PLATE ROLL BENDING MACHINE

Typical Tools and Geometry Produced

Many shapes can be formed by roll bending. Typical examples are round and tapered cylinders and different curved panels.

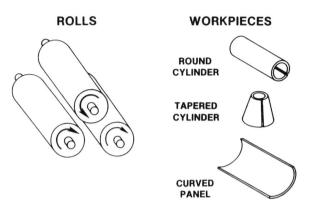

Geometrical Possibilities

The final geometry of the workpiece is created by size, arrangement, and style of the rolls. Typical workpiece sizes range from 0.25 in. to 2.5 in. thick, 15 in. to 120 in. wide (or final workpiece length), and 24 in. to 80 in. in diameter. However, some machines can produce workpieces that are between 0.030 in. and 10 in. thick, 1 in. and 400 in. wide, and 4 in. and 144 in. in diameter.

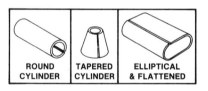

PLATE ROLL BENDING CAPABILITIES

	.01	.1	1	10	100	1000
THICKNESS (IN)						
WIDTH (IN)						
DIAMETER (IN)						

▓ TYPICAL RANGE   ▨ FEASIBLE RANGE

Tolerances and Surface Finish

For plate roll bending, the tolerance of the form diameter can typically be held to within ±0.06 in. However, it is feasible to improve the tolerance. The surface finish may range from 10 to 150 microinches, depending upon the severity of the forming and the original surface finish of the material. Typically, the surface finish ranges from 16 to 50 microinches for gold finished sheet stock.

TOLERANCES		
ROLL BENDING	TYPICAL	FEASIBLE
FORM DIAMETER (IN)	±0.06	±0.01

SURFACE FINISH								
PROCESS	MICROINCHES (A.A.)							
	2	4	8	16	32	64	125	250
ROLL BENDING/FORMING								

▓ TYPICAL RANGE   ▨ FEASIBLE RANGE

Tool Style

Plate roll bending tools are available in a variety of styles and sizes. Arrangement of the rolls and their sizes depend on material thickness and desired workpiece geometry. The most common roll style is the pinch roll style. Here an upper and a lower roll pinch the workpiece, forcing it against a rear forming roll. Shoe-type pinch rolls are often used on automated machines, whereas pyramid rolls are used for heavier gage materials with higher bending forces. Usually, a plate would have an edge that remains straight and would not bend in the rolls. Thus, in most cases, the ends of a workpiece are prebent in a press brake.

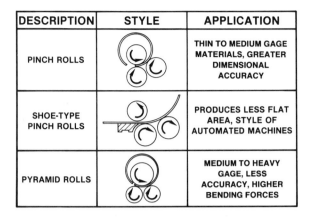

DESCRIPTION	STYLE	APPLICATION
PINCH ROLLS		THIN TO MEDIUM GAGE MATERIALS, GREATER DIMENSIONAL ACCURACY
SHOE-TYPE PINCH ROLLS		PRODUCES LESS FLAT AREA, STYLE OF AUTOMATED MACHINES
PYRAMID ROLLS		MEDIUM TO HEAVY GAGE, LESS ACCURACY, HIGHER BENDING FORCES

Workholding Methods

Small, lightweight parts may be handled manually as they are fed into a roll bending machine. Large, heavy workpieces are handled by cranes, forklifts, or hoists.

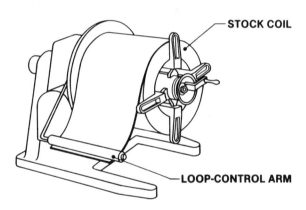

STOCK COIL

LOOP-CONTROL ARM

Effects on Work Material Properties

Because of workhardening, there is a possibility of microcracks at bend areas. Changes to dimensional properties include possible thinning at bends. Physical properties and chemical properties are only slightly influenced.

Work material properties	Effects of roll bending
Mechanical	* May cause workhardening * May cause microcracks * Thinning at bends
Physical	* Little effect
Chemical	* Little effect

Typical Workpiece Materials

Generally, most alloys of aluminum and copper have good to excellent ratings. Nickel and magnesium have fair to good ratings, whereas mild steel has a poor to excellent rating and stainless steel has a fair to excellent rating.

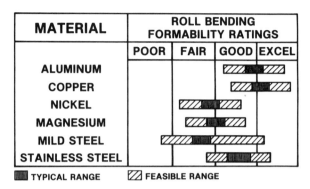

MATERIAL	ROLL BENDING FORMABILITY RATINGS			
	POOR	FAIR	GOOD	EXCEL
ALUMINUM			▨▨	▨
COPPER			▨	▨
NICKEL		▨▨		
MAGNESIUM		▨	▨	
MILD STEEL		▨▨	▨▨	
STAINLESS STEEL			▨▨	▨

▮ TYPICAL RANGE   ▨ FEASIBLE RANGE

Tool Materials

Plate roll forming machines with tool steel rolls are commonly used. When processing involves severe alterations, sharp radii bends, or long production runs, then rolls manufactured from high alloy tool steel or carbide are used. Such rolls are also used to process heavy gage materials or hard workpieces. When the final surface finish is critical, or when a workpiece is abrasive, generally chrome-plated tool steel rolls must be used. In thin-walled tube production, when surface scratches must be avoided, the machine rolls will have synthetic coverings such as nylon, teflon, or urethane.

Tool materials	Applications
Alloyed steel	* Most common * Medium severe forming
High alloy tool steels	* Long production runs * Hard workpiece materials * Heavy gage materials
Chrome-plated	* Critical surface finishes * Abrasive workpiece surfaces
Nylon Teflon Urethane	* For thin-walled tube production

Factors Affecting Process Results

Tolerance and surface finish depend upon the following:

* Roll material and finish
* Workpiece material and finish
* Severity of bending
* Forming speed
* Lubrication effectiveness

Tool Geometry

Rolls used in this process are designed specifically for each particular application. However, there are a few basic shop rules that usually apply to their design and geometry. Bending rolls are generally designed to form the material in an upward direction, with equal-diameter upper and lower rolls for shallow contours and larger upper rolls for deep sections. Bending rolls are generally designed with equal diameters of upper and lower rolls on shallow contours and larger upper rolls for deep sections.

Production Rates

Suggested forming rates are given for selected workpiece materials. These figures are based on specified material thicknesses, where the bend radius is greater than 50 times the material thickness.

Material thickness (in.)	Minimum rolling diameter (in.)	Rolling speed (fpm)
1/4	11	22
1/2	13	20
1	21	15

Values are for a mild steel pipe 6 ft long.

Lubrication and Cooling

Lubricants provide a protective barrier between the roll dies and the workpiece surface. Lubricants also reduce tool wear and allow for higher forming rates.

Work material	Roll lubricants	Application
Nonferrous	Chlorinated oils or waxes, mineral oils	Spray, wiping roller
Ferrous	Water-soluble oils	Wiping, drip, spray
Stainless steels	Chlorinated oils or waxes	Wiping roller
Polished surfaces	Plastic film	Calendaring, covering, spraying
Precoated materials	Film/forced air	

Power Requirements

Power requirements depend primarily on the yield strength of the material being rolled and the degree of deformation the workpiece undergoes during rolling. Typical values for workpieces of various thicknesses are given in this table.

Material thickness (in.)	Minimum rolling dia (in.)	Rolling speed (fpm)	Horsepower required
1/4	11	22	5
1/2	13	20	10
1	21	15	20

Values are for a mild steel pipe 6 ft long.

Cost Elements

Cost elements include the following:

* Setup time
* Equipment and tool costs
* Load/unload time
* Bending/forming time
* Direct labor rate
* Overhead and maintenance
* Amortization of equipment and tooling

Time Calculation

Roll forming time is a function of rolling speed, the length of the workpiece section to be rolled, and the number of passes required. Only on relatively large radius bends can the workpiece be completed in one pass. The basic variables that influence roll bending process times follow. A pictorial representation of these variables are identified. These variables are used in computing bending times.

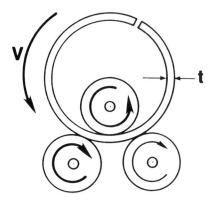

Material thickness (in.) = t
Workpiece unbent length (ft) = L
Bending speed (fpm) = V
Number of passes required = N

$$\text{Bending time} = \frac{L \times N}{V}$$

Safety Factors

The primary hazard in this process is from moving workpieces because they process at speeds up to 800 fpm. The following hazards should also be taken into consideration:

* Personal
  - Moving workpieces
  - Getting caught in rolls
  - Skin and eye irritation from lubricant

Progressive Die Drawing

Progressive die drawing is a cold forming process utilizing a series of stations to perform two or more simultaneous operations on sheet metal. During each stroke of the machine, the final workpiece is developed as the strip of sheet metal moves through the die.

Process Characteristics

* Utilizes two or more cutting and/or forming operations simultaneously
* Typically produces small workpieces at a rapid rate
* Typically requires expensive die sets
* Saves time and money by combining operations
* Is capable of maintaining close tolerances

Process Schematic

A workpiece in the form of strip stock or coil is moved through a series of die stations that produce the finished workpiece, leaving some scrap material. In this process, the workpiece is created in stages. The sheet metal advances a predetermined distance, and the die descends, either piercing, drawing, or blanking the workpiece, all at different locations. The metal continues to advance until each operation has been performed to produce the finished part.

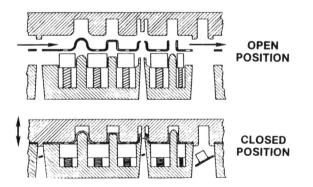

OPEN POSITION

CLOSED POSITION

Workpiece Geometry

In progressive die drawing, the workpiece is usually in the form of a strip or coil. It progresses through a series of stages during which time various operations are performed, including shallow drawing, notching, piercing, louvering, coining, dimpling, and, finally, blanking out the finished part.

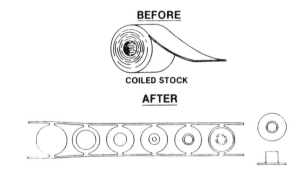

BEFORE

COILED STOCK

AFTER

Setup and Equipment

Hydraulic powered presses are used most often for progressive die drawing, but mechanical presses may also be used. Processes are single-, double-, or triple-action according to the number of slides the machine has. The most common press has a *straight-side*-type frame, but *gap* frame presses are also used. A punch holder is attached to the ram, and a die shoe is attached to the bolster plate. A stripper holds the workpiece in place while the punch is withdrawn. Multistage dies require a larger press with a wider bed than single-stage dies.

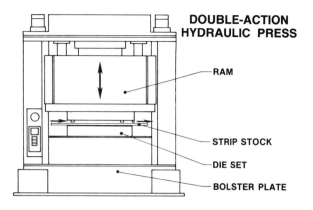

DOUBLE-ACTION HYDRAULIC PRESS

RAM

STRIP STOCK

DIE SET

BOLSTER PLATE

Typical Tools and Geometry Produced

Punches in the die perform cutting, forming, and drawing operations. Forming dies can be used for drawing a variety of regular or irregular shapes. Blanking dies, on the other hand, are used for removing the finished part from the strip stock. It must be emphasized that this process is limited to shallow draws. Deep draws cause the metal strip to be stretched excessively, which could cause wrinkling. This could also jam the machine or disturb the other parts.

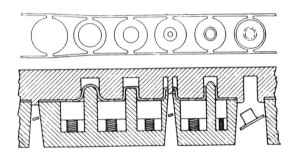

Geometrical Possibilities

Geometrical possibilities are governed by the workstrip width and the press size. Some possible drawn shapes are featured with cup diameters ranging from 0.03 in. to 8 in., and cup heights from 0.25 in. to 7.5 in. Parts are usually small and can be produced at high production rates.

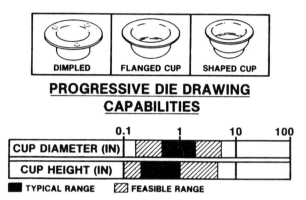

DIMPLED   FLANGED CUP   SHAPED CUP

PROGRESSIVE DIE DRAWING CAPABILITIES

	0.1	1	10	100
CUP DIAMETER (IN)				
CUP HEIGHT (IN)				

■ TYPICAL RANGE   ▨ FEASIBLE RANGE

Tolerances and Surface Finish

Tolerances in progressive die drawing depend upon such factors as press and die conditions, tooling rigidity, and workpiece material. Tolerances are usually around ±0.010 in. The final surface finish depends on the initial material finish, tooling, and lubrication.

Tolerances		
Progressive die drawing	Average	Precision
Diameter (in.)	±0.010	±0.001

Surface finish
The resultant surface finish depends on the proper use of tooling, lubricants, and initial sheet condition

Tool Style

The progressive drawing operation is typically a sequence of forming operations involving piercing, drawing, and blanking. As the workpiece moves from station to station, it is held by a carrier strip until the final operation, at which time the part is blanked from the workpiece, and the carrier strip becomes scrap.

DESCRIPTION	STYLE	APPLICATION
PIERCING PUNCH/DIE		HOLE FORMING OPERATION PRODUCED BY REMOVING A SLUG OF METAL
DRAWING PUNCH/DIE		FORMING OPERATION UTILIZING BENDING, STRAIGHTENING, AND THE FORCES OF FRICTION, COMPRESSION, AND TENSION
BLANKING PUNCH/DIE		CUTTING ACTION ABOUT A COMPLETE OR ENCLOSED CONTOUR

Workholding Methods

Workholding methods for progressive drawing involve a pressure plate, such as the dynamic blankholder, or a static stripper. The dynamic blankholder prevents wrinkling of the blank outer edge while the drawing operation is performed. A stripper plate aids in removing the workpiece from the punch upon punch retraction.

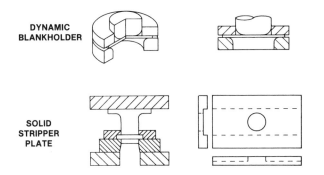

DYNAMIC BLANKHOLDER

SOLID STRIPPER PLATE

Effects on Work Material Properties

Mechanical effects of this process are a thickening of the metal in areas of compression and thinning of the metal at tension regions. The material may also be burnished or workhardened. Physically, the material may wrinkle, tear, or form an orange-peel surface. Scratching, scoring, and galling may occur if lubrication is deficient.

Work material properties	Effects of progressive die drawing
Mechanical	* Sheet metal thickens in the compression region and thins in the tensile region * The top of the cup wall may be burnished or workhardened
Physical	* Sheet metal may form wrinkles, tears, or an orange-peel surface * Scratching, scoring, and galling may occur
Chemical	* Little effect

Typical Workpiece Materials

Aluminum, brass, and cold-rolled steel (CRS) have fair to good formability ratings. Stainless steel has a good to excellent formability rating. CRS also has a good to excellent rating for its ability to be drawn.

MATERIAL	PROGRESSIVE DIE DRAWING FORMABILITY RATINGS			
	POOR	FAIR	GOOD	EXCEL
ALUMINUM				
BRASS				
CRS-COMMERCIAL				
CRS-DRAW				
STAINLESS STEEL				

▓ TYPICAL RANGE   ▨ FEASIBLE RANGE

Tool Materials

Listed here are several different materials used in the tooling along with their applications.

Tool materials	Applications
Tool steel	* The most widely used punch and die material
Carbon steel	* Strong and weldable punch and dies * Used for blankholder
Alloy steel	* Used for ejector * Durable and heat-resistant blankholder
Cemented carbides	* Wear-resistant punch and dies

Factors Affecting Process Results

Tolerance and surface finish depend upon the following:

* Radii of punch and die
* Size of clearance
* Blank material
* Lubricant
* Rigidity of die set
* Proper alignment system
* Correct feed and positioning
* Proper design of tooling

Tool Geometry

The following considerations should be made when designing progressive dies:

* The workpiece must remain attached to the scrap skeleton until the final station, without hindering the drawing operation
* Drawing operations must be completed before the final station is reached
* Sometimes, in deep drawing, it is difficult to move the workpiece to the next station
* If the draw is relatively deep, stripping is often a problem
* The length of the press stroke must be more than twice the depth of the draw

Process Conditions

Process parameters include punch diameter, drawing ratio, blank thickness, and tensile strength of the work material. The punch force equation is shown as a function of these parameters. The blankholder force should be limited to roughly one third of the punching force to avoid tearing the workpiece.

| d = 5 in. | t = 0.040 in. | R = 39% | K = 0.8 |

Workpiece material	Workpiece tensile strength (psi)	Punch force (lbs)
Aluminum	45,000	22,600
Brass	60,000	30,200
Cold rolled steel	65,000	32,700
Stainless steel	85,000	42,700

Punch force $P_{max} = \sigma_T \times \pi \times d \times t \times K$ Blankholder force p = 1/2 P_{max}							
R	50%	45%	43%	39%	36%	34%	31%
K	1	0.9	0.8	0.7	0.6	0.5	0.4
R = drawing ratio d = punch ratio t = blank thickness σ_T = tensile strength K = correction factor							

Lubrication

Lubricants for progressive die drawing include lard oil, wax, soap, and pigmented compounds of heavy oil. These lubricants are most frequently applied by brush or roller.

Work material	Cutting fluid	Application
Aluminum	Lard oil	Brush on
Brass	Wet wax	Brush on
CRS-commercial	Chlorinated wax	Brush on
CRS-draw	Soap	Brush on
Stainless steel	Pigmented compound heavy oil	Brush on

Power Requirements

The power requirements are determined by the draw ratio (blank diameter/punch diameter) and the necessary punch force for a given material. The draw ratio expresses the maximum blank size capable of being drawn by a punch of known size. The conditions for this chart were taken from the section on Process Conditions.

Drawing Force (lbs)

Material	Reduction			
	39%	43%	47%	50%
Aluminum	19,800	22,600	25,400	28,300
Brass	26,400	30,200	34,000	37,700
Cold-rolled steel	28,600	32,700	36,800	40,800
Stainless steel	37,400	42,700	48,100	53,400

Cost Elements

Cost elements include the following:

* Setup time
* Forming operation time
* Die set change/replace time
* Die set costs
* Direct labor rate
* Overhead rate
* Amortization of equipment and tooling

Time Calculation

Given here are the equations for drawing, ejection, and positioning times. Feed rate, punch velocity, ejector velocity, depth of draw, and length of coil feed/stroke are the factors that are used to calculate process time.

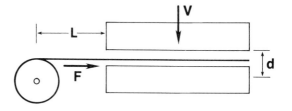

Coil feed rate (in./sec) = F
Punch velocity (in./sec) = V
Ejector velocity (in./sec) = E
Depth of draw (in.) = d
Length of coil feed/stroke (in.) = L

$$\text{Drawing time} = \frac{d}{V}$$

$$\text{Ejection time} = \frac{d}{E}$$

$$\text{Positioning time} = \frac{L}{F}$$

$$\text{Process time} = \frac{d}{V} + \frac{d}{E} + \frac{L}{F}$$

Safety Factors

The following risks should be taken into consideration:

* Personal
  - Rapid die closing
  - Elevated noise level
  - Sharp scrap disposal
* Environmental
  - Noise
  - Vibration

Progressive Roll Forming

In progressive roll forming, a metal strip is cold formed by passing it through shaped rolls to achieve a desired shape. The stock is fed longitudinally through successive pairs of contoured rolls that progressively form the workpiece into the desired shape.

Process Characteristics

* Is used for mass production of long pieces with relatively close tolerances
* Utilizes ductile workpiece materials
* Material is usually less than 1/8 in. thick and 20 in. wide
* Produces workpieces at a typical forming speed of 100 fpm
* Is suitable for decorative and structural products

Process Schematic

Progressive roll forming involves shaped, mating rolls, which form the metal strip stock. Each roll pair is termed a roll stage. There are usually about 14 roll stages used to produce a finished workpiece, depending on the workpiece complexity and size.

PROGRESSIVE ROLL FORMING

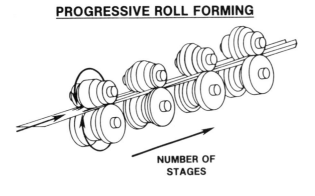

NUMBER OF STAGES

Workpiece Geometry

Different workpiece geometries include simple trough-shaped parts, cylindrically shaped parts, and complex parts having many angles. Workpiece thicknesses range from 0.004 in. to 0.125 in. Workpieces strip widths can be as wide as 20 in. or more.

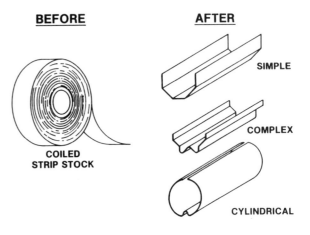

BEFORE

COILED STRIP STOCK

AFTER

SIMPLE

COMPLEX

CYLINDRICAL

Setup and Equipment

The typical progressive roll forming machine consists of a base with two or more roll sets containing a pair of forming rolls. Machines vary in size and power according to the workpiece material, stock feeders, straighteners, and cutoff equipment available.

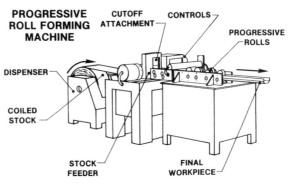

PROGRESSIVE ROLL FORMING MACHINE

CUTOFF ATTACHMENT

CONTROLS

PROGRESSIVE ROLLS

DISPENSER

COILED STOCK

STOCK FEEDER

FINAL WORKPIECE

Typical Tools and Geometry Produced

Mating die-set rolls are constructed to form the desired shape in stages sequentially by means of various shaped rolls. Side rolls can also be used. Roll size depends on material formability and thickness.

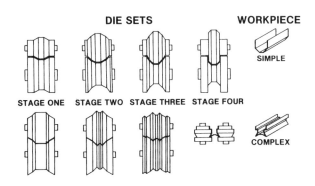

Geometrical Possibilities

The final geometry of the workpiece is a result of the size, arrangement, and style of the rolls. Typical finished workpiece sizes range from 0.025 in. to 0.25 in. thick and just about any length when cut off. Part widths can be from 1.0 in. to 20 in. or more.

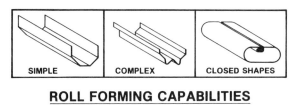

Tolerances and Surface Finish

Tolerances can typically be held to within ±0.015 in. for the width of the cross-sectional form, and 0.060 in. for its depth. However, better tolerances are feasible. Surface finish may range from 10 to 150 microinches, depending upon the severity of the forming required. Typically, surface finishes range from 16 to 50 microinches for cold-finished sheet stock.

TOLERANCES		
ROLL FORMING	**TYPICAL**	**FEASIBLE**
WIDTH OF FORM (IN)	±0.015	±0.004
DEPTH OF FORM (IN)	±0.060	±0.030

SURFACE FINISH	
PROCESS	**MICROINCHES (A.A.)**
ROLL FORMING	2 4 8 16 32 64 125 250

▓ TYPICAL RANGE   ▨ FEASIBLE RANGE

Tool Style

Progressive roll forming tools are available in a variety of styles and sizes. The arrangement of the rolls and their sizes depend on material thickness and desired workpiece geometry. The most common roll style is the pinch roll style. Here, the upper and lower roll pinch the workpiece, forcing it against the rear forming roll. Shoe-type pinch rolls are often used on automated machines. Pyramid rolls are used for heavier gage materials requiring higher bending forces. Workpieces requiring higher bending forces, in most cases, must be prebent on a press brake to insure proper forming.

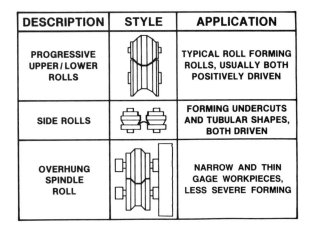

DESCRIPTION	STYLE	APPLICATION
PROGRESSIVE UPPER/LOWER ROLLS		TYPICAL ROLL FORMING ROLLS, USUALLY BOTH POSITIVELY DRIVEN
SIDE ROLLS		FORMING UNDERCUTS AND TUBULAR SHAPES, BOTH DRIVEN
OVERHUNG SPINDLE ROLL		NARROW AND THIN GAGE WORKPIECES, LESS SEVERE FORMING

Workpiece Methods

Small, lightweight parts may be handled manually as they are fed into the roll bending machine. Large, heavy workpieces are handled by cranes, forklifts, or hoists.

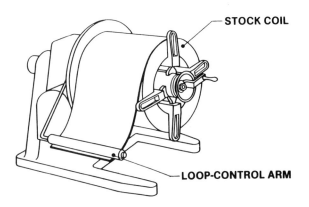

STOCK COIL

LOOP-CONTROL ARM

Tool Geometry

Each roll is designed specifically for a particular application. However, there are a few basic shop rules that usually apply to their design and geometry. Bending rolls are generally designed to form the material in an upward direction, and upper and lower rolls usually have nearly the same diameters.

Forming rolls are generally designed with the following considerations:

* To form the material in an upward direction
* With horizontal reference level at the lowest point of the form
* To begin forming near the center of the workpiece
* With each successive roll set being 0.5% to 1% larger to prevent buckling

Production Rates

Suggested rolling speeds (in feet per minute) are noted below. These figures are based on workpieces where the bend radius is greater than 50 times the material thickness.

WORKPIECE SHAPE	NUMBER OF ROLL STATIONS	ROLLING SPEED (FPM)
	8	85
	12	55
	22	50

•FOR LOW CARBON STEEL .07 IN. THICK
•PART HEIGHT AND WIDTH MEASUREMENTS RANGE FROM ¾ in. to 3 in.

Lubrication and Cooling

The chart shows the typical roll form lubricants used for a variety of work materials. Lubricants serve a variety of purposes, one of the most important functions being to provide a protective barrier between the roll dies and the workpiece surface. Lubricants also reduce tool wear and allow for higher forming rates.

Work material	Roll lubricants	Application
Nonferrous	Chlorinated oils or waxes, mineral oils	Spray, wiping roller
Ferrous	Water-soluble oils	Wiping, drip, spray
Stainless steels	Chlorinated oils or waxes	Wiping roller
Polished surfaces	Plastic film	Calendaring, covering, spraying
Precoated materials	Film/forced air	

Power Requirements

Power requirements for roll bending depend primarily on the yield strength of the material being rolled and the degree of deformation the workpiece undergoes during rolling. Typical values for workpieces of various thicknesses are shown below for mild steel at medium deformation and moderate forming speeds.

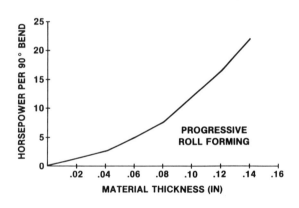

Effects on Work Material Properties

Roll forming primarily affects the mechanical properties of the workpiece material. There is some degree of workhardening, which allows for the possibility of microcracks at bend areas. Possible changes to dimensional properties include thinning at bends. Physical properties and chemical properties are only slightly influenced.

Work material properties	Effects of roll forming
Mechanical	* May cause workhardening * May cause microcracks * Thinning at bends
Physical	* Little effect
Chemical	* Little effect

Typical Workpiece Materials

Listed are relative formability ratings for common workpiece materials used in roll forming. Generally, most alloys of aluminum and copper have good to excellent ratings. Nickel and magnesium have fair to good ratings, whereas mild steel and stainless steel have fair to excellent ratings, depending on the specific steel used.

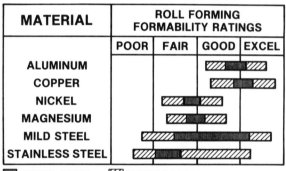

Tool Materials

Materials for rolls generally consist of tool steels, with various alloys and/or platings depending upon process requirements. Plain alloyed steel is the most common tool material for plate roll forming machines, used for moderately severe forming. High alloy tool steels are used for long production runs, hard workpiece materials, heavy gage material, excessively sharp radii bends, and severe forming. Chrome-plated tool steels are used where the final surface finish is very critical or where the workpiece is very abrasive. Rolls with synthetic coverings, such as nylon, teflon, or urethane, are used occasionally for thin-walled tube production, where surface scratches must be avoided.

Tool materials	Applications
Alloyed steel	* Most common * Medium to severe forming
High alloy tool steels	* Long production runs * Hard workpiece materials * Excessively sharp radii * Severe forming
Chrome-plated	* Critical surface finishes * Abrasive workpiece surfaces
Nylon	* For thin-wall pipe production
Teflon	
Urethane	

Factors Affecting Process Results

Tolerance and surface finish depend upon the following:

* Die finish and material
* Workpiece material and finish
* Severity of forming
* Forming speed
* Lubrication effectiveness
* Number of roll stages

Cost Elements

Cost elements include the following:

* Setup time
* Equipment and tool costs
* Load/unload time
* Direct labor rate
* Overhead rate
* Amortization of equipment and tooling

Time Calculation

Roll forming is a function of rolling speed, the length of the workpiece section to be rolled, and the number of passes required. Relatively large radius bends can be done with a few stages, whereas small radius bends require more stages.

PROGRESSIVE ROLL FORMING

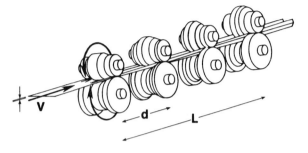

Velocity of strip through rolls (fpm) = V
Length of workpiece (ft) = L
Distance between forming stands (ft) = d
Number of forming stands = n

$$\text{Forming time} = \frac{L + n(d)}{V}$$

Safety Factors

In roll forming, the safety hazards to the operator include the moving workpieces (up to 800 fpm), the high pressure rolls, and skin and eye irritation from the various lubricants. Environmental risks are minimal.

The following hazards should be taken into consideration:

* Personal
  - Moving workpieces
  - High pressure rolls
  - Sharp, sheared metal edges

Stretch-Draw Forming

Stretch-draw forming is the process of cold forming sheet stock over a forming block by a mating die of the desired shape while the workpiece is held in tension. The workpiece material is extended just beyond the yield point to retain the desired shape. Bar, rolled, and extruded stock may also be used.

Process Characteristics

* Uses flexible low cost tooling
* Produces large parts at a reduced weight
* Reduces material thickness by 5% to 7%
* Increases yield stress up to 10%
* Requires less forming pressure

Process Schematic

Stretch-draw forming maintains a workpiece in tension while lowering it onto a stationary die block. A descending upper die stretches the workpiece over the lower die block, forming details and contours on the workpiece. Tension in the workpiece coupled with drawing stresses produce accurate parts with relatively low forming pressure.

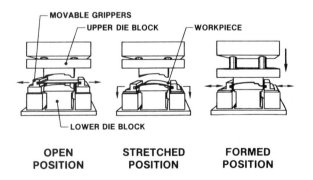

Workpiece Geometry

This process produces a wide variety of workpiece shapes from flat sheet stock. It is used for architectural shapes and aerospace forms requiring a number of contours in lightweight parts. For contoured parts, a trim press may be used to shear the part from the parent sheet. During the drawing process, material may be reduced in thickness by 5% to 7%.

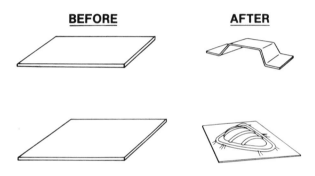

Setup and Equipment

The mating die method of stretch draw forming involves an upper and lower die block mounted in a hydraulic press. The upper die block is mounted in the descending ram. The lower die block is mounted in the press bed. The workpiece is securely held in tension by movable grippers. Yield stress of the finished part may be increased as much as 10% by the stretching and cold-working operations.

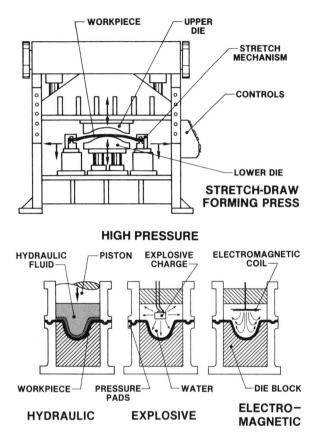

STRETCH-DRAW FORMING PRESS

HIGH PRESSURE

HYDRAULIC   EXPLOSIVE   ELECTRO–
                                MAGNETIC

Typical Tools and Geometry Produced

Shown are two examples of mating die sets with the upper and lower dies and the resulting workpieces. Large stretch-draw forming dies are often made of inexpensive materials, such as kirksite or filled plastics, because they are not subject to severe forces. The best stretch-draw forming results are obtained when forming shallow or nearly flat contours.

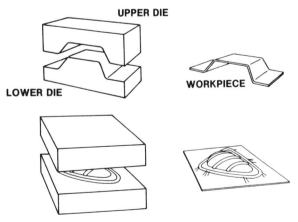

Geometrical Possibilities

Typical shapes are illustrated here. The size, complexity, and material determine which of the various stretch forming processes to use. Typical workpiece dimensions range from 20 in. to 70 in. in length and 0.2 in. to 0.5 in. in thickness. However, it is possible to stretch workpieces that are between 7 in. and 200 in. in length and 0.007 in. to 1 in. thick.

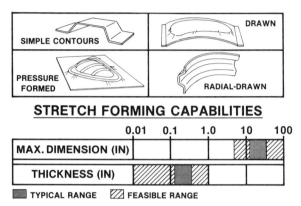

STRETCH FORMING CAPABILITIES

	0.01	0.1	1.0	10	100
MAX. DIMENSION (IN)				▨	▨▨
THICKNESS (IN)	▨▨	▨			

▨ TYPICAL RANGE   ▨ FEASIBLE RANGE

Tool Style

Stretch forming dies can be categorized by their basic construction. Single dies are used for simple, two-dimensional contours and curves, whereas closed punch and dies are used in presses where complex, two-dimensional contours and deep drawn parts are made. The wiping die or rolls are used in radial-draw forming to apply some compressive forces to the workpiece and aid in the forming process. Confined dies are used in the high pressure stretch forming process and are generally self-contained, requiring no additional equipment.

DESCRIPTION	STYLE	APPLICATION
SINGLE-DIE		SIMPLE TWO DIMEN. CONTOURS AND CURVES
CLOSED PUNCH & DIE		COMPLEX TWO DIMEN. CONTOURS DEEP DRAWN PARTS
WIPING DIE/ ROLLS		CHANNEL SHAPED STRUCTURAL PARTS
CONFINED DIE		HIGH PRESSURE FORMING OF SIMPLE GEOMETRIES

Workholding Methods

Shown are three of the most common workholding methods. The workpiece must be gripped or held in such a way to prevent it from slipping out of the holding mechanism under processing. Typically, in mechanical stretch forming, the workpiece is gripped by mechanical or hydraulic jaws. In high pressure forming processes, a drawn bead is typically built into the die faces, so the workpiece is held in position when the die faces are closed. (Note the "pressure formed" geometry.)

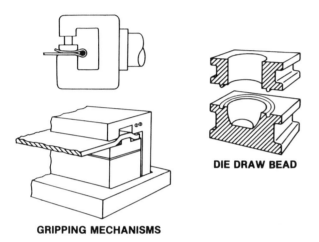

DIE DRAW BEAD

GRIPPING MECHANISMS

Tolerances and Surface Finish

For stretch forming operations, tolerances are typically ±0.015 to ±0.020 in. but tolerances of ±0.002 to ±0.005 in. are feasible in some cases. These tolerances are primarily determined by the accuracy of the die being used. The surface finish generally depends on the original surface finish quality of the sheet stock material. When using the wiping die, some surface adhesion or welding with the workpiece may occur. Also, if stretching beyond 2% to 5%, some surface graining (stretch marks) may appear.

TOLERANCES		
STRETCH FORMING	TYPICAL	FEASIBLE
MAX. DIMENSION (IN)	± 0.015	± 0.002
DEPTH (IN)	± 0.020	± 0.005

SURFACE FINISH	
PROCESS	MICROINCHES (A.A.)
STRETCH FORMING	2 4 8 16 32 63 125 250 500

■ TYPICAL RANGE ▨ FEASIBLE RANGE

Typical Workpiece Materials

Listed are some of the typical workpiece materials and their relative formability ratings. Most aluminum alloys generally have excellent formability ratings. Mild steel ranges from fair to excellent, while nickel alloys and stainless steels have fair to good ratings. Titanium alloys are very difficult to form by any other process, but, with this process, they rate at least poor to fair.

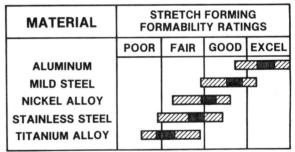

MATERIAL	STRETCH FORMING FORMABILITY RATINGS			
	POOR	FAIR	GOOD	EXCEL
ALUMINUM				
MILD STEEL				
NICKEL ALLOY				
STAINLESS STEEL				
TITANIUM ALLOY				

■ TYPICAL RANGE ▨ FEASIBLE RANGE

Tool Materials

In stretch-draw forming and radial-draw forming, zinc alloys and tool steels are used for male and female dies and wiping dies. Zinc dies are used for low production, while the harder tool steel dies are commonly used for high production. For single-die, straight stretch forming, less costly and less permanent die materials are used, such as phenolic resin, plastic-faced aluminum, and kirksite. (Kirksite is used because it is easily cast and machined.) For high pressure stretch forming, tool steel is normally used because it resists shock and wear.

Tool/die materials	Applications
Tool steel	* Stretch-draw dies, radial-draw and wiping dies
Zinc alloys	
Cast phenolic resins	* Die block in single-die stretch forming
Kirksite	
Plastic-faced aluminum	
Steel-faced concrete	* Electromagnetic, hydraulic, and explosive forming dies
Steel	

Effects on Work Material Properties

Stretch forming mainly affects the mechanical properties of the workpiece material. There may be workhardening, possible tearing or cracking of improperly designed stress points, increasing of yield stress up to 10%, reduction in cross-sectional thickness up to 15%, and, when improperly stretched, visible surface graining (or stretch marks). Improperly produced stretch marks may reduce the corrosive resistance of the workpiece material.

Work material properties	Effects of stretch forming
Mechanical	* Possible tears or cracks at stress points * Increased yield stress * Decreased ductility * Up to 15% reduction in thickness * Visible stretch marks
Physical	* Little effect
Chemical	* Little effect

Tool Geometry

Stretch forming tool geometries are subject to the following limitations:

* Sharp contours, deep draws, and re-entrant angles are not practical
* Progressive or transfer-type methods of die forming are not suitable (draw depth and deformation are limited)
* Draft angles must allow the part to be removed

Process Conditions

Typical maximum elongation values that can be achieved are shown for typical workpiece materials. Also shown are maximum tensile strength ranges for these materials. Soft, lower strength aluminum- and nickel-based alloys can undergo the greatest elongation and deformation ranging up to 40%.

Work material	Tensile strength ($\times$1000 psi)	Max. elongation (% for 2-in. specimen)
Aluminum	12 to 80	10 to 35
Mild steel	60 to 100	15 to 22
Nickel alloys	50 to 160	15 to 40
Stainless steel	70 to 200	10 to 20
Titanium alloys	80 to 150	2 to 10

Factors Affecting Process Results

Tolerance and part quality depend upon the following:

* Part design and absence of square corners or sharp details
* Uniformity and quality of the workpiece stock material
* Proper mating of die halves
* Proper prestretch of the workpiece (not typically used in high pressure stretch forming processes)
* Proper processing conditions

Power Requirements

The power requirements for stretch forming depend on the method used. Shown are the force requirements for the conventional press compared to the stretch press. The amount of force required to form different workpieces is significantly less for the stretch press.

Workpiece	Tool	Conventional press (tons)	Stretch press (tons)
Auto roofs	Punch	900	250
	Gripper/blank-holder	600	85
	Tooling weight	22	7
Quarter panels	Punch	1000	800

These numbers are based on equal production rates.

Lubrication and Cooling

This chart shows typical die lubricants used for stretch forming. When stretch forming polished aluminum (or any other polished workpieces), a layer of poly (vinyl chloride) is often placed over the finished surface to reduce friction and to prevent adhesion/cold welding to the die. For mild and stainless steels, either molybdenum disulfide or a mixture of white lead and 30-weight oil is used. These lubricants are usually brushed on the surface of the workpiece.

Work material	Die lubricants	Application
(Polished) aluminum	Sheet of poly (vinyl chloride)	Laid over material
Mild steel, stainless steel	Molybdenum disulfide, white lead with oil	Brushed on workpiece

Cost Elements

Cost elements include the following:
* Setup time
* Load/unload time
* Tool and die costs
* Overhead rate
* Direct labor rate
* Amortization of equipment and tooling

Time Calculation

Time required to stretch the workpiece the desired amount is a function of the amount of elongation desired and the stretching rate. The work material is stretched so much that it is near its yield point and thus requires little additional force to form it to the desired shape. Forming time is a function of draw distance and die velocity. Shown is the velocity of the forming die and the forming die travel distance. Stretch forming is a relatively fast process, and many pieces per hour can usually be produced.

MECHANICAL

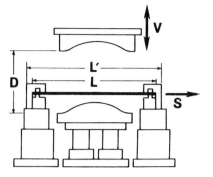

Stretching rate (in./sec) = S
Velocity of forming die (in./sec) = V
Forming die travel distance (in.) = D
Length of workpiece (in.) = L
Stretched out length (in.) = L'

$$\text{Stretch time} = \frac{L' - L}{S}$$

$$\text{Forming time} = \frac{D}{V}$$

Safety Factors

Safety hazards are due to moving machine parts (e.g., die halves closing), noise of operations, especially in explosive forming, and explosive and high pressure dangers.

The following hazards should be taken into consideration:

- * Personal
  - – Contact with moving machine parts
  - – Noise
  - – Explosive danger
  - – High pressure dangers
- * Environmental
  - – Minimal

Swaging

Swaging is a forming process in which a metal workpiece is reduced to a desired size or shape by a succession of rapid hammer blows from a rotating or stationary die.

Process Characteristics

* Involves rapidly repeated hammer blows
* Workpieces usually have a symmetrical cross-section
* Workpiece may be hot or cold
* Produces a smooth surface finish

Process Schematic

Shown is a two-die rotary swager used to put tapers on round stock or reduce overall workpiece diameter. In either case, a workpiece is slowly fed into a swaging die opening. As the spindle assembly rotates, dies and backers are forced outward against rollers by centrifugal force. As backers ride up over the rollers, they push against dies, which, in turn, deliver a blow to the workpiece. The dies continue rotating and rapidly delivering powerful blows to form the workpiece quickly.

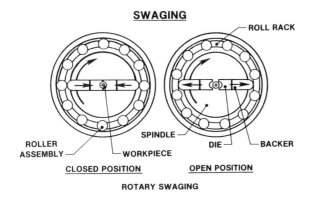

SWAGING

CLOSED POSITION   OPEN POSITION

ROTARY SWAGING

Workpiece Geometry

Typical operations include pointing, forging, and attaching. Swaged parts may be either solid or hollow, with rotational or nonrotational cross-sections. The operation is fast, accurate, and can produce excellent surface finishes. The part to be swaged may be hot or cold, depending on the amount of material to be deformed.

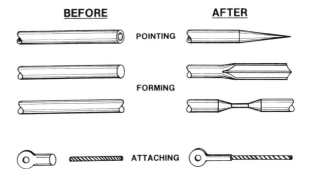

BEFORE   AFTER

POINTING

FORMING

ATTACHING

Setup and Equipment

A swaging machine typically consists of a die cavity, where the swaging dies rotate. A swaging machine is generally simple in design. A part may be subjected to 1000 to 3000 blows per minute. The machine size capacity for solid stock is generally about one half that of hollow or tube stock.

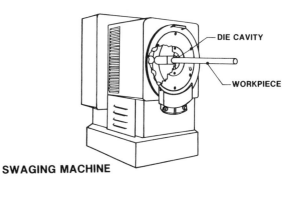

SWAGING MACHINE

Typical Tools and Geometry Produced

Most swagers have two or four dies. These dies either rotate or remain stationary, depending on the design of the machine. A four-die head almost doubles the working surface, speeds material reduction, and increases the size of work that can be handled. Stationary die swaging permits the reduction of a variety of cross-sectional shapes, including nonsymmetrical ones. Stationary dies have the same radial reciprocating action as rotary dies. However, there is no rotation of either the dies or workpiece.

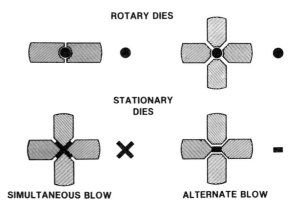

ROTARY DIES

STATIONARY DIES

SIMULTANEOUS BLOW   ALTERNATE BLOW

Thread Rolling _____

Thread rolling is a cold-forming process in which external threads are formed by rolling workpieces between shaped hardened dies. As the dies and workpiece rotate and advance relative to one another, metal flows radially to achieve the desired thread form. The die head or workpiece is automatically retracted after producing the proper thread form for fitness and fasteners.

Process Characteristics

* Produces external rolled threads
* Material is plastically deformed and cold-worked
* Produces rolled threads with excellent strength and surface finish
* Forms major thread diameters greater than blank diameter
* Is a high production process

Process Schematic

The thread rolling method shown is called "in-feeding." With the thread rollhead locked in the closed position, the thread rolls engage the workpiece and roll the threads as the workpiece is fed into the rollhead. When the desired length of thread has been rolled, the thread rollhead opens, and the work is withdrawn from the head. When using continuous rolling, the head remains closed at all times.

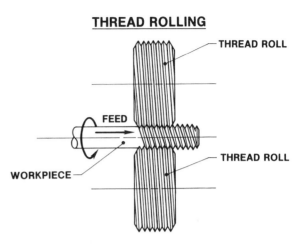

THREAD ROLLING

Workpiece Geometry

Several types of workpieces can be thread rolled. The process is faster than conventional thread cutting, provides a better part, and uses less material. Thread rolling is usually three to five times faster than thread cutting and is highly repeatable. Not only is thread rolling economical, but the threads produced are very accurate and strong because they were cold worked.

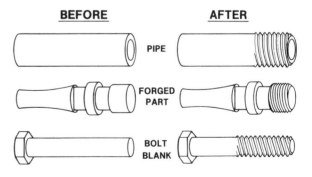

Setup and Equipment

Shown is a typical horizontal thread rolling machine. The workpiece is held in a chuck or other workpiece holding device while the thread rolling diehead traverses the workpiece. Several sizes of thread rolling heads can be used over a wide range of workpiece diameters to produce threads that are superior in strength and finish to threads produced by other machining processes. Vertical thread rolling machines are also produced.

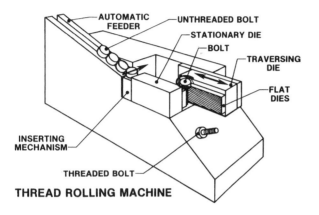

THREAD ROLLING MACHINE

Typical Tools and Geometry Produced

Several styles of thread rolling dies are available. The rotating type is usually mounted in a headstock while the workpiece is advanced into the die. A stationary-type die forms thread on a rotating workpiece, whereas a rotating-type die acts on a stationary workpiece. The straddle-type die is mounted to a cross slide that advances laterally onto the rotating workpiece, forming the desired thread.

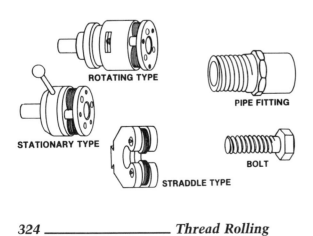

Geometrical Possibilities

A variety of geometrical possibilities in thread types are available with thread rolling. Thread types include UNC, UNF, ISO, ACME, Whitworth, worm, buttress, ball-screw, wood-screw, tapping-screw, bag-screw, and many others. Typical diameters for rolled threads range from 0.075 in. to 2 in. Typical lengths range from 2 in. to 10 in., but, with through-feed thread forming, lengths of 100 in., or more, can be readily produced.

THREAD ROLLING CAPABILITIES

	.01	.1	1	10	100
DIAMETER (IN)					
LENGTH (IN)					

■ TYPICAL RANGE ▨ FEASIBLE RANGE

Tolerances and Surface Finish

Tolerances and surface finish depend on several variables, such as type of material to be formed, quality of the workpiece material, condition and type, and the rigidity of the equipment. In general, the tolerances and surface conditions are very good. Tolerances range from ±0.001 in. to ±0.0006 in. for the die major diameter and pitch. The surface finishes range from 6 to 32 microinches.

TOLERANCES		
THREAD ROLLING	TYPICAL	FEASIBLE
MAJOR DIAMETER	±0.001	±0.0006
PITCH DIAMETER	±0.001	±0.0006

SURFACE FINISH									
PROCESS	MICROINCHES (A.A.)								
THREAD ROLLING	2	4	8	16	32	63	125	250	500

■ TYPICAL RANGE ▨ FEASIBLE RANGE

Tool Style

Thread rolling can be done on a variety of thread rolling machines, including reciprocating flat-die machines, cylindrical-die machines, and rotary-planetary machines. Shown are tools used for each machine. Thread rolling may also be done on turret lathes and on screw machines by using special attachments, such as stationary-type or straddle-type die heads.

DESCRIPTION	STYLE	APPLICATION
FLAT DIES		MACHINE, TAPING AND WOODSCREW
CYLINDRICAL THRU-FEED 2 DIES		LARGE OR UNBALANCED SCREWS
CYLINDRICAL IN-FEED 3 DIES		TUBE FITTING, SPARK PLUGS
PLANETARY DIES		HIGH VOLUME MACHINE, SHEET METAL, DRIVE SCREWS

Workholding Methods

Through-feed die machines involve a work support to position the center line of the workpiece in the same plane as the center line of the dies. In-feed and planetary die machines require only loading and unloading mechanisms.

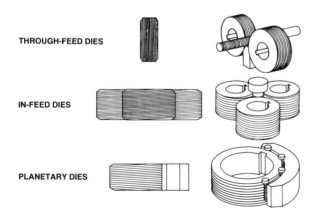

THROUGH-FEED DIES

IN-FEED DIES

PLANETARY DIES

Effects on Work Material Properties

Cold working of the surface during rolling significantly increases yield stress of the work material. Thread rolling also increases resistance to fatigue failure because the surface layer of the thread is left with smooth burnished roots and flanks and is left stressed in compression, especially in the root section of the threads.

Work material properties	Effects of thread rolling
Mechanical	* Yield and shear yield stress are increased * Fatigue resistance is increased * Ductility is decreased (corresponding to strain hardening)
Physical	* Little effect
Chemical	* Little effect

Typical Workpiece Materials

Rolling rates well for soft ductile materials but rates poorly for low alloy steels. Steels containing free-machining elements, such as sulphur and lead, should be avoided.

MATERIAL	THREAD ROLLING FORMABILITY RATINGS			
	POOR	FAIR	GOOD	EXCEL
ALUMINUM				▨▨
BRASS				▨▨
CARBON STEEL		▨▨		
LOW ALLOY STEEL	▨	▨▨		
STAINLESS STEEL			▨▨	
MONEL		▨▨		

■ TYPICAL RANGE   ▨ FEASIBLE RANGE

Tool Materials

Different workpiece materials dictate the use of different tool materials. Die hardness depends on die material and type of die.

Workpiece materials	Tool steel die material	Flat die hardness, RC	Clyindrical die hardness, RC
Aluminum, brass	A2	57 to 60	56 to 58
Carbon steel, low alloy steel	D2	60 to 62	58 to 60
Stainless steel, monel	M2	58 to 60	58 to 60

Factors Affecting Process Results

Tolerance and surface finish depend upon the following:

* Dimensional accuracy of the blank
* Tolerances, surface finish, and wear of the dies
* Workpiece material
* Machine rigidity
* Thread form and size

Tool Geometry

There are very few limitations on the size or geometry of the dies and on the size of the material used. Given are the ranges for die widths, lengths, pitches, etc.

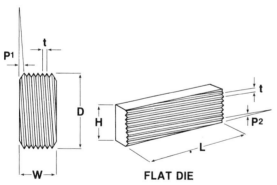

CYLINDRICAL DIE

FLAT DIE

Dimensional ranges (in.)

Cylindrical dies				
D	W	P¹	t	Workpiece diameter
0.75 to 2.75	0.25 to 0.75	0.1° to 90°	0.0125 to 0.2	0.0625 to 4.0

Flat dies				
L	H	P²	t	Workpiece diameter
4 to 28	1 to 10.5	1° to 45°	0.0125 to 0.1	0.125 to 1.5

Production Rates

Production rates depend on thread diameters, as well as die type. Production rates for flat dies range from 15 pieces per minute to as much as 500 pieces per minute. The rate for cylindrical dies ranges from 1 to 300 pieces per minute. Planetary die rate ranges from 100 to as much as 2000 pieces per production minute.

Thread dia (in.)	Flat dies (pieces/min)	Cylindrical (pieces/min)	Planetary (pieces/min)
1/8	40 to 500	75 to 300	450 to 2000
1/4	40 to 400	60 to 150	250 to 1200
1/2	25 to 90	50 to 100	100 to 400
3/4	20 to 60	5 to 10	—
1	15 to 50	1 to 50	—

Process Conditions

One important factor in thread rolling is the number of times the blank needs to revolve through the dies in order to achieve the desired material flow and to obtain a satisfactory finish. Given here are some general figures for both cylindrical dies and for flat dies. These figures can be used to calculate the process time.

Thread rolling travel

Cylindrical dies (two roll)			
Threads per inch	Revolutions of blank		
	Brass or aluminum	Steel	Stainless steel
32	10 to 15	11 to 18	14 to 23
18	12 to 19	15 to 22	18 to 28
8	21 to 27	25 to 31	29 to 37

Flat dies			
Type of thread	Revolutions of blank		
Machine screw	5 to 6	6 to 7	7 to 8
Tapping screw	5 to 6	5 to 6	6 to 7

Lubrication and Cooling

The purpose of lubrication is to reduce friction and wear between the dies and workpiece, thereby improving surface finish. Because high temperatures are generated during the drastic mechanical forming of thread rolling, the cutting fluid also acts as a cooling agent.

Type of roll	Cutting fluid	Diameter of thread
Flat	None Soluble oil	Up to 1/4 in. 1/4 in. to 1-1/4 in.
Cylindrical	Soluble oil Mineral oil	Up to 1/2 in. 1/2 in. to 4-1/2 in.
Planetary	Soluble oil Mineral oil	1/4 in. to 1-1/8 in.

Power Requirements

Power requirements for thread rolling are generally less than those for conventional thread cutting. Given here are average horsepower requirements for producing 1/4-in., 3/8-in., and 1/2-in. threads. Horsepower requirements range from 1.75 hp to 8.0 hp in the small sizes illustrated and are correspondingly higher for larger blank diameters.

Material	Blank size (in.)	Threads per inch*	HP required
Brass or aluminum	1/4	20	1.75
	3/8	16	3.0
	1/2	13	4.0
Alloy steel	1/4	20	3.5
	3/8	16	6.0
	1/2	13	8.0

* For A 3-in. long workpiece with a standard thread feed rate of 0.0005 to 0.002 in./revolution.

Cost Elements

Cost elements include the following:

* Setup time
* Thread rolling time
* Positioning time
* Tool change time
* Direct labor rate
* Tool costs
* Overhead rate
* Amortization of equipment and tooling

Productivity Tips

To increase productivity, the blank diameter should equal the pitch diameter. Ends of the blank should be beveled approximately 60°.

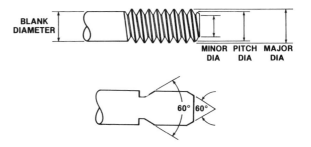

When considering the blank diameter tolerance, a change in blank diameter will affect the major diameter by an approximate ratio of 3 to 1.

BLANK DIAMETER TOLERANCE

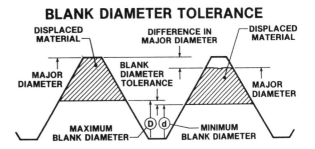

Choice of equipment and tooling depends largely upon the following:

* Workpiece material
* Size of workpiece
* Type of thread
* Equipment capacity (speed, power range, and automation)
* Production capacity
* Dimensional accuracy

Time Calculations

Cylindrical-Die Thread Rolling

The variables for cylindrical-die thread rolling are the blank diameter, workpiece length, workpiece feed rate, number of revolutions, and retract time. Forming time is workpiece length divided by feed rate.

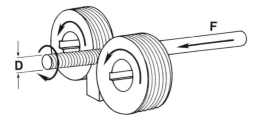

Blank diameter (in.) = D
Die/workpiece length (in.) = L
Die/workpiece feed rate (ipm) = F
Number of blank revolutions = N
Retract time (min) = T

$$\text{Forming time} = \frac{L}{F}$$

Typical feed rates:

1/8-in. blank = 85 ipm

3/8-in. blank = 80 ipm

Flat-Die Thread Rolling

Calculating process time for flat-die thread rolling is similar to that of cylindrical-die thread rolling except that different feed rates apply.

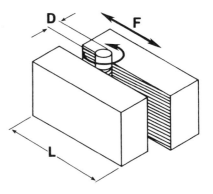

Blank diameter (in.) = D
Die/workpiece length (in.) = L
Die/workpiece feed rate (ipm) = F
Number of blank revolutions = N
Retract time (min) = T

$$\text{Forming time} = \frac{L}{F}$$

$$\text{Die length} = D \times N \times \pi$$

$$\text{Total time} = \frac{L}{F} + T$$

Feed rates:

1/8-in. blank = 16 to 48 ipm

3/8-in. blank = 8 to 24 ipm

Safety Factors

The following risks should be taken into consideration:

* * Personal
  - – Noise
  - – Skin irritation
* * Environmental
  - – Minimal risk involved

Tube Bending

Tube bending is a cold forming process that permanently bends metal tubing to the shape of a die while retaining the original cross-sectional shape.

Process Characteristics

* Reduces the tubular outer wall thickness while increasing inner wall thickness
* Generally requires internal and external supports
* Produces complex shapes from ductile metal tubing

Process Schematic

Tube bending requires the workpiece to be rigidly held to a die by a clamping block. As the die rotates, a flexible mandrel of some sort is used to prevent workpiece collapse. A wiper die helps keep the tube in a state of tension, thus preventing wrinkling.

TUBE BENDING

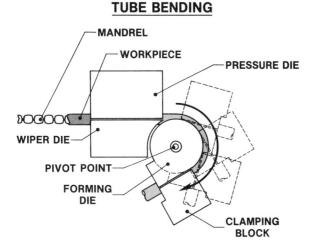

Workpiece Geometry

Straight tube stock can be processed into a variety of single or multiple bends. Powered rotary benders can quickly process workpieces as needed, thus greatly reducing the need for storage space.

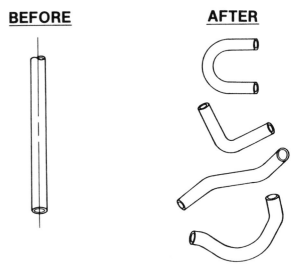

BEFORE · AFTER

Setup and Equipment

Shown is a hydraulically powered tube bending machine, which shapes the rigidly held tube about a die. Internal and external supports may be used to prevent crushing the workpiece or significantly changing its cross-section. Numerically controlled bending machines are available for medium production work. Automated and Computer Numerical Control (CNC) equipment are used for higher levels of production.

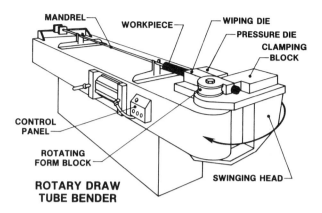

ROTARY DRAW TUBE BENDER

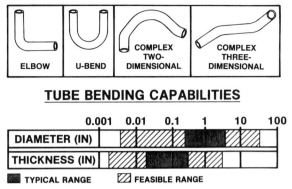

TUBE BENDING CAPABILITIES

	0.001	0.01	0.1	1	10	100
DIAMETER (IN)						
THICKNESS (IN)						

■ TYPICAL RANGE　▨ FEASIBLE RANGE

Typical Tools and Geometry Produced

A single die can produce an almost unlimited number of shapes with individual bends from approximately 2° to 180°. A different die is required for each tube size and bend radius. With a special round helical die, coiled bends of 360° or greater can also be produced.

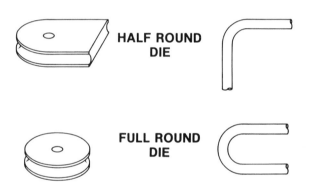

HALF ROUND DIE

FULL ROUND DIE

Tolerances and Surface Finish

For most bending applications, radial tolerances are held around ±0.031 in. for tubing and ±0.062 in. for pipes. Surface finish depends on the original tube finish. Wrinkles on the inner radius may affect surface finish, but, generally, surface finish is unaltered by this process.

TOLERANCES		
TUBE BENDING	**TYPICAL**	**FEASIBLE**
TUBE	±0.031	±0.005
PIPE	±0.062	±0.016

SURFACE FINISH							
PROCESS	**MICROINCHES (A.A.)**						
TUBE BENDING	4	8	16	32	64	128	250

■ TYPICAL RANGE　▨ FEASIBLE RANGE

Geometrical Possibilities

Common shapes produced by this process include elbows and U-bends, as well as complex two-dimensional bends. Typical tube diameters range from 0.25 in. to 8 in. Tubing wall thickness ranging from 0.031 in. to 0.375 in. can be handled by power-driven tube bending machines.

Tool Style

A plug mandrel is used where a bend is not very critical. The form mandrel is similar to the plug mandrel but has a radius formed at the end to give the tube more support during a bending. A ball mandrel and a ball mandrel with steel cable are used where bends are critical. Sand is also used, mainly for very small production sizes.

DESCRIPTION	STYLE	APPLICATION
PLUG MANDREL		NORMAL BENDS
FORM MANDREL		BENDS WHERE SUPPORT IS NEEDED
BALL MANDREL		CRITICAL BENDS
BALL MANDREL WITH STEEL CABLE		CRITICAL BENDS
SAND, CERROBEND		LOW PRODUCTION INEXPENSIVE

MATERIAL	TUBE BENDING FORMABILITY RATINGS			
	POOR	FAIR	GOOD	EXCEL
MILD STEEL			▨▨	
STAINLESS STEEL			▨	▨
COPPER			▨	
TITANIUM		▨		
BRASS			▨	▨
ALUMINUM			▨	▨

▨ TYPICAL RANGE   ▨ FEASIBLE RANGE

Workholding Methods

A common workholding method is shown here. A pressure die and wiping die act as stationary blocks, and a rotating form block and clamping block hold the workpiece through a rotational bending action. The workpiece is drawn under tension between the wiping die and the pressure die.

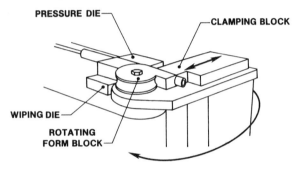

PRESSURE DIE — CLAMPING BLOCK

WIPING DIE —

ROTATING FORM BLOCK —

Effects on Work Material Properties

This process may cause work hardening and uneven wall thicknesses during bending. There are very few physical or chemical effects on the workpiece due to tube bending.

Work material properties	Effects of tube bending
Mechanical	* Workhardening * Uneven wall thickness
Physical	* Little effect
Chemical	* Little effect

Typical Workpiece Materials

Stainless steel, mild steel, copper, brass, and aluminum have good to excellent ratings, whereas titanium has slightly lower ratings.

Tool Materials

The mandrel, pressure die, forming block, and clamping blocks are made of hardened tool steel to prolong life. The wiping die is made of aluminum or bronze to reduce the possibility of scratching the workpiece.

Tool materials	Applications
Hardened steel and tool steel	* Mandrel * Pressure die * Clamping block * Rotating form block
Aluminum and bronze	* Pressure dies * Wiper die * Mandrel

Factors Affecting Process Results

Tolerance and surface finish depend upon the following:

* Workpiece material
* Bend geometry
* Type and quantity of lubrication
* Rigidity of pipe holding and bending equipment
* Surface finish of holding tools

Tool Geometry

Shown is a basic clamping block and its position in relation to a rotating form die. The length of the clamping block is determined by radius of the bend and wall thickness of the workpiece, as indicated in the chart.

RADIUS OF BEND · CLAMP LENGTH · CLAMPING BLOCK

RADIUS OF BEND CENTER LINE	WALL THICKNESS (IN)	APPROXIMATE CLAMP LENGTH
1×OD	< .035	4 to 5 OD
	.035 to .065	3 to 4 OD
	> .065	2 to 3 OD
2×OD	< .035	3 to 4 OD
	.035 to .065	2 to 3 OD
	> .065	1.5 to 2.5 OD
3×OD	< .065	2 to 3 OD
	> .065	1 to 2 OD

Bending Speed

Tube bending speed is shown for a steel pipe with a 0.875 in. outside diameter and 0.06 in. wall thickness. Typically, a U-bend will take 4 seconds, a 70° bend, 2 seconds, and a 15° bend, 1 second.

Steel workpiece with 0.875 (in.) outside diameter and 0.06 (in.) wall thickness

Bend radius	Speed (sec)
U-bend	4
70° bend	2
15° bend	1

Lubrication

The chart shown lists several work materials, their respective die lubricants, and their form of application. For steels, aluminum, and bronze, mineral oils and organic fats are best suited as lubricants. Stainless steel, copper, and titanium use similar lubricants (which must not contain sulfur or chlorides). Magnesium uses these lubricants as well as tallow and soap.

Work material	Die lubricants	Application
Steels, aluminum, and bronze	Mineral oils, organic fats	Spray, brush, or wipers
Stainless steel, copper, and titanium	Mineral oils, organic fats, no sulfur or chlorides	Spray, brush, or wipers
Magnesium	Mineral oils, organic fats, tallow, soap	Spray, brush, or wipers

Power Requirements

Shown in this chart are the power requirements (in foot-pounds) for bending aluminum and stainless steel pipes, in a variety of diameters, with a wall thickness of 0.035 in. The power requirements typically range from 4 to 245 ft-lbs, as indicated.

Material	Outside diameter (in.)	Power (ft-lbs)
Aluminum (6061-T6)	1/4	4
	1/2	20
	3/4	45
	1	80
Stainless steel (21-6-9)	1/4	12
	1/2	55
	3/4	135
	1	245

Cost Elements

Cost elements include the following:

* Setup time
* Load/unload time
* Bend geometry
* Direct labor rate
* Overhead rate
* Amortization of equipment and tooling

Time Calculation

The primary factors include bend angle, bending rate, and bend radius. Bending rates can be obtained from a handbook of rates based upon the pipe's wall thickness, outer diameter, and material type.

Bending time for a single bend is a function of bend angle and rate. In some instances, it may be necessary to reset the equipment several times to produce complex bends. The major variables that influence tube bending time are bend angle and bending rate. Bending rate, in degrees per second, is a function of material type, diameter, and wall thickness. Bending rates of 45°/sec are not uncommon for small diameter tubing.

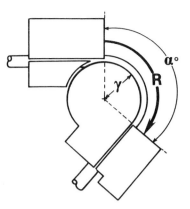

Bending time (sec) = T
Bend angle (°) = α
Bending rate (°/sec) = R
Bend radius to centerline (in.) = γ

$$\text{Bending time} = \frac{\alpha}{R}$$

Safety Factors

The following hazards should be taken into consideration:

* Personal
  - Rotating and swinging workpieces
  - Eye and skin contact with lubricants
  - Pinch points

Tube Drawing

Tube drawing is a reduction process in which one end of a tube is grasped and pulled through a die that is smaller than the tube diameter. To obtain the desired size, a series of successive reductions, or passes, may be necessary. Because of its versatility, tube drawing is suited for both small and large production runs.

Process Characteristics

* Stock is pulled through a die, reducing its diameter
* Increases length as diameter decreases
* Typically requires several passes
* Results in improved material properties through cold working
* Is suitable for small production runs of long workpieces

Process Schematic

Tube drawing involves the reduction of a workpiece by entry into a smaller cross-sectional die with a floating mandrel to prevent workpiece buckling or wrinkling. The workpiece end must be pointed or reduced to a smaller size than the die so that it may be inserted through the die and into the gripper. Tube drawing results in improved workpiece mechanical properties.

Workpiece Geometry

Tube stock is reduced by the drawing process, which elongates and slightly hardens the workpiece. Generally, more than one reducing pass is required. Typical stock lengths may be up to 100 ft in length and up to 1 in. in diameter. For very severe reductions in diameter, annealing between passes may be necessary.

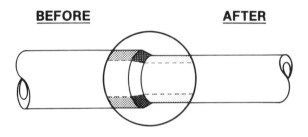

BEFORE   **AFTER**

Setup and Equipment

A multiple-die or draw bench is composed of a die block with appropriate dies, hydraulically powered grip, and rotating pull chain unit. A grip clamps the end of the workpiece and pulls it through a reducing die. The finished workpiece is then transferred to the side holder. Automated handling of the workpiece is usual in tube drawing operations.

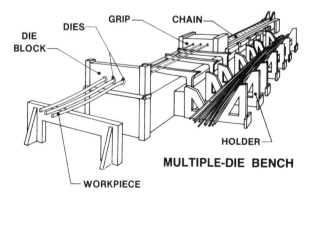

MULTIPLE-DIE BENCH

Typical Tools and Geometry Produced

Carbide dies and floating mandrels are available in a wide range of sizes and shapes. Drawing dies can be easily changed in the die block to permit drawing a variety of different-sized tubes.

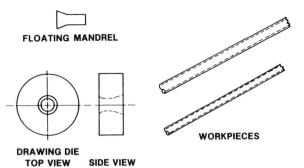

FLOATING MANDREL

DRAWING DIE
TOP VIEW   SIDE VIEW

WORKPIECES

Upset Forging

Upset forging is a metal shaping process in which a heated workpiece of uniform thickness is gripped between split female dies while a heading die (punch) is forced against the workpiece, deforming and enlarging the end of the workpiece. A sequence of die cavities may be used to control the workpiece geometry gradually until it achieves its final shape. This is a rapid "cold forming" process.

Process Characteristics

* Increases the diameter of the end of the central portion of a workpiece by compressing its length
* Complex parts are usually formed gradually in a sequence of separate die cavities
* May produce diameters up to three times the original diameter
* Impressions may be in the punch die, gripping die, or both
* Usually requires no trimming
* May be performed cold on ductile materials

Process Schematic

Upsetting involves placing heated metal stock in a stationary die. The amount of stock to be upset is set by the stop gage. A moving-grip die closes against a stationary die, gripping the stock tightly. As a punch, which is attached to a punch heading die, moves forward and presses against a workpiece, it forces the stock to take the shape of a die cavity. The workpiece is ejected, and the cycle is repeated.

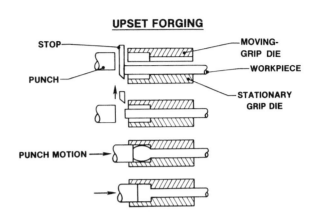

UPSET FORGING

Workpiece Geometry

Many complex parts are formed in stages with a sequence of cavities. Upset forging machines are used to forge heads on bolts, fasteners, valves, couplings, axles, and many other components. The force of forgings may range from less than 1 lb to as much as 500 lbs. Upset forgings are uniform in size and weight, and very little machining is usually required to finish the part. Thread rolling may be performed on fasteners.

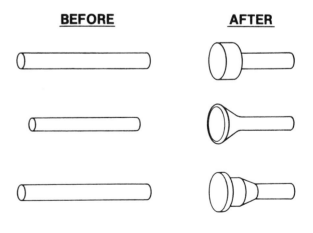

BEFORE   **AFTER**

Setup and Equipment

Shown is a typical upset forging machine with the punch heading die, grip die, workpiece, and stop. These machines are mechanically operated from a main shaft. An eccentric drive operates a header slide and moves a punch heading die against the workpiece head (held in a grip die). A die opening determines the maximum diameter of upset that can be formed.

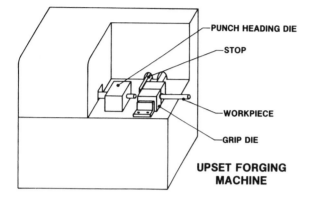

UPSET FORGING MACHINE

Typical Tools and Geometry Produced

Shown is a typical punch and die set. Although most upset forgings are produced by a single stroke of the ram, some shapes require several passes. In this case, the die set will have several different impressions, and the stock will be moved from one impression to the next, until the final shape is obtained. The desired impressions may be in the gripping die, punch die, or both.

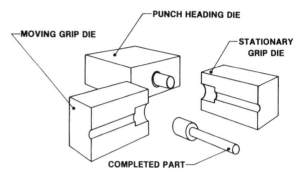

Wire Drawing

Wire drawing is a metal-reducing process in which a wire rod is pulled or drawn through a single die or a series of continuous dies, thereby reducing its diameter. Because the volume of the wire remains the same, the length of the wire changes according to its new diameter ($l_2 = D_1^2 l_1 / D_2^2$). Various wire *tempers* can be produced by a series of drawing and annealing operations. (*Temper* refers to toughness.)

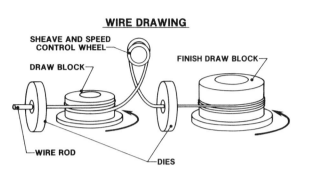

WIRE DRAWING

SHEAVE AND SPEED CONTROL WHEEL

DRAW BLOCK

FINISH DRAW BLOCK

WIRE ROD

DIES

Process Characteristics

* Pulls a wire rod through a die, reducing its diameter
* Increases the length of the wire as its diameter decreases
* May use several dies in succession (tandem) for small diameter wire
* Improves material properties due to cold working
* Wire temper can be controlled by swaging, drawing, and annealing treatments

Process Schematic

A coil of wire rod is placed on a reel. A portion of the wire has a reduced diameter so that it can easily pass through the die and be wound around a draw block. This smaller end is grasped by a clamp, and the remaining wire is mechanically pulled through the die, thereby reducing its diameter. A rotating draw block or reel continues to pull the elongated wire through the die and forms it into a coil. Drawn wire is stronger and less ductile than the parent stock.

Workpiece Geometry

Wire stock is typically coiled both before and after drawing for ease in handling. If the desired wire reduction is over 50% in area, the wire stock may require annealing between successive drawing operations.

BEFORE   **AFTER**

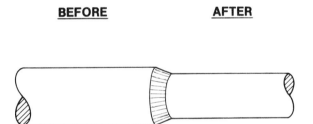

ENLARGED VIEW OF WIRE REDUCTION

Setup and Equipment

This process uses a powered draw block around which a few turns of wire are wound to provide sufficient friction for pulling wire through the dies. The sizing die and feed rolls maintain proper tension on incoming wire. The die box may contain a lubricant through which the wire passes before going through the die. The wire is pulled through a die box and coiled again by a powered cylindrical block. Continuous drawing machines generally have one draw die and draw block for each reduction pass.

Typical Tools and Geometry Produced

The most common die is a circular steel case with a carbide insert drawing die nib. Hardened tool steel nibs are sometimes used for large diameter wire. Diamond or ceramic nibs are commonly used for small diameter wire to resist wear and to maintain wire diameter accuracy.

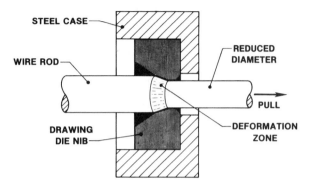

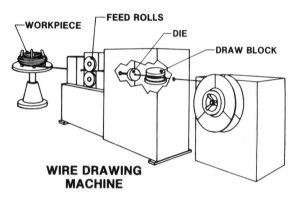

WIRE DRAWING
MACHINE

Thermal Joining

Electron Beam Welding

Electron beam welding is a metal-joining process wherein melting is produced by the heat of a concentrated stream of high velocity electrons. The kinetic energy of the electrons is changed into heat upon impact with the workpiece. Filler material is not generally used in this process.

Process Characteristics

* Requires a vacuum chamber
* Uses no filler metal
* Causes little workpiece distortion
* Produces excellent penetration
* Is easily automated
* Permits joining of dissimilar alloys
* Weld zone is narrow
* Temperatures are very high, but heat-affected zone is very small

Process Schematic

An electron beam gun emits electrons, accelerates the beam of electrons, and focuses it on the workpiece. The principle is similar to that in a television picture tube. The electrons are emitted by a heated filament and then accelerated by a positively charged anode. Magnetic focusing coils deflect and focus the electron beam at the weld point. The workpiece is automated under the electron beam. This process is done in a vacuum chamber because air not only contaminates the weld but also can deflect the beam of electrons.

ELECTRON BEAM WELDING

DC FILAMENT SUPPLY

CONTROL ELECTRODE

ANODE

FOCUS COIL

WORKPIECE

Weld Beam and Geometry Produced

An electron beam distinguishes this process from laser beam welding. The electron beam is generated by a filament and focused by magnetism. Occasionally, a filler material is used when welding mild steel. Typical geometries produced include L-joints, T-joints, and lap joints.

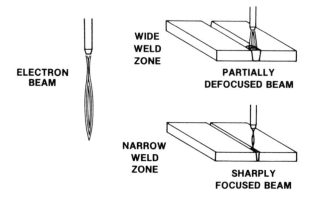

ELECTRON BEAM

WIDE WELD ZONE

PARTIALLY DEFOCUSED BEAM

NARROW WELD ZONE

SHARPLY FOCUSED BEAM

Setup and Equipment

The electron beam welder is almost identical to the laser beam except that the work is done in a vacuum, and the beam generated from the filament is focused by magnetism.

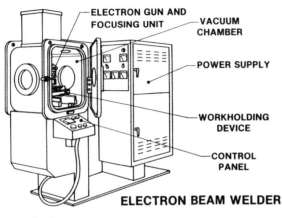

ELECTRON BEAM WELDER

Workpiece Geometry

Because of the awkwardness of working within a vacuum chamber of limited size, workpiece geometry is limited. Three factors are involved. One limiting factor is that the workpiece has to be able to fit into the chamber. Another factor is the difficulty in properly aligning the workpiece under the electron beam. The final factor is that it is sometimes difficult to control the motion of the workholding device to insure a good weld.

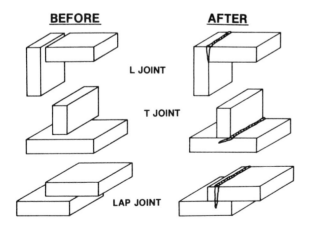

Geometrical Possibilities

The welding capabilities of this process are tremendous. No large gaps are produced, and no filler metals are required. The typical welding thickness is between 0.3 in. and 2 in. Gap widths typically range between 0.001 in. and 0.01 in.

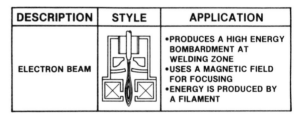

Tool Style

The electron beam is generated by a filament, and the electrons accelerate, forming a beam of electrons. This beam is focused by a magnetic field.

DESCRIPTION	STYLE	APPLICATION
ELECTRON BEAM		• PRODUCES A HIGH ENERGY BOMBARDMENT AT WELDING ZONE • USES A MAGNETIC FIELD FOR FOCUSING • ENERGY IS PRODUCED BY A FILAMENT

Workholding Methods

Workholding methods consist of a table or bed on which the workpiece is positioned. Special fixtures usually are used to hold the workpieces. Positional tolerances of 0.0001 in. or better are possible. Electron beam welding is for relatively small components. Vacuum chambers have been developed to accommodate a continuous flow of components.

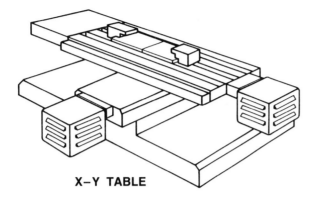

X-Y TABLE

Electron Beam

Shown in this frame is an electron beam and its different stages. The lightly shaded region is where the welding takes place; the darker area is either not affected or is just heated.

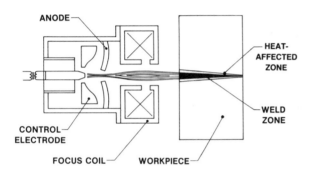

ANODE

HEAT-AFFECTED ZONE

WELD ZONE

CONTROL ELECTRODE

FOCUS COIL

WORKPIECE

Weld and Finish Quality

Cavities and internal stresses are sometimes created, but the heat-affected zone is normally very small, so difficulties are minimized.

Defect	Possible causes	Effect
Cavities	* Excessive welding speed * Poor focus * Insufficient power supply * Excessive gap in joint	* Reduction of strength
Internal stresses	* Insufficient welding speed * Poor focus * Excessive power supply	* Reduction of strength * Distortion

Effects on Work Material Properties

The effects on a workpiece are minimal.

Work material properties	Effects of electron beam welding
Mechanical	* Hardening or softening may occur
Physical	* None
Chemical	* None

Typical Workpiece Materials

Shown is a list of the more popular metals and their weldability ratings.

MATERIAL	ELECTRON BEAM WELDABILITY RATINGS			
	POOR	FAIR	GOOD	EXCEL
ALUMINUM		▨▨	■■■	▨▨
STEEL		▨▨	■■■	▨
STAINLESS STEEL		▨▨	■■■	
TANTALUM			▨▨	■■■ ▨
TUNGSTEN		▨▨	■■■	
COPPER		▨▨	■■■	

■ TYPICAL RANGE   ▨ FEASIBLE RANGE

Process Conditions

Process conditions depend largely on the vacuum pressure. Generally, the better the vacuum, the deeper the penetration becomes. When deeper penetration is required, lower welding speeds are used.

Type of system	Vacuum pressure (torr)	Penetration (in.)	Welding speed (ipm)
High vacuum	10^{-4} to 10^{-5}	3.5	4
Medium vacuum	10^{-1}	2.5	6
Nonvacuum	760	0.5	20

Factors Affecting Process Results

Variables that influence weldability include the following:

* Workpiece surface condition
* Workpiece composition
* Workpiece thickness
* Focusing of electron beam
* Power applied
* Workpiece melting point
* Joint design

Cooling Method

Coolant is pumped through cooling coils that are located at the top and bottom of the focusing coil assembly.

COIL ASSEMBLY

COOLING COILS

Power Requirements

The power requirements are related to vacuum pressure, machine capacity, and joint type.

Joint type	Vacuum (torr)	Power
Circumferential butt	10^{-4}	130 at 24 mÅ 26 at 54 mÅ
Circular butt	0.1	175 at 45 mÅ
Corner	10^{-4}	110 at 35 mÅ
"T"	10^{-4}	110 at 35 mÅ
Crack repair	20^{-4}	140 at 40 mÅ

kV = kilovolts; mÅ = milliamps.

Time Calculation

Welding time is equal to the length of the workpiece, L, divided by the welding rate, R.

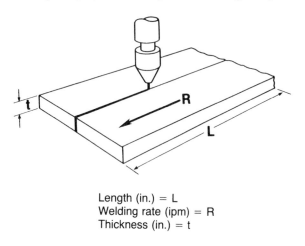

Length (in.) = L
Welding rate (ipm) = R
Thickness (in.) = t

$$\text{Welding time} = \frac{L}{R}$$

Cost Elements

Cost elements include the following:

* Setup time
* Load/unload time
* Pump down time
* Welding time
* Direct labor rate
* Overhead rate
* Amortization of equipment and tooling

Safety Factors

The following risks should be taken into consideration:

* Personal
  - Eye contact with welding beam
  - Heat from tool and workpiece
  - Skin exposure to welding beam
  - Exposure to X-rays
* Environmental
  - Smoke
  - Very high voltage

Furnace Brazing

Furnace brazing is a joining process that uses filler metal to join two contacting workpieces. Heat generated by the furnace melts the filler metal and distributes it throughout the joint by capillary action.

Process Characteristics

* Uses a furnace to heat the workpiece and melt the filler metal
* Furnaces are heated only to a temperature sufficient to melt the filler metal
* May braze multiple joints simultaneously

Process Schematic

A filler metal is applied in contact with, or adjacent to, the joint(s) of the workpieces to be joined. The workpieces are conveyed into the heating chamber where sufficient heat is applied to melt the filler metal. The melted filler is then drawn into the joint by capillary action with the aid of the cleaning and flowing additives contained in the flux. Temperatures used are between 800°F and 2100°F.

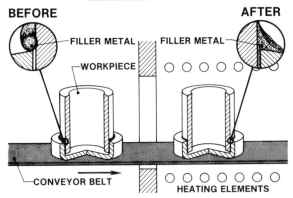

FURNACE BRAZING

Workpiece Geometry

It is best to use some form of lap joint (such as a modified T-joint, corner joint, and cap to tube). This produces an excellent bond and increases joint strength. Clearance between workpieces ranges around 0.008 in., depending on the type of filler metal used. Workpieces are normally small- to medium-sized assemblies.

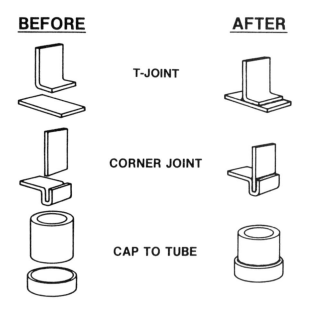

Setup and Equipment

Furnace brazing equipment normally consists of a heating chamber, cooling chamber, and a conveyor system used to move the workpiece. An inert gas, such as helium, may be used throughout the furnace as a protective atmosphere. Three types of furnaces commonly used are the box, mesh-belt conveyor, and the roller-hearth conveyor furnace.

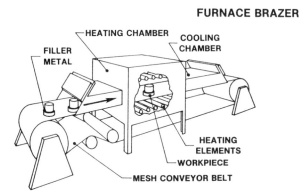

FURNACE BRAZER

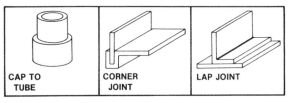

FURNACE BRAZING CAPABILITIES

	.01	.1	1	10	100	1000
SIZE (IN³)						
THICKNESS (IN)						

▓ TYPICAL RANGE   ▨ FEASIBLE RANGE

Typical Tools and Geometry Produced

Filler metals are available in a wide variety of shapes, sizes, and consistencies (such as pastes, preforms, and strips). Paste filler metals are easily adapted to automated operations and can conform to a variety of joining designs. The temperature to which the workpiece is heated varies according to the type of filler metal used. Alloys of copper, silver, and brass are most commonly used.

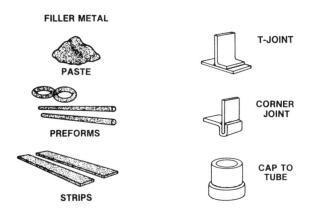

Geometrical Possibilities

Flux and filler metal need to be properly distributed throughout the joint. Self-supported workpieces are desirable but not mandatory. Workpiece size is limited by furnace sizes, which typically range from 8 in.³ to 125 in.³ The workpiece thickness ranges typically from 0.2 in. to 0.8 in.

Workholding Methods

Special jigs and fixtures can be designed to support the workpiece as it moves through the furnace. Shown is a typical type of fixture designed to hold three separate pieces together.

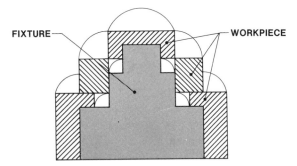

Tool Style

Three types of furnaces are the vacuum furnace, the continuous flow brazing furnace, and the batch furnace.

Description	Style	Application
Vacuum furnace	Cold or hot wall-type furnace	For stainless steel and other super alloys, for temperatures up to 2000°F
Continuous flow brazing furnace	Air or protective atmosphere	General use parts protected by flux
Batch furnace	Controlled atmosphere	For parts sensitive to oxide formation

Toolholding Methods

Workpieces are typically self-supported and are carried on a conveyor belt. A dish is often placed under the workpiece, and, in special cases, small jigs or fixtures can be used to support the workpiece.

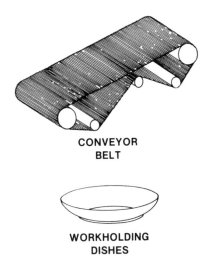

CONVEYOR BELT

WORKHOLDING DISHES

Braze Finish and Quality

Defects include incomplete fusion, porosity, and a rough or dirty finish. These can be caused by a contaminated surface, insufficient heating, poor flux coverage, insufficient filler metal, too much heat, or an improper filler metal. The effects can be a reduction in strength or bad appearance. In the case of pipes, leaks may occur.

Defect	Possible causes	Effect
Incomplete fusion Porosity Rough and dirty	* Contaminated surfaces * Insufficient heating * Poor flux coverage * Insufficient filler metal * Too much heat * Improper filler metal	* Leaks in pipes or lines * Bad appearance * Reduction in strength

Effects on Work Material Properties

The mechanical consequences of furnace brazing are hardening, embrittlement, and loss of strength. A structural change is the only possible physical effect of this process, and chemical properties are not affected.

Work material properties	Effects of furnace brazing
Mechanical	* Cause hardening * Cause embrittlement * Loss of strength
Physical	* Possible structural change
Chemical	* Little effect

Brazing Materials

Filler materials are nonconsumable. Listed are the different types of alloying elements with the American Welding Society (AWS) classification number and the working temperature range.

Principal brazing element	AWS classification	Brazing temperature range (°F)
Aluminum	BAlSi-3	1060 to 1120
Copper	BCuP-3	1300 to 1500
Silver	BAg-6	1425 to 1600
Gold	BAu-2	1635 to 1850
Nickel	BNi-4	1850 to 2150

The flux used in furnace brazing is the consumable material. Listed are the different types of flux combinations for different types of workpiece materials and the working temperature range.

Flux type	Workpiece material	Temperature range (°F)	Form
Chlorides	Magnesium	900 to 1200	Powder
Florides	Aluminum	700 to 1190	
Boric acid Borates Fluoborates	Steels Most common metals	700 to 2000	Powder Paste liquid
Chlorides Florides Borates	Aluminum-bronze Aluminum-brass and alloys	1050 to 1800	Paste powder

Another type of consumable material is protective gas. Given here is the AWS number, along with the chemical composition and the workpiece materials for which it is used.

AWS atmosphere type number	Compositions %				Workpiece materials
	H_2	N_2	CO	CO_2	
1	1	87	1	11	Copper Brass
2	15	70	10	5	Copper Brass Low-carbon steel Nickel Monel
3	15	75	10	—	Low- to high-carbon steels Nickel Copper Brass

Factors Affecting Process Results

Tolerance and surface finish depend upon the following:

* Cleanliness of workpiece
* Type of material
* Type of filler material
* Type of flux and coverage
* Temperature
* Type of joint
* Cooling rate and uniformity

Typical Workpiece Materials

Brazability ratings vary depending on pre-surface treatment, type of flux, and type of filler metal used. Best results are achieved when brazing copper and brass, whereas aluminum is a more difficult material to furnace braze.

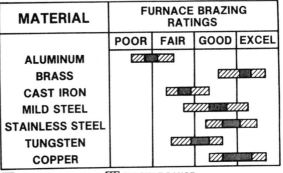

TYPICAL RANGE   FEASIBLE RANGE

Heating Time—Brazing

Variables that affect heating time are type and size of workpiece, type of furnace, and protective atmosphere. Indicated heating times are based on relatively small workpieces (between 0.25 lb and 1 lb). Heating times can be as long as 20 min for larger workpieces.

Workpiece materials	Type of furnace	Protective atmosphere	Heating time (min)
1010 to 1012 steel	Conveyor	Endothermic	11
Tungsten to steel	Conveyor	Endothermic	8
Brass to steel	Box	Endothermic	10
Low carbon steel	Conveyor	Exothermic	12
1008 to 1020 steel	Conveyor	Dissoc. ammonia	12
1017 to 1117 steel	Box	Exothermic	12

Cooling

Cooling is achieved by passing workpieces through a water-cooled jacket or a cool atmosphere. The goal is to cool the workpieces at an even rate to prevent disfiguration and discoloration. The water does not come in direct contact with the workpiece.

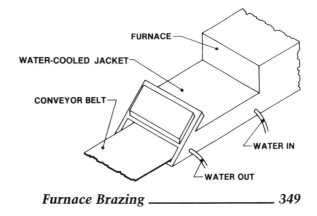

Furnace Brazing _____ **349**

Power Requirements

Power requirements are given in Btu's. The power required for brazing an average size workpiece depends on the type of workpiece material, the type of filler material, and the process temperature range.

Workpiece materials	AWS filler number	Temperature (F°)	Specific heat (Btu/ lb-°F)*
1010 to 1012 steel	BCu-1	2040	0.117
Tungsten to steel	BCu-1	2050	0.120
Brass to steel	BAg-la	1350	0.117
Low-carbon steel	BCu-1	2010	0.117
1008 to 1020 steel	Silver	1600	0.118
1017 to 1117 steel	BCu-1	2050	0.117

* For a workpiece weighing 1 lb.

Cost Elements

Cost elements include the following:

* Setup time
* Load and unload times
* Preheating and heating time
* Cleaning time
* Cooling time
* Direct labor rate
* Overhead rate
* Amortization of equipment and tooling

Time Calculation

The operational elements are setup time, T_s, load/unload time, T_{lu}, workpiece heating time, T_h, cooling time, T_c, conveyor capacity, C, length of the oven (conveyor length), L, conveyor velocity, V, and production size, N. The time calculations can be determined through the indicated formula. The heating time is calculated first. The heating time is controlled by conveyor velocity (which is variable) and by the length of the heating zone. The cooling time is determined by the length of the cooling zone and the conveyor velocity.

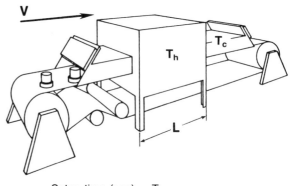

Setup time (sec) = T_s
Load/unload time (sec) = T_{lu}
Workpiece heating time (sec) = T_h
Cooling time (sec) = T_c
Conveyor capacity (parts) = C
Conveyor length (oven length) (ft) = L
Conveyor velocity (ft/min) = V
Production size (parts) = N

$$\text{Process heating time} = \frac{L}{V}$$

Safety Factors

The following hazards should be taken into consideration:

* Personal
    - Toxic fumes
    - Hot working environment
    - Explosion hazards

Gas Metal Arc Welding (MIG)

Gas metal arc welding, commonly referred to as MIG, is an electric arc welding process in which an arc is struck between a consumable wire electrode and a workpiece. Shielding is provided by an inert gas, hence MIG for metal inert gas.

Process Characteristics

* Uses a consumable wire electrode
* Uses shielding gas
* Results in a uniform weld bead
* Produces a slag-free weld bead
* Is commonly used for automatic welding

Process Schematic

An arc is formed between the wire electrode and the workpiece. The electrode melts within the arc and is deposited as filler material. A shielding gas is provided throughout the process to protect the weld from atmospheric contamination during solidification.

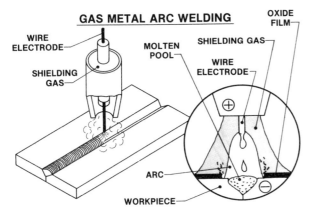

GAS METAL ARC WELDING

Workpiece Geometry

MIG is generally considered to be the most versatile of all arc welding processes. It can be used to join most types of metals. Welding by gas metal arc can be done in most positions; however, it is done most efficiently in flat, horizontal positions.

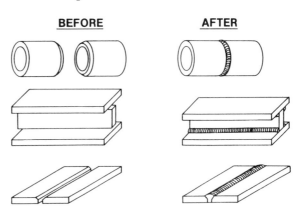

BEFORE   AFTER

Setup and Equipment

Essential equipment includes a welding machine, which provides a regulated source of power for melting electrodes; a wire feeder, which advances the electrode; a control system, which regulates gas flow and electrode feed; and a welding gun, which holds the electrode and carries the gas to the workpiece.

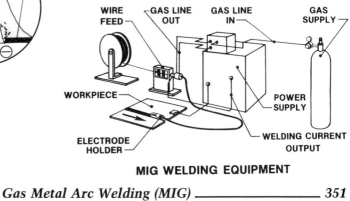

MIG WELDING EQUIPMENT

Geometrical Possibilities

Assuming adequate space for an electrode holder to move about the weld area, we find that most geometries are possible. Typically, metal from 0.125 in. to 1 in. thick is welded, but the feasible workpiece thickness range is from 0.02 in. to more than 2.5 in.

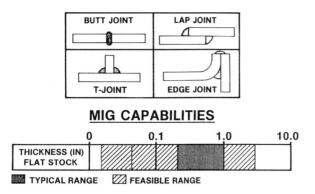

Tool Style

The top electrode holder is a semiautomatic air-cooled holder. Compressed air is circulated through the electrode holder in order to maintain moderate temperatures. The second example is a water-cooled electrode holder. Here water is circulated in place of air. The third typical holder is an automatic electrode holder. This is mounted on a welding machine and is water-cooled.

DESCRIPTION	STYLE	APPLICATION
AIR-COOLED SEMIAUTOMATIC ELECTRODE HOLDER		USED WITH LOWER CURRENT LEVELS USED FOR WELDING LAP OR BUTT JOINTS
WATER-COOLED SEMIAUTOMATIC ELECTRODE HOLDER		USED WITH HIGHER CURRENT LEVELS USED FOR WELDING T- OR CORNER JOINTS
WATER-COOLED AUTOMATIC ELECTRODE HOLDER		USED IN AUTOMATED EQUIPMENT

Toolholding Methods

There are two main toolholding methods. In automated welding, the electrode holder is mounted (usually on an arm), and the workpiece is moved in a manner to permit proper welding. With semiautomatic welding, the electrode holder is manually positioned. The manual method is more flexible and is used for small batches of parts. The automatic method is used for high production.

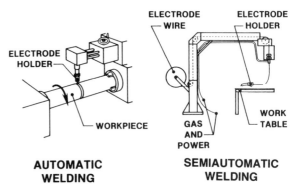

Typical Tools and Geometry Produced

Electrode wires are available in a wide range of materials for welding common metals and alloys. The shielding gas is commonly argon, helium, carbon dioxide, or a gas mixture. Gas mixtures are used to improve arc action and metal transfer.

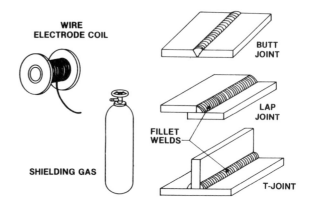

Workholding Methods

Various devices, such as clamps and vices, are used to secure workpieces on positioning tables, such as those shown. Positioning tables make it easy to manipulate heavy or awkward workpieces.

ROTATING AND TILTING POSITIONER

HEADSTOCK-TAILSTOCK POSITIONER

Shielded Gas

Argon and helium are among the most popular shielding gases. These gases are often mixed with carbon dioxide to reduce cost.

Shielding gas	Applications
Argon	* Superior for welding metals * Operates at a lower arc voltage * Arc starts easier, is smoother, more stable * Mild steel, aluminum, titanium, alloy steels
Helium	* Gas flow 1-½–3 times greater than argon * High speed welding of mild steel, titanium * Small heat-affected zone * Stainless steel, copper
Carbon dioxide	* Used as a component of gas mixtures * Produces good penetration and welding speed, low cost * Carbon and low alloy steels

Weld Finish and Quality

Weld finish and quality for gas metal arc welding are similar to those of other welding processes. Shown here are typical weld defects, along with their possible causes and their resulting effects.

Defect	Possible causes	Effects
Cracks	* Mechanical stress * Thermal stress * Impurities in alloying elements	* Reduction in strength * Unsatisfactory weld
Cavities (porosity)	* Poor gas coverage * Entrapped gas	* Reduced weld strength
Rough and dirty bead	* High weld current	* Bad appearance
Incomplete fusion	* Dirty surfaces * Weld speed too fast	* Reduced static strength * Produces stress risers * Unsatisfactory weld

Effects on Work Material Properties

Work material properties	Effects of MIG welding*
Mechanical	* May harden material * May reduce fatigue strength * Shrinkage may occur * Metal may be annealed * Warpage
Physical	* Bad appearance * May create cracks or porosity
Chemical	* May reduce corrosion resistance in welding zone

* MIG = gas metal arc welding.

Typical Workpiece Materials

Most common metals may be joined by MIG welding. Weldability ratings are good to excellent for mild steel, aluminum, stainless steel, and magnesium. Composition of the electrode

MATERIAL	MIG WELDING WELDABILITY RATINGS			
	POOR	FAIR	GOOD	EXCEL
MILD STEEL			▨▨ ▨▨	
CAST IRON		▨▨▨	▨	
STAINLESS STEEL			▨▨	▨▨
ALUMINUM				▨▨
COPPER			▨▨▨ ▨▨	
MAGNESIUM			▨▨ ▨	
TITANIUM		▨▨▨	▨▨	

▪ TYPICAL RANGE   ▨ FEASIBLE RANGE

may be altered to permit higher weldability ratings for different types of metals.

Filler Material

Common filler metal used in MIG welding includes both solid and cored wires. Various alloys are used for the electrode wire. Argon and helium are among the most popular shielding gases. These gases are often mixed with carbon dioxide to reduce cost.

Nonconsumable

Filler materials	Applications
Manganese and silicon	* Best suited for use with carbon dioxide shielding gas
Aluminum	* Helium and/or argon shielding gas
Low alloy steel	* Used on critical welds in low-carbon steel with carbon dioxide shielding
Synthetic powder core	* Used when special compositions are desired and for specialized welding applications
Traces of oxidizers	* Used on rusty surfaces and on semikilled or rimmed steel

Process Conditions

Welding speed is governed by the type of metal in the electrode wire and workpiece, the current used, and the size of the electrode wire. In a chart based on low-carbon steel joints, typical values are given for electrode feed rates (in inches per minute), electrode diameter (in inches), and welding speed (in inches per minute).

Low-carbon steel workpiece and low-carbon steel electrode

Joint type	Electrode feed rate (ipm)	Electrode diameter (in.)	Welding speed (ipm)
Circumferential corner	500	0.045	45
Circumferential modified butt	340 to 380	0.030	46.6
Corner	30 to 160	0.030	10
Lap	—	0.0625	60
T	300	0.045	16
Butt	400	0.045	7

Factors Affecting Process Results

Variables that influence weldability include the following:

* Welding current
* Type of current used (AC, DC)
* Heat impact/arc voltage/preheat
* Gas coverage of weld zone
* Cleanliness of welding area
* Welding speed
* Electrode composition and end profile
* Metal conductivity
* Joint design
* Workpiece material composition and properties

Cooling Methods

The electrode holder is commonly cooled with water and air. The cooling air or water is forced through channels in the holder, causing temperatures to remain at a somewhat moderate level. It is important to maintain the electrode's seal to prevent coolant leakage into the shielding gas, which would produce poor finish and weld quality.

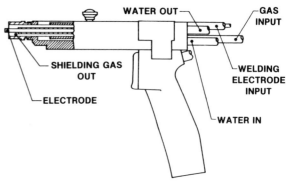

Power Requirements

The power requirements for MIG welding depend on several variables, including desired electrode feed rate and electrode diameter. Given are typical joint types, electrode feed rates, electrode diameters, and power requirements for welding low-carbon steel.

Low-carbon steel workpiece and low-carbon steel electrode

Joint type	Electrode feed rate (ipm)	Electrode diameter (in.)	Power supply (amp)
Circumferential corner	500	0.045	280 to 300 (DCEP)
Circumferential modified butt	340 to 380	0.030	170 to 190 (DCEP)
Corner	30 to 100	0.030	80 to 85 (DCEN)
Lap	234	0.0625	380 (DCEN)
T	300	0.045	200 (DCEP)
Butt	400	0.045	275 (DCEP)

amp = amperes.

Cost Elements

Cost elements include the following.

* Setup time
* Load/unload time
* Welding time
* Tool costs
* Direct labor rate
* Overhead rate
* Amortization of equipment and tooling

Time Calculation

Welding time is governed by workpiece length, number of passes required, and welding rate. The number of passes is directly related to the width of the gap produced by the bevel. If the gap is quite wide, it is sometimes necessary to oscillate or weave the electrode as it progresses along the weld gap. Total welding time, therefore, equals weld length divided by welding rate times the number of passes required.

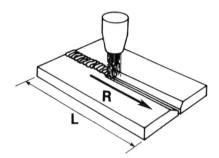

Length (in.) = L
Welding rate (ipm) = R
Number of passes = N

$$\text{Welding time} = \frac{L}{R}$$

$$\text{Total welding time} = \frac{L}{R} \times N$$

Safety Factors

The following risks should be taken into consideration:

* Personal
  - Eye contact with welding arc
  - Heat from tool and workpiece
  - Skin exposure to welding arc
* Environmental
  - Smoke

Gas Torch Braze Welding

Gas torch braze welding is a joining process that uses a filler metal to join two workpieces. Heat generated by a gas flame melts the filler metal. Metal is then distributed throughout the joint by a depositing action.

Process Characteristics

* Uses flame from a gas torch to heat the workpiece and melt the filler metal
* Typically involves workpieces of steel and cast iron
* Uses a filler metal, generally of a copper alloy in the shape of a rod or wire
* Distributes the filler metal by deposition
* Generally requires the use of a flux

Process Schematic

In gas torch braze welding, the filler metal is brought into contact with the joint between the workpieces. The joint is heated until the filler metal begins to melt. The filler metal has a lower melting point than that of the workpieces. The melted filler metal then fills in the joint, causing a strong bond between the workpieces. Generally a flux material is used to facilitate "wetting" of the workpieces with the molten filler material.

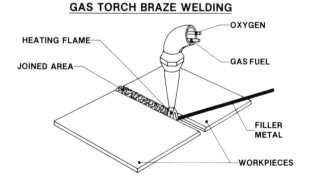

GAS TORCH BRAZE WELDING

Workpiece Geometry

The joint to be welded does have limits on its gap width. It is best if the joining surfaces have a small bevel at the surface joint.

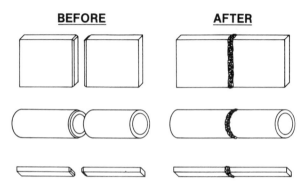

Setup and Equipment

Supplies for this process consist of a fuel gas (such as acetylene, natural gas, or propane), oxygen or air, filler metal (usually a copper alloy), and a regulator.

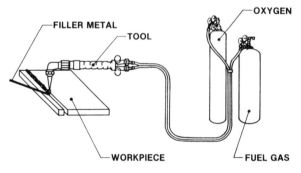

Typical Tools and Geometry Produced

Filler metals are available in a variety of sizes, depending on the size of the gap that needs to be filled. Filler metals are generally copper alloys.

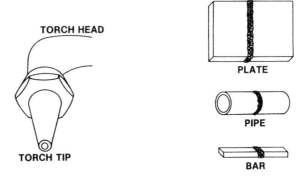

TORCH HEAD

TORCH TIP

PLATE

PIPE

BAR

Gas Tungsten Arc Welding (TIG)

In tungsten inert gas arc welding (TIG), an arc is struck between a nonconsumable tungsten electrode and the workpiece. Gas shielding (to protect the molten metal from contamination) and constant amperage are supplied during the welding operation.

Process Characteristics

* Uses nonconsumable tungsten electrode
* Uses shielding gases (Ar, He, or CO_2)
* Produces very high quality welds
* Produces no slag or spatter
* Is readily applied to thin materials

Process Schematic

An electric arc is formed between a nonconsumable tungsten tip and workpiece, causing the workpiece to melt. A shielding gas is provided at low pressure and forms a protective shield around the arc and the molten pool of metal. The tungsten electrode is not consumed because of its extremely high melting point. The operator must maintain a constant arc length to insure a stable arc. Filler material is supplied as a wire or rod.

GAS TUNGSTEN ARC WELDING

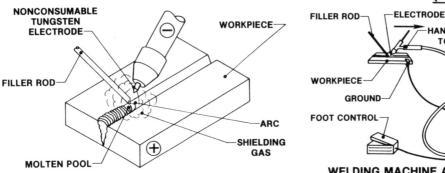

Typical Tools and Geometry Produced

TIG produces extremely high quality welds on metals such as aluminum, magnesium, stainless steel, titanium and nickel alloys. Filler metal is not usually required for thin materials, but it is used for thicker material. Filler material should be matched to the base metal.

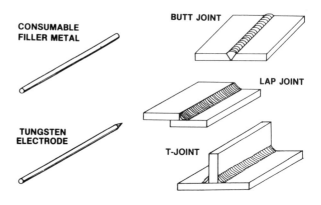

Setup and Equipment

Equipment consists of a power supply, a water supply to cool the torch, shielding gas, a hand-held gun, filler rod, and foot control. This control allows the operator to adjust the current easily to match welding conditions.

(TIG)

WELDING MACHINE AND CONTROL SYSTEM

Tool Style

Electrodes have a variety of end profiles. The two most common electrodes are the pointed and the hemispherical types. The hemispherical profile is produced by first grinding the electrode to a sharp point with a taper three to six times as long as the electrode diameter. The end becomes rounded as the point melts back onto itself until the ideal diameter for the job has been achieved.

DESCRIPTION	STYLE	APPLICATION
POINTED		• USED PRIMARILY WITH DCSP* • DIRECTS THE ARC • USED FOR NARROW JOINTS AND RESTRICTED LOCATIONS
FULL HEMISPHERE		• USED PRIMARILY WITH DCRP† • HAS THE GREATEST CURRENT DENSITY

*DCSP = direct current straight polarity.
†DCRP = direct current reverse polarity.

Toolholding Methods

Manual electrode holders are classified as either air- or water-cooled. Air-cooled holders are designed for light duty welding (up to 150 amps), whereas water-cooled holders are used for heavy duty welding (up to 600 amps). Holders are rated by their capacity to carry current at 100% duty cycles.

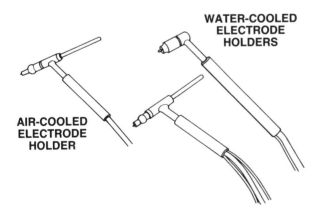

WATER-COOLED
ELECTRODE
HOLDERS

AIR-COOLED
ELECTRODE
HOLDER

Electrode Materials

Electrode size must be selected with care and is related to the type of work material, current, amperage required, and material thickness. Pure tungsten, which melts at 6170°F, provides a less expensive electrode. Thorium and zirconium additives are included in some electrodes. These additives improve the electrode's performance for some applications.

Nonconsumable

Material	Applications
Pure tungsten	* Uses a lower current * Used on less critical and general purpose work
Thoriated tungsten	* Arc starting is free from sputtering and flashing * Greater arc stability * Maintains good end profile during welding
Zirconiated tungsten	* Resistant to contamination of the weld * Maintains good end profile during welding * Used with low amperage AC

Shielding Gas

Argon is used more extensively than other shielding gases. Characteristics of each shielding gas are shown.

Consumable

Shielding gas	Applications
Argon	* Superior for welding metals, low cost * Operates at a lower arc voltage * Arc starts easier, is more smooth, stable * Used on mild steel, aluminum, titanium
Helium	* Gas flow 1-½–3 times greater than argon * High speed welding of mild steel, titanium * Small heat-affected zone, more penetration * Hotter more intense arc increases welding speed 30–40% * Used on stainless steel, copper
Argon–helium mix	* Used for a hotter arc in welding aluminum * Widely used in automatic welding * Used on aluminum alloys

Geometrical Possibilities

Common joint types are butt, lap, corner, and edge joints. Good welds depend on proper edge preparation, especially welds for square groove butt joints and other joints requiring filler metal.

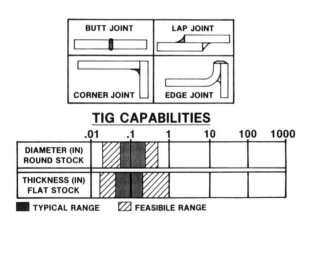

Workholding Methods

There are manual and automatic workholding methods. Manual systems include jigs, tack welds, and fixtures; automatic methods include fixtures, longitudinal seamers, pedestal boom manipulators, and rotating positioners. Automatic clamping methods are used for high production welding.

* Manual
  – Jigs
  – Tack welds
  – Fixtures
* Automatic
  – Longitudinal seamers
  – Pedestal boom manipulators
  – Rotating positioners

Weld Finish and Quality

Weld defects, such as cracks, porosity, round or dirty weld bead, or incomplete fusion, may be avoided by the following methods: keeping the welding area and equipment clean; making sure that the equipment used is in proper working condition; and using the proper melting technique and material for the job.

Defect	Possible causes	Effects
Cracks	* Mechanical stress * Thermal stress * Metallurgical problems	* Reduced weld strength
Cavities (porosity)	* Poor gas coverage * Entrapped gas * Wrong electrodes	* Reduced weld strength
Rough and dirty bead	* Air or water leak in shielding gas * High weld current	* Bad appearance * Brittle welds * Cracks/porosity
Incomplete fusion	* Dirty surfaces * Weld speed too fast * Insufficient amperage	* Reduced static strength * Produces stress risers

Effects on Work Material Properties

Some of the adverse effects on work material properties are given below.

Work material properties	Effects of TIG welding*
Mechanical	* May harden material * May reduce fatigue strength * Shrinkage may occur * Metal may be annealed/normalized * Warpage
Physical	* Bad appearance * May create cracks or porosity
Chemical	* May reduce corrosion resistance

*TIG = gas tungsten arc welding.

Typical Workpiece Materials

Primarily nonferrous metals, such as aluminum, magnesium, titanium, and refractory metals, are welded with the TIG process. However, stainless steel, cast iron, and mild steel can also be welded.

MATERIAL	TIG WELDABILITY RATINGS			
	POOR	FAIR	GOOD	EXCEL
MILD STEEL			▨▨ ▨	
CAST IRON			▨▨ ▨	
STAINLESS STEEL			▨▨ ▨	
ALUMINUM				▨▨ ▨
COPPER		▨▨ ▨		
TITANIUM			▨▨ ▨	

▨ TYPICAL RANGE   ▨ FEASIBLE RANGE

Workpiece Geometry

Almost any type of geometry can be welded using TIG. Before a surface can be welded, however, it must be clean and free from oil, grease, paint, and rust because the inert gas does not provide any cleaning or fluxing action. Welds are usually clean.

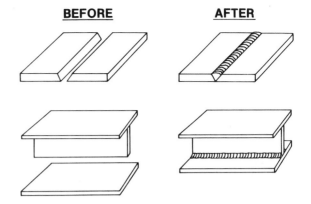

BEFORE   **AFTER**

Factors Affecting Process Results

Skillful operators can create welds that are smooth and uniform. TIG is often used for welding thin materials.

Workpiece variables that influence weldability include the following:

* Workpiece shape
* Workpiece thickness
* Alloy content
* Electrical resistivity
* Thermal conductivity
* Thermal expansion
* Hardness and strength
* Oxide forming characteristics

Variables that influence weld quality include the following:

* Welding current and voltage
* Type of current used
* Gas coverage of weld zone
* Cleanliness of welding area
* Welding speed or deposition rate
* Electrode composition and end profile
* Metal conductivity
* Joint design
* Workpiece material composition and properties

Process Conditions

Shown are factors required when welding various thicknesses of aluminum. Generally, an electrode size should be chosen so that it will operate at near maximum current carrying capacity. This insures maximum penetration, a stable arc, higher welding speed, and minimum width and convexity of the weld bead.

Aluminum, single-bevel joint

Material thickness (in.)	Bevel angle	AC current (amp)	Electrode dia (in.)	Filler rod dia (in.)
1/16	90°	70–100	1/16	1/16–3/32
1/8	90°	115–150	3/32	1/16–1/8
1/4	60°–100°	210–275	5/32	3/32–3/16
3/8	60°	325–400	1/4	1/4
	100°	275–350	3/16	1/4
1/2	60°	375–400	1/4	1/4
	100°	275–340	3/16	1/4
1	60°	500–600	5/16–3/8	1/4–3/8

dia = diameter.

Cooling Methods

Water cooling of metal nozzles makes it possible to use a welding current from 200 amps to 600 amps. The maximum practical current for manual welding with TIG is 500 amps. Automatic TIG machines may be used for higher amps if necessary.

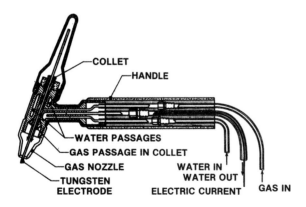

COLLET
HANDLE
WATER PASSAGES
GAS PASSAGE IN COLLET
GAS NOZZLE
TUNGSTEN ELECTRODE
WATER IN
WATER OUT
ELECTRIC CURRENT
GAS IN

Power Requirements

Shown are the welding current ranges that may be used for various sizes of electrodes. Electrode size must be selected with care and is related to the type of work material, current required, and material thickness. Thorium and zirconium additives are included in these electrodes.

Argon is used as a shielding gas

Electrode size	Direct current (amp)		Alternating current (amp)	
	Straight polarity	Reverse polarity	Tungsten electrode	Tungsten plus Th, Zr additives
0.010	Up to 15	NR	Up to 15	Up to 15
0.020	5–20	NR	5–15	5–20
1/16	70–150	10–20	50–100	70–150
3/32	150–250	15–30	100–160	140–235
1/8	250–400	25–40	150–210	225–325
3/16	500–750	55–80	250–350	400–500
1/4	750–1000	80–125	325–450	500–630

NR = not recomended.

Cost Elements

Cost elements include the following:

* Setup time
* Load/unload time
* Welding time
* Cost of consumable electrodes
* Direct labor rate
* Overhead rate

Productivity Tip

When it is impractical to move the work to the welding equipment, the equipment may be moved to the workpiece by means of a pedestal boom (as shown), a portable cart, or an overhead track-type cart.

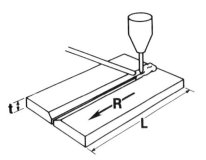

Length (in.) = L
Welding rate (ipm) = R
Number of passes required = N
Thickness (in.) = t

$$\text{Welding time} = \frac{L}{R} \times N$$

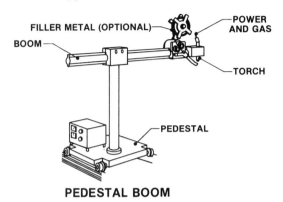

PEDESTAL BOOM

Time Calculation

Welding speed is governed by current density and the amount of material to be melted. Welding time per piece equals weld length divided by welding speed, times the required number of passes. Welding speed is measured in inches per minute. The number of passes depends on weld size, filler rod size, and weld current.

Safety Factors

The following hazards should be taken into consideration:

* Personal
  - Radiation to the eyes
  - Burn from tool and workpiece
  - Skin exposure to welding arc
  - Inhalation of toxic fumes
* Environmental
  - Smoke
  - Ozone gas
  - Fire

Laser Beam Welding

In laser beam welding, materials melt under the heat obtained from a narrow beam of coherent, monochromatic light (a laser beam). Typically, no filler metal is used.

Process Characteristics

* Is used for thin-gage workpieces
* Is used for welding areas that are not readily accessible
* Provides excellent welding precision
* Permits joining of dissimilar alloys
* Uses no electrodes
* Causes little or no thermal damage to the workpiece
* Is easily automated

Process Schematic

Frequently a CO_2 laser or a solid state neodymium–yttrium–aluminum–garnet (Nd-YAG) laser is used. Laser output can be pulsed or continuous. A focal point is selected just below the surfaces to be welded. The workpiece is then manipulated under the beam. Shielding gas blown through the nozzle coaxially with the laser beam protects the weld.

LASER BEAM WELDING

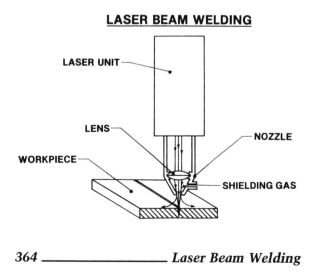

Typical Tools and Geometry Produced

Although laser welding is a relatively new technology in the manufacturing industry, it is seeing wider and wider use because of its pinpoint heat and penetration capabilities. The power intensity at its focal point can be greater than a billion watts per square inch, enough to melt many metals. Typical weld joints are shown for foil and heavy-gage material.

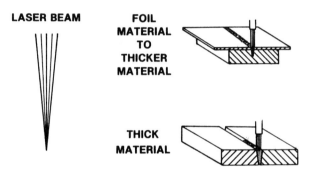

Setup and Equipment

The equipment is comparatively expensive. A laser beam welder consists of a power supply, a control unit, a laser beam unit and lens/nozzle section, and a workholding fixture on a numerically controlled X–Y table. An X–Y table is used to manipulate the workpiece for more complicated welds. A laser control unit keeps the beam focused with the right intensity.

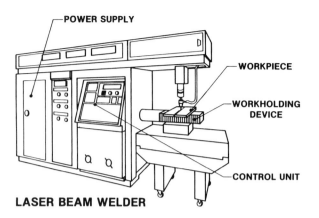

LASER BEAM WELDER

Workpiece Geometry

Typical workpiece shapes include butt joints, edge flange joints, and various lap joints, where penetration is essential to the success of the weld. Because of the types of joints used and penetration capabilities of a laser beam, generally no filler metal is required. Most laser welding is done on thin materials.

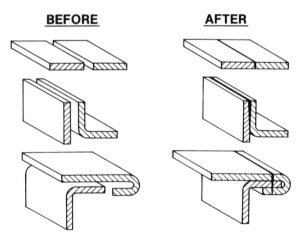

BEFORE   **AFTER**

Tool Style

A laser beam is generated in either a gas (CO_2) or solid (Nd-YAG) medium and focused through lenses or mirrors. CO_2 laser beams produce light with wavelengths of 10.6 micrometers; Nd-YAG lasers, 1.06 micrometers. Typically, a beam diameter can be focused down to about ten times the wavelength.

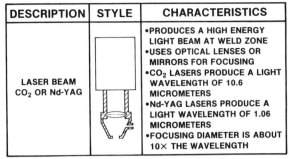

DESCRIPTION	STYLE	CHARACTERISTICS
LASER BEAM CO_2 OR Nd-YAG		• PRODUCES A HIGH ENERGY LIGHT BEAM AT WELD ZONE • USES OPTICAL LENSES OR MIRRORS FOR FOCUSING • CO_2 LASERS PRODUCE A LIGHT WAVELENGTH OF 10.6 MICROMETERS • Nd-YAG LASERS PRODUCE A LIGHT WAVELENGTH OF 1.06 MICROMETERS • FOCUSING DIAMETER IS ABOUT 10× THE WAVELENGTH

Tool Materials

A laser beam may be finely focused to produce a thin beam or defocused to produce a broad beam. In some instances, it is possible to direct a beam through the surface of a transparent object without affecting it and weld a feature inside the object. The most used industrial lasers are the CO_2 laser and the Nd-YAG laser.

Tool materials	Applications
CO_2	* Large industrial laser for cutting and welding
Neodymium (Nd)	* For small or medium size lasers
Nd-YAG (neodymium-yttrium-aluminum-garnet)	* For high power solid-state industrial lasers

Geometrical Possibilities

No large gaps or filler metals are required. Typical welding thickness is between 0.01 in. and 0.75 in., but, under laboratory conditions, up to 9 in. of aluminum have been welded. The melt-through weld shown is a good example of the capabilities of laser beam welding.

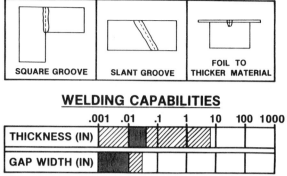

| | SQUARE GROOVE | SLANT GROOVE | FOIL TO THICKER MATERIAL |

WELDING CAPABILITIES

	.001	.01	.1	1	10	100	1000
THICKNESS (IN)							
GAP WIDTH (IN)							

■ TYPICAL RANGE   ▨ FEASIBLE RANGE

Workholding Methods

Special fixtures placed on a numerically controlled or computer numerically controlled (NC/CNC) X–Y table help position workpieces. A high degree of accuracy is required of both components and fixtures because of the small dimensions of the laser beam.

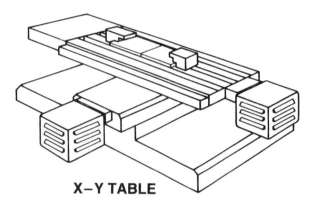

X–Y TABLE

Surface Finish

Shown is a table depicting possible causes for major types of weld defects associated with this process.

Defect	Possible causes	Effects
Cavities	* Excessive welding speed * Poor focus * Insufficient power supply * Excessive gap in joint	Reduction of strength
Internal stresses	* Insufficient welding speed * Poor focus * Excessive power supply	Reduction of strength, distortion

Effects on Work Material Properties

The effects of this process are minimal.

Work material properties	Effects of laser beam welding
Mechanical	* Hardening or softening may occur
Physical	* None
Chemical	* None

Typical Workpiece Materials

Weldability ratings depend on low thermal conductivity and low reflectivity of the material. Metals, such as gold, silver, and copper, cannot be laser welded. Nickel and titanium have good weldability ratings. The most suitable materials are steel and stainless steel.

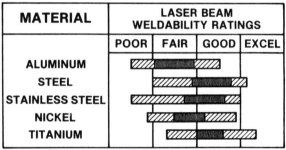

MATERIAL	LASER BEAM WELDABILITY RATINGS			
	POOR	FAIR	GOOD	EXCEL
ALUMINUM				
STEEL				
STAINLESS STEEL				
NICKEL				
TITANIUM				

■ TYPICAL RANGE   ▨ FEASIBLE RANGE

Process Conditions

The following are important process conditions:

* Type of material (conductivity, reflectivity)
* Joint design
* Type and size of laser
* Position of focal point
* Focusing
* Welding speed
* Pulse frequency
* Duration of pulses
* Type and amount of shielding gas
* Nozzle design

Cooling

Cooling is sometimes needed for the laser material, and sometimes the lens needs cooling.

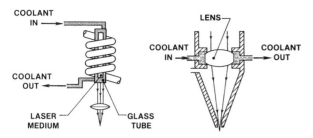

Factors Affecting Process Results

Variables that influence weldability include the following:

* Workpiece surface condition
* Workpiece composition
* Workpiece thickness
* Focusing of laser beam
* Power applied
* Workpiece melting point
* Joint design

Power Requirements

Power requirements depend largely upon the following:

* Type of material (conductivity, reflectivity)
* Joint design
* Welding speed
* Type and size of laser

* Commercial CO_2 lasers are available in the range of 0.1 through 10 kilowatts.
* Commercial Nd-YAG lasers are available in the range of 50 through 500 kilowatts.

Cost Elements

Cost elements include the following:

* Setup time
* Load/unload time
* Welding time
* Direct labor rate
* Overhead rate
* Amortization of equipment and tooling

Time Calculation

Length and thickness of the workpiece and welding velocity are process time parameters. Welding time is equal to length of the workpiece divided by welding rate.

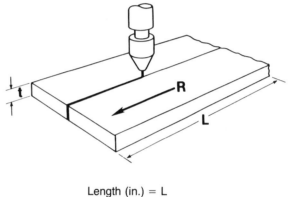

Length (in.) = L
Welding rate (ipm) = R
Thickness (in.) = t

$$\text{Welding time} = \frac{L}{R}$$

Safety Factors

The following risks should be taken into consideration:

* Personal
  - Eye contact with welding beam
  - Heat from tool and workpiece
  - Skin exposure to welding beam
* Environmental
  - Smoke
  - Very high voltage

Metal Bath Dip Soldering

Metal bath dip soldering is defined as a metal-joining process where the workpieces to be joined are immersed in a pot of molten solder. Because of the relatively low melting temperature of the solder (between 350 and 600°F), only adhesion between the solder and the workpieces results. A flux or metal cleaner is used to prepare the workpiece for bonding with the solder. Typically, dip soldering is an automated process used extensively in the electronics assembly industry.

Process Characteristics

* Uses a pot of molten solder to join workpieces
* Requires an application of flux
* Is a relatively low temperature joining process where no fusion takes place
* Requires an immersion period of between 2 and 12 seconds
* Can be successfully automated
* Is used extensively in the electronics assembly industry

Process Schematic

The dip soldering process is as follows. First, the workpieces to be joined are treated with a cleaning flux. The workpiece is then mounted in the workholding device and immersed in the molten pool of solder for a period of between 2 and 12 seconds. Often the workpiece is agitated to aid the flow of the solder. The workholder must allow the workpiece to be inclined 3 to 5 degrees to insure solder runoff and a smooth finish.

METAL BATH DIP SOLDERING

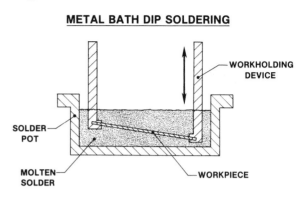

WORKHOLDING DEVICE

SOLDER POT

MOLTEN SOLDER

WORKPIECE

Workpiece Geometry

Dip soldering is a process that is generally limited to only all-metal workpieces, although parts such as circuit boards can tolerate momentary contact with the hot molten solder without damage.

BEFORE

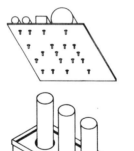

AFTER

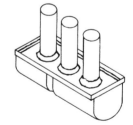

Setup and Equipment

The essential equipment includes the solder pot with its temperature control panel, the pool or bath of molten solder, and the workholding device. The workholding device is usually custom designed for each respective workpiece for either manual or automated dipping.

Typical Tools and Geometry Produced

The primary elements of the dip soldering process are the molten pool of solder and the cleaning flux. Traditionally, solder has been an alloy of 50% tin, 49.5% lead, and 0.5% antimony. The lead gives the solder its low-melting-point characteristics, and the tin provides the wetting or flowability of the solder. Typical products include circuit boards, wire tinning or precoating, wire joining, and products such as automobile radiators.

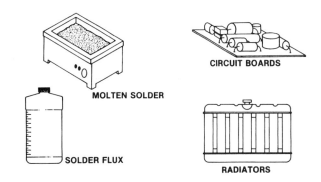

Metal Bath Dip Soldering

Plasma Arc Welding

In plasma arc welding, a shielded arc is struck between a nonconsumable electrode and the torch body, and this arc transforms an inert gas into plasma. This plasma is then used to melt the workpiece and filler metal.

Process Characteristics

* Uses nonconsumable tungsten electrodes
* Requires inert gases, such as argon, helium, and nitrogen, for the shielding gas and to form the plasma
* Produces a high temperature arc of 30,000°F.

Process Schematic

An arc between the torch body and a nonconsumable electrode is used to convert an inert gas into plasma. Plasma exists only at very high temperatures and thus can be used to melt and weld materials.

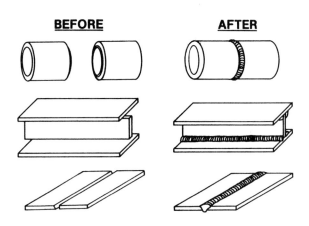

BEFORE   **AFTER**

Setup and Equipment

Essential equipment consists of a welding machine, which provides a regulated source of current; the shielding gas; the water coolant supply; a control console; an optional foot control, which serves as an on–off switch; a hand-held or automated welding gun; sources of plasma; and a source of shielding gas.

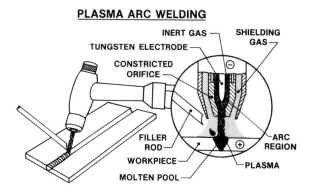

PLASMA ARC WELDING

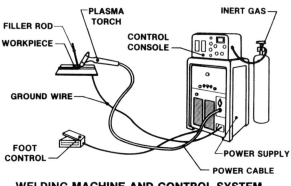

WELDING MACHINE AND CONTROL SYSTEM

Workpiece Geometry

Because of the preciseness of the plasma jet stream, plasma arc welding is highly suitable for thin-gage workpieces, providing a superior penetration without burn-through.

Typical Tools and Geometry Produced

The principal elements of this process are the filler metal (sometimes used), the nonconsumable tungsten electrode, and a source of plasma gas.

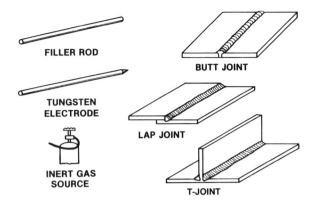

DESCRIPTION	STYLE	APPLICATION
SINGLE PLASMA PORT	○	SIMPLE MOST COMMON GENERAL PURPOSE
MULTIPLE PLASMA PORT	○ ○ ○	ELONGATES THE ARC INCREASES WELDING SPEED UP TO 50% LARGER WELD GAPS

Geometrical Possibilities

Most geometries, such as butt, lap, T-, and edge joints can be welded using this process. Geometric capabilities are limited only because it is a manually operated system. Feasible thickness ranges start at 0.0085 in. and run up to several inches.

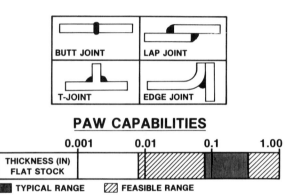

Toolholding Methods

Generally, the welding gun is hand-held. Automated equipment has been developed but is not common.

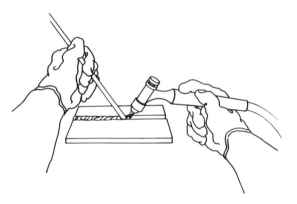

Tool Style

Shown are two types of orifice tips. The single port is used for more detailed and general purpose welds, whereas the multiple port is used for elongated arcs, larger welds, and increased welding speeds.

Electrode Current Capacity

Here are current ranges for different-size electrodes. Diameters range from 0.010 in. to 5/32 in. The current ranges from 5 amps to 500 amps.

Tungsten electrode diameter (in.)	Current capacity (amp)
0.010	Up to 15
0.020	5 to 20
0.040	15 to 80
1/16	70 to 150
3/32	150 to 250
1/8	250 to 400
5/32	400 to 500

Shielding Gas

Argon is used for low-carbon steel, copper, and zinc. Argon and oxygen mixes are used for stainless steel, keyhole welds, and root passes. Helium is used for melt-in welds in copper and stainless steel, and for keyhole welds.

Shielding gas	Applications
Argon	* Carbon and low alloy steels * Copper and zinc
Argon and oxygen	* For stainless steel * Keyhole weld * Root passes
Helium	* Melt-in welds * Copper, stainless steel * Keyhole welds

Workholding Methods

Shown are two possible methods. A rotating and tilting positioner is used for large, heavy, and awkward workpieces; a balanced positioner is used for small workpieces. The workpiece is clamped to the positioner by bolts or small vises.

ROTATING AND TILTING POSITIONER

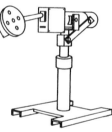

BALANCED POSITIONER

Weld and Finish Quality

Given are possible defects and their possible causes.

Defect	Possible causes	Effects
Cracks	* Mechanical stress * Thermal stress * Impurities in alloying elements	* Reduced weld strength
Cavities (porosity)	* Poor gas coverage * Entrapped gas	* Reduced weld strength
Rough and dirty bead	* High weld current	* Bad appearance
Incomplete fusion	* Dirty surfaces * Weld speed too fast	* Reduced static strength * Produces stress risers

Effects on Work Material Properties

Work material properties	Effects of plasma arc welding
Mechanical	* May harden material * May reduce fatigue strength * Shrinkage may occur * Metal may be annealed * Warpage
Physical	* Bad appearance * May create cracks or porosity
Chemical	* May reduce corrosion resistance

Typical Workpiece Materials

Mild steel, stainless steel, aluminum, copper, and magnesium have good to excellent ratings, whereas cast iron and titanium have fair to excellent ratings.

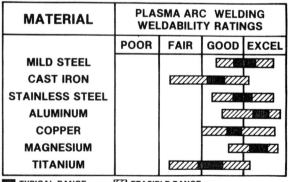

MATERIAL	PLASMA ARC WELDING WELDABILITY RATINGS			
	POOR	FAIR	GOOD	EXCEL
MILD STEEL			▨	▨
CAST IRON		▨	▨	
STAINLESS STEEL			▨	▨
ALUMINUM			▨	▨
COPPER			▨	▨
MAGNESIUM			▨	▨
TITANIUM		▨	▨	

■ TYPICAL RANGE   ▨ FEASIBLE RANGE

Process Conditions

Welding speeds depend on the type of workpiece material, the thickness of the material, and the welding current. Thickness ranges from 0.093 in. to 0.250 in. Welding speeds range from 10 in./min to 30 in./min (ipm).

Workpiece metal	Welding speed (ipm)			
	Workpiece thickness (in.)			
	0.093	0.125	0.187	0.250
Low alloy steel	—	12	11	14
Stainless steel	24	30	16	14
Titanium alloy	—	20	13	12
Copper or brass	10	10	15	20

Factors Affecting Process Results

Variables that influence weldability include the following:

* Plasma temperature
* Gas coverage of weld zone
* Cleanliness of welding area
* Welding speed
* Joint design
* Workpiece material composition and properties

Cooling Methods

The torch must be cooled in order to be handled. Pumping liquid (usually water) through the torch and around the tungsten electrode cools it.

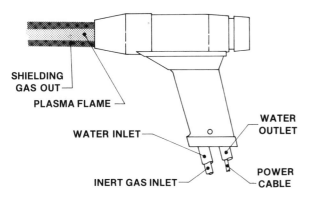

SHIELDING GAS OUT
PLASMA FLAME
WATER INLET
INERT GAS INLET
WATER OUTLET
POWER CABLE

Power Requirements

Power requirements depend on workpiece material and thickness. They range from 115 amps for thin stainless steel to 670 amps for 1/4 in. copper or brass.

Workpiece material	Power requirements (amp)			
	Workpiece thickness (in.)			
	0.093	0.125	0.187	0.250
Low alloy steel	145	185	215	275
Stainless steel	115	145	165	240
Titanium alloy	165	185	175	200
Copper or brass	180	300	460	670

Cost Elements

Cost elements include the following:

* Setup time
* Load/unload time
* Welding time
* Direct labor rate
* Overhead rate
* Amortization of equipment and tooling costs

Time Calculation

The time parameters for this process include workpiece length and thickness; the latter determines the welding rate. To calculate welding time, the length of the workpiece is divided by the welding rate.

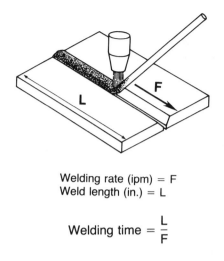

Welding rate (ipm) = F
Weld length (in.) = L

$$\text{Welding time} = \frac{L}{F}$$

Safety Factors

The following risks should be taken into consideration:

* Personal
  - Heat from welding arc
  - Skin exposure to radiation
  - Possibility of burns from hot workpieces
* Environmental
  - Smoke, fumes

Projection Welding

Projection welding is a metal-joining process where heat obtained from an electrical current passed between projections or embossments causes the two mating workpieces to fuse together. Workpieces are held together under pressure exerted by the nonconsumable electrodes.

Process Characteristics

* Requires no filler metal or fluxes
* Requires a preformed projection, embossment, or intersection on one of the workpieces
* Can produce multiple spot welds in one cycle
* Uses nonconsumable, low resistance copper alloy electrodes
* Is used primarily on sheet metals or on relatively low-mass workpieces
* Is easily automated

Process Schematic

In this process, a specially designed projection is formed on one of two conforming workpieces, which serves to concentrate an AC current flow between the electrodes. After a weld is formed, the electrodes return to their original position.

PROJECTION WELDING

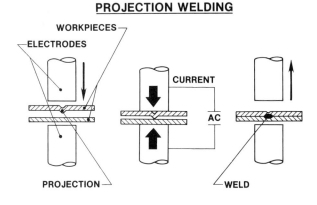

Workpiece Geometry

The workpiece geometry varies in shape, size, and number of welds. Low-carbon steel material is commonly used, although most electrically conductive materials strong enough to support a projection or the process pressure may be projection welded.

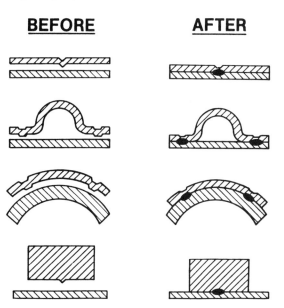

Setup and Equipment

A press-type projection welding machine is normally used. The machine is composed of a welding head containing the electrodes and a current control panel. The welding head stroke is powered by hydraulics, pneumatics, springs, or electromagnets.

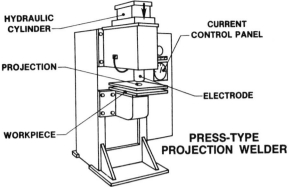

PRESS-TYPE PROJECTION WELDER

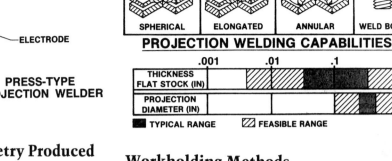

PROJECTION WELDING CAPABILITIES

	.001	.01	.1	1
THICKNESS FLAT STOCK (IN)				
PROJECTION DIAMETER (IN)				

■ TYPICAL RANGE　▨ FEASIBLE RANGE

Typical Tools and Geometry Produced

Flat-faced electrodes provide the necessary pressure (up to 1200 psi) and current to perform projection welding. Necessary electrode material properties include electrical conductivity, strength, hardness, and temperature resistance. Projection shapes include dimples, ridges, and solid cones, which provide filler material alignment points and concentrate the current flow.

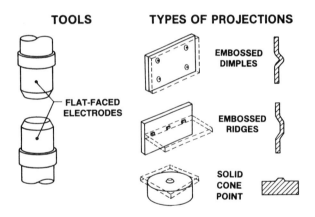

TOOLS

FLAT-FACED ELECTRODES

TYPES OF PROJECTIONS

EMBOSSED DIMPLES

EMBOSSED RIDGES

SOLID CONE POINT

Geometrical Possibilities

Spherical and elongated projections are most often used, whereas weld bolts and annular projections are more specialized. Feasible stock thickness ranges from 0.006 in. to around 1.0 in. for most projection welds. The typical range is from 0.05 in. to 0.5 in. A weld bolt can be welded to any thickness of material. The feasible range for projection diameters is from 0.1 in. to around 1.0 in.

Workholding Methods

For projection welding the workpiece is normally hand-held.

Tool Style

Large area electrodes are used to cover the projections and are able to withstand high pressures. Die electrodes are used for larger workpieces and are usually designed to fit the workpiece.

DESCRIPTION	STYLE	APPLICATION
LARGE AREA ELECTRODE	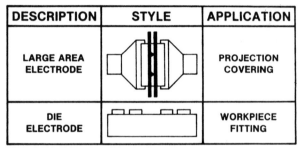	PROJECTION COVERING
DIE ELECTRODE		WORKPIECE FITTING

Toolholding Methods

There are many types of electrode holders available. Shown are three types: universal offset, light duty offset, and paddle-type offset electrode holders.

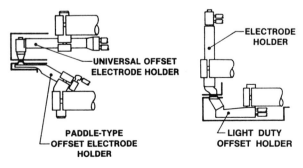

Weld Finish and Quality

Generally, defects occur when the electrode current or pressure is at a wrong setting. Bad welds cause a bad appearance, reduce fatigue strength, and increase the likelihood of corrosion.

Defects	Possible causes	Effects
Indentation	* Improper electrode force * High heat rate	* Loss of weld strength * Bad appearance
Surface fusion	* Low electrode force * High weld current	* Undersized weld * Weld zone extends through surface
Irregularly shaped weld	* Worn electrode * Improper projection size	* Reduced weld strength
Cracks, cavities, pinholes	* Excessive heat * Removing electrode force prematurely * Dirty weld surface	* Reduced fatigue strength * Increased corrosion

Effects on Work Material Properties

Mechanical effects include hardening of the workpiece, reduction of fatigue strength, warpage, and annealing. Physical effects include bad appearance, internal cracks, porosity, and surface cracks. The only chemical effect is a loss in corrosion resistance.

Work material properties	Effects of projection welding
Mechanical	* May harden material * May reduce fatigue strength * Warpage may occur * Metal may be annealed * Notch (stress riser)
Physical	* Bad appearance * May create internal cracks, porosity, or spongy metal * May create surface cracks
Chemical	* May reduce corrosion resistance

Electrode Materials

Class 1 electrodes with a copper base alloy are used for coated low-carbon steel and for non-ferrous materials. Class 2 copper base electrodes are used for cold- or hot-rolled, low-carbon steels, stainless steels, nickel alloys, silicon, bronze, and nickel silver. Class 3 electrodes are for high pressures, high electrical resistance workpiece materials, and thick workpieces. Stainless steel, monel, and inconel require class 3 electrodes. Refractory metals in classes 10 through 14 require high heat, long weld time, adequate cooling, and high pressures.

Materials	Applications
Copper base alloys, class 1	* Low-carbon steel coated with tin or galvanized (poor weld chemistry) * Chromium, zinc or aluminum
Copper base alloys, class 2	* Cold- and hot-rolled low-carbon steel * Stainless steels * Nickel alloys * Silicon, bronze, and nickel silver
Copper base alloys, class 3	* High pressures and workpiece resistance * Thickness sections of low-carbon steel * Stainless steel * Monel and inconel
Refractory metal compositions, classes 10–14	* High heat * Long weld time * Adequate cooling * High pressure

Factors Affecting Process Results

Variables that affect welds are projection geometry, thermal and electrical conductivity of electrodes, welding current, electrode composition and geometry, electrode force, welding time, weld spacing, workpiece material, and cleanliness of work material.

Variables that affect welds include the following:

* Projection size and geometry
* Thermal and electrical conductivities of electrodes
* Welding current

* Electrode composition and design
* Welding force or electrode force
* Time (squeeze, weld, hold, off)
* Weld spacing
* Workpiece material
* Cleanliness

Typical Workpiece Materials

Aluminum, brass, mild steel, and stainless steel are the best materials for use with this process. Cast iron and magnesium do rather poorly under projection welding.

MATERIAL	PROJECTION WELDING CAPABILITY RATINGS			
	POOR	FAIR	GOOD	EXCEL
ALUMINUM				▨▦▨
BRASS		▨▦▨		
CAST IRON	▨▦			
MILD STEEL		▨▦▨		
STAINLESS STEEL	▨▦▨			
MAGNESIUM	▨▦			

▦ TYPICAL RANGE   ▨ FEASIBLE RANGE

Electrode Force

Process conditions are functions of steel thickness, electrode force, projection diameter, and weld current. Welding times range from 1 second for 0.053-in. thick metal to 2.42 seconds for 0.245-in. thick metal.

Low-carbon steel workpiece

Steel thickness (in.)	Electrode force (lbs)	Projection diameter (in.)	Weld time (sec)
0.153	2000	0.33	1
0.164	2300	0.35	1.16
0.179	2630	0.39	1.36
0.195	2930	0.41	1.63
0.210	3180	0.44	1.86
0.225	3610	0.47	2.10
0.245	3900	0.53	2.42

Cooling of Electrodes

Electrodes are cooled by pumping water or a brine solution through the electrode. Cooling prevents deterioration of the electrode.

METHOD	COOLING MEDIUM
FACE COOLANT HOLE BORED INTO ELECTRODE	• WATER 60 to 80°F • BRINE SOLUTIONS FOR REFRIGERATED SYSTEMS 0 to 40°F

Power Requirements

Power requirements are a function of the steel thickness, the electrode force, the projection diameter, and the welding time, as shown in the electrode force table below. Welding currents range from 15,400 amps for metal 0.153 in. thick to 23,300 amps for metal 0.245 in. thick.

Low-carbon steel workpiece

Steel thickness (in.)	Electrode force (lbs)	Projection diameter (in.)	Welding current (amp)
0.153	2000	0.33	15,400
0.164	2300	0.35	16,100
0.179	2630	0.39	17,400
0.195	2930	0.41	18,800
0.210	3180	0.44	20,200
0.225	3610	0.47	21,500
0.245	3900	0.53	23,300

Cost Elements

Cost elements include the following:

* Setup time
* Load/unload time
* Welding time
* Tool change time
* Tool costs
* Direct labor rate
* Overhead rate
* Amortization of equipment and tooling

Time Calculation

Operation elements used to calculate total welding time are setup time, load and unload times, hold time, weld time, and the number of welds to be performed. To calculate the total time, add load time, weld time, hold time, and unload time; multiply this sum by the number of welds; and add the setup time to this product.

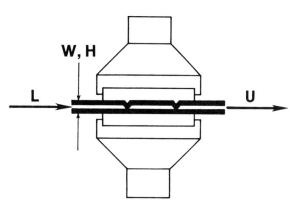

Setup time (sec) = T_s
Load time (sec) = L
Unload time (sec) = U
Hold time (sec) = H
Weld time (sec) = W
Number of welds = N

Total time = T_s + [(L + W + H + U) × N]

Safety Factors

The following risks should be taken into consideration:

* Personal
  – Clamping electrodes
  – Eye irritation from flash
  – Hot workpiece

Shielded Metal Arc Welding (SMAW)

In shielded metal arc welding (SMAW), an electric arc is established between a flux-coated consumable rod electrode and the workpiece. A gaseous shield is provided by vaporization of the flux coating.

Process Characteristics

* Uses a consumable rod electrode
* Deposits slag on the weld bead
* Provides shielding by vaporization of the flux coating on the electrode
* Supplies constant welding current
* Weld appearance and quality depend on operator skill in maintaining a constant arc length and travel speed

Process Schematic

SMAW involves the use of a flux-coated consumable electrode. As the arc is established, the flux coating on the rod disintegrates and forms a gas shield that protects the weld from the atmosphere. In addition to providing a gas shield, scavengers contained in the flux clean impurities on the surface of the work. The slag produced by the flux coating prevents the weld metal from oxidizing.

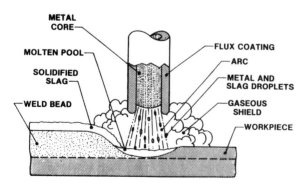

METAL CORE
MOLTEN POOL
SOLIDIFIED SLAG
WELD BEAD
FLUX COATING
ARC
METAL AND SLAG DROPLETS
GASEOUS SHIELD
WORKPIECE

Workpiece Geometry

Shown are some typical examples of types of workpieces. The top illustration shows two geometric possibilities, cylindrical and flat, that can be easily attached to each other. The bottom illustration shows how two pieces of the same geometric configuration can be welded to each other.

BEFORE

AFTER

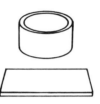

Setup and Equipment

SMAW equipment consists of a constant current power source that supplies power to the consumable rod electrode. The power source is determined by the current required to operate any rod diameter efficiently. Usually, 1 amp is used for every 0.001 in. of rod diameter; however, the type of flux coating and electrode material may cause differences between electrodes. The workpiece needs to be grounded to prevent shock to the operator and also to complete the welding circuit.

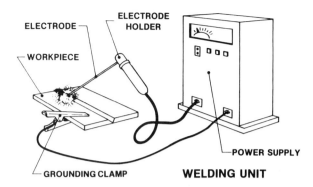

WELDING UNIT

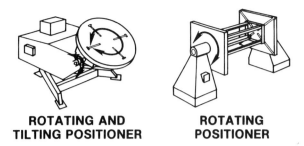

ROTATING AND TILTING POSITIONER

ROTATING POSITIONER

Typical Tools and Geometry Produced

Flux-coated electrodes come in various core wire diameters from 1/16 in. to 5/16 in., and also come in lengths of 9 in., 14 in., and 18 in. The illustration shows three types of joints suitable to SMAW: the butt joint, lap joint, and T-joint. Matching electrode properties to the base materials is a general rule for choosing an electrode. Examples of electrode types are mild steel, stainless steel, nickel, bronze, and aluminum bronze.

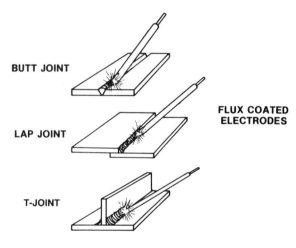

BUTT JOINT

LAP JOINT

FLUX COATED ELECTRODES

T-JOINT

Workholding Methods

Various devices, such as clamps and vises, are used to secure the workpieces on positioning tables, such as those shown. Position tables make it easier to manipulate heavy or awkward workpieces.

Tool Style

Given are several types of electrodes and their application. E6010 is an example of a code for an electrode. E stands for electrode; 60 means 60,000 lbs. of tensile strength (60 ksi); 1 stands for the welding position; and 0 stands for the type of current to be used.

DESCRIPTION	STYLE	APPLICATION
E6010 E6011		DEEP PENETRATION, FORCEFUL, SPRAY TYPE ARC
E6012 E6013		MEDIUM PENETRATION GLOBULAR VISCOUS SLAG, WELDING SHEET METAL
E6020		SPRAY TYPE ARC, MEDIUM PENETRATION HEAVY HONEYCOMBED SLAG

Toolholding Methods

A large clamp holds the electrode and conducts electricity through to the electrode.

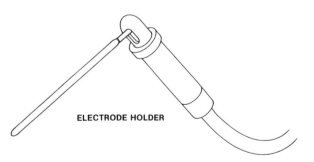

ELECTRODE HOLDER

Tool Materials

Shown are several types of fluxing elements and their corresponding applications that are coated on a mild steel electrode.

Coating materials	Applications
Calcium carbonate	* Shielding gas * Fluxing agent
Titanium dioxide	* Slag former * Arc stabilizer
Mica	* Arc stabilizer
Iron powder	* Contact welding * High deposition rate
Ferromanganese	* Alloying * Deoxidizer

Mild steel electrode.

Geometrical Possibilities

Most joint types are possible. The butt joint, lap joint, T-joint, and edge joint are common examples. The geometrical capabilities are limited only because this is a manual process. Feasible workpiece thicknesses range from 0.125 in. to around 12 in.

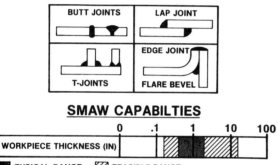

SMAW CAPABILTIES

Weld and Finish Quality

Shown here are five of the most common defects, along with their possible causes and their effects.

Defect	Possible causes	Effects
Cracks	* Mechanical stress * Thermal stress * Impurities in alloying elements	* Unacceptable weld requires rework
Subsurface porosity	* Dirty surfaces * Entrapped gas	* Reduced weld strength
Rough and ropey bead	* High weld current	* Bad appearance
Incomplete fusion	* Dirty surfaces * Weld speed too fast	* Reduced weld strength * Produces stress risers
Inclusions	* Trapped slag or dirt	* Reduced weld strength

Effects on Work Material Properties

Work material properties	Effects of shielded metal arc welding
Mechanical	* May harden material * May reduce fatigue strength * Shrinkage may occur * Metal may be annealed * May warp material
Physical	* Poor appearance * May create cracks or porosity
Chemical	* May reduce corrosion resistance in welding zone

Typical Workpiece Materials

Mild steel, cast iron, and stainless steel have fair to excellent weldability ratings.

MATERIAL	SHIELDED METAL-ARC WELDABILITY RATINGS			
	POOR	FAIR	GOOD	EXCEL
MILD STEEL			▓	▓
CAST IRON		▓	▓	
STAINLESS STEEL		▓	▓	

■ TYPICAL RANGE   ▨ FEASIBLE RANGE

Process Conditions

The process conditions are based on a low-carbon workpiece and electrode. Electrode classifications, deposition rate, and size (in diameter) determine welding speed. Welding speeds range from 9 in./min to 14 in./min.

Low-carbon steel workpiece and low-carbon steel electrode

Electrode class	Electrode deposition rate (oz./hr)	Electrode diameter (in.)	Welding speed (ipm)
E6011	88	3/16	9 to 13
E6012	66	3/16	13 to 14
E7014	87	3/16	11 to 13
E7018	83	3/16	10 to 14
E7024	134	3/16	11 to 13

Flat weld position.

Factors Affecting Process Results

Variables that influence weld quality include the following:

* Welding current and voltage
* Type of current used (AC or DC)
* Flux coverage of weld zone
* Cleanliness of welding area
* Welding speed or deposition rate
* Electrode composition and end profile
* Metal conductivity
* Joint design
* Workpiece material composition and properties

Power Requirements

The power requirements are based upon having a low-carbon steel workpiece and electrode. The power requirement (in amps) is a function of deposition rate, electrode diameter, and electrode class.

Low-carbon steel workpiece and low-carbon steel electrode

Electrode class	Electrode deposition rate (oz./hr)	Electrode diameter (in.)	Power supply (amp)
E6011	88	3/16	200
E6012	66	3/16	225
E7014	87	3/16	260
E7018	83	3/16	240
E7024	134	3/16	270

Flat weld position.

Cost Elements

Cost elements include the following:

* Setup time
* Load/unload time
* Welding time
* Electrode costs
* Direct labor rate
* Overhead rate
* Amortization of equipment and tooling

Time Calculation

Process time parameters include preheating time, workpiece length, number of passes required, setup time, welding rate, and welding time. Welding time is obtained by dividing length by the welding rate and then multiplying by the number of passes.

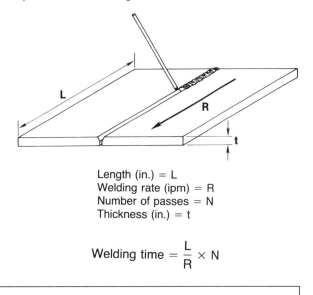

Length (in.) = L
Welding rate (ipm) = R
Number of passes = N
Thickness (in.) = t

$$\text{Welding time} = \frac{L}{R} \times N$$

Safety Factors

The following hazards should be taken into consideration:

* Personal
  - Eye and skin exposure to welding arc rays
  - Heat from electrode and workpiece
  - Skin exposure to welding arc
  - Respiratory irritation
* Environmental
  - Smoke

Spot Welding

Spot welding is a process in which contacting metal surfaces are joined by the heat obtained from resistance to electric current flow. Workpieces are held together under pressure exerted by electrodes.

Process Characteristics

* Requires no filler metal or fluxes
* Can achieve high speed production
* Is easily automated
* Does not require skilled operators
* Is used primarily on sheet metal
* Uses nonconsumable, low resistance, copper alloy electrodes

Process Schematic

There are three stages in making a spot weld. During the first stage, electrodes are brought together against the metal, and pressure is applied. Next, the current is turned on momentarily. This is followed by the third, or hold, stage, in which the current is turned off, but the pressure is continued. This hold time forges the metal while it is cooling.

SPOT WELDING

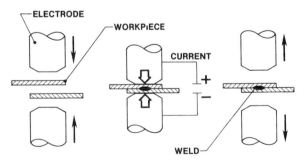

Workpiece Geometry

Good design practice must always allow for adequate accessibility. Connecting surfaces should be free of contaminants, such as scale, oil, and dirt, for quality welds. Metal thickness is generally not a factor in determining good welds.

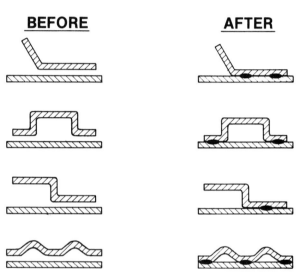

Setup and Equipment

A spot welding machine normally consists of tool holders and electrodes, which are mechanisms for making and holding contact at the weld. Tool holders have two functions: to hold the electrode firmly in place and to support water hoses that provide cooling of the electrodes.

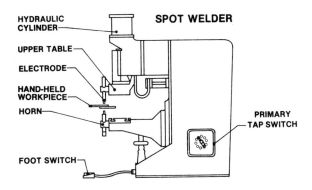

SPOT WELDER

- HYDRAULIC CYLINDER
- UPPER TABLE
- ELECTRODE
- HAND-HELD WORKPIECE
- HORN
- FOOT SWITCH
- PRIMARY TAP SWITCH

Typical Tools and Geometry Produced

The electrodes, which conduct the current and apply the pressure, are made of low resistance copper alloy and are usually hollow to facilitate water cooling. Electrode tip size directly affects the size and shear strength of the weld.

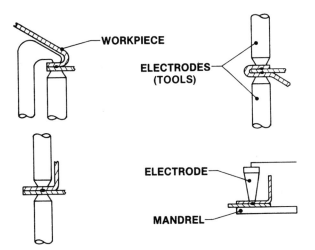

WORKPIECE

ELECTRODES (TOOLS)

ELECTRODE

MANDREL

Geometrical Possibilities

The two workpieces that are to be joined should be placed to allow for easy electrode accessibility. The spot welding capabilities chart describes the range of workpiece sizes that can be joined. The width (as limited by the throat length of the spot welder) is typically 5 in. to 50 in. Workpiece thickness can range from 0.008 in. to 1.25 in.

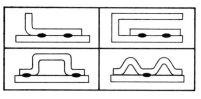

SPOT WELDING CAPABILITIES

	0	0.01	0.1	1.0	10	100
WIDTH (IN)						
THICKNESS (IN)						

▓ TYPICAL RANGE ▨ FEASIBLE RANGE

Workholding Methods

For stationary spot welding equipment, the workpieces are normally hand-held. Larger workpieces are either self-supported or fed by conveyors. Portable spot welders are in the form of robots or hand-held electrodes. Magnetic fixturing cannot normally be used in the spot welding process.

Tool Style

Electrodes in spot welding perform three major functions: they conduct current to the workpiece, transmit enough force to produce a satisfactory weld, and dissipate the heat from the weld zone rapidly. Each electrode has a hollow shank that allows coolant circulation.

DESCRIPTION	STYLE	APPLICATION
RADIUS		HIGH HEAT
TRUNCATED		HIGH PRESSURE
ECCENTRIC		FOR CORNERS
OFFSET ECCENTRIC		REACHING INTO CORNERS
OFFSET TRUNCATED		REACHING INTO WORKPIECE

Toolholding Methods

Electrode holders are made in a wide variety of shapes and sizes that are adjustable to suit the workpieces. The most common types include universal, paddle-type, light duty, and regular offset electrode holders.

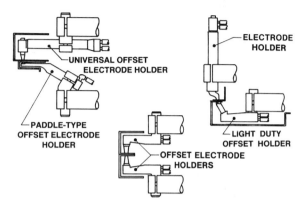

Weld Finish and Quality

Correct machine settings are necessary for good weld results. Electrode force, electrode temperature, electrode composition, weld current, and cleanliness of both the electrode and the workpiece constitute the main factors in weld quality.

Defects	Possible causes	Effects
Indentation	* Improper electrode force/shape * High heat rate	* Loss of weld strength * Bad appearance
Surface fusion	* Low electrode force * High weld current	* Undersized welds * Weld zone extends through surface
Irregularly shaped weld	* Worn electrode * Dirty electrode	* Reduced weld strength
Electrode fouling	* Low electrode force * Improper electrode material	* Reduced weld strength * Reduced electrode life
Cracks, cavities, pinholes	* Excessive heat * Removing electrode force prematurely	* Reduced fatigue strength * Increased corrosion

Effects on Work Material Properties

Spot welding may harden the material, reduce fatigue strength, and cause warpage. Annealing may occur and some stretching of material may occur. The physical effects of spot welding include a bad appearance and internal or external cracks. The material may have a reduction in corrosion resistance, which is classified as a chemical effect.

Work material properties	Effects of spot welding
Mechanical	* May harden material * May reduce fatigue strength * Warpage may occur * Metal may be annealed * Stretching of material
Physical	* Bad appearance * May create internal cracks * Porosity, or spongy metal * May create surface cracks
Chemical	* May reduce corrosion * Resistance

Electrode Materials

Most electrodes are made from copper base alloys because of their high thermal and electrical conductivities. Other electrodes that are used for special applications are made from refractory metal compositions. Class 2 electrodes are the best general purpose electrodes.

Materials	Applications
Copper base alloys, class 1	* Low-carbon steel coated with tin or galvanized * Chromium, zinc, and aluminum
Copper base alloys, class 2	* Cold- and hot-rolled low-carbon steel * Stainless steel * Nickel alloys * Silicon, bronze, and nickel silver
Copper base alloys, class 3	* High pressures and workpiece resistance * Thick sections of low-carbon steel * Stainless steel * Monel and inconel
Refractory metal compositions, classes 10–14	* High heat * Long weld time * Inadequate cooling * High pressure

Factors Affecting Process Results

Variables that affect welds include the following:

* Thermal and electrical conductivities of electrodes
* Welding current
* Electrode composition and design
* Welding force or electrode force
* Time (squeeze, weld, hold, off)
* Weld spacing
* Workpiece material

Typical Workpiece Materials

Workpiece materials must conduct electricity.

MATERIAL	SPOT WELDING CAPABILITY RATINGS			
	POOR	FAIR	GOOD	EXCEL
ALUMINUM				
BRASS				
CAST IRON				
MILD STEEL				
STAINLESS STEEL				
MAGNESIUM				

▓ TYPICAL RANGE   ▨ FEASIBLE RANGE

Weld Time

Process conditions that affect welding time are metal thickness, electrode face diameter, and electrode force. Welding times range from 0.01 sec to 0.63 sec.

Metal thickness (in.)	Electrode face diameter (in.)	Electrode force (lbs)	Weld time (sec)
1/8	0.010	200	0.01
0.021	3/16	300	0.10
0.40	1/4	500	0.16
0.78	5/16	1100	0.28
0.125	3/8	1800	0.43
0.250	7/16	1950	0.50
0.375	1/2	2100	0.56
0.625	5/8	2400	0.63

For mild steel workpiece and class 1 electrode.

Cooling

Coolant holes in electrodes are either round or fluted. Fluted holes offer more cooling surface than do round holes. Coolant holes should extend as close to the face of the electrode as possible without endangering electrode strength. Water or a coolant brine solution may be used.

METHOD	COOLING MEDIUM
—FACE —COOLANT HOLE BORED INTO ELECTRODE	• WATER 60 to 80°F • BRINE SOLUTIONS FOR REFRIGERATED SYSTEMS 0 to 40°F

Weld Current

Weld current requirements range from 4000 amps to 24,000 amps for different thicknesses of 1010 mild steel.

Metal thickness (in.)	Electrode face diameter (in.)	Electrode force (lbs)	Weld current (amp)
0.010	1/8	200	4000
0.021	3/16	300	6500
0.040	1/4	500	9500
0.078	5/16	1100	14,000
0.125	3/8	1800	19,000
0.250	7/16	1950	20,000
0.375	1/2	2100	21,500
0.625	5/8	2400	24,000

For mild steel workpiece and class 1 electrode.

Cost Elements

Cost elements include the following:

* Setup time
* Load/unload time
* Idle time
* Welding time
* Tool change time
* Tool costs
* Direct labor rate
* Overhead rate
* Amortization of equipment and tooling

Productivity Tip

Portable and robot welders increase maneuverability and productivity. Robot welders are able to produce excellent welds with exacting precision.

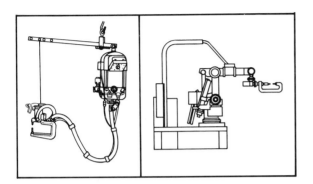

Time Calculation

Time parameters depend on load time, number of welds, unload time, and weld time, including squeeze, hold, and off-line welding. First, the workpieces are loaded. The weld is then performed, and the workpieces are unloaded. Load and unload times are added to the weld time. This sum is then multiplied by the number of welds.

SPOT WELDING SETUP

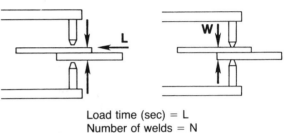

Load time (sec) = L
Number of welds = N
Unload time (sec) = U
Weld time (sec) = W

$$\text{Total time} = (L + W + U) \times N$$

Safety Factors

The following risks should be taken into consideration:

* Personal
  - Clamping electrodes
  - Eye irritation from flash
  - Hot workpiece

Submerged Arc Welding (SAW)

In submerged arc welding, or SAW, the heat for coalescence is provided by an electric arc struck between the workpiece and the consumable electrode. Shielding is provided by a blanket of granular flux, deposited over the area to be welded.

Process Characteristics

* Uses a consumable wire-fed electrode
* Is shielded by a granular flux that partially vaporizes
* Has a slag deposit on the weld bead
* Is capable of high welding speeds and deposition rates
* Produces high quality welds on thick workpieces
* Can be semi- or fully automated

Process Schematic

Current is fed through a consumable wire electrode to produce an arc between the electrode and the workpiece. A granular flux blanket is deposited ahead of the arc as the electrode advances along the weld joint. After the weld pool solidifies, the unfused flux is removed, screened, and recycled.

Workpiece Geometry

Workpiece materials commonly used include low-carbon, medium-carbon, high strength, or stainless steels. The process is usually not suited to materials less than 3/16 in. thick. Workpieces requiring flat butt welds or horizontal fillets are the most common joints produced.

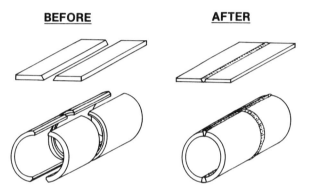

BEFORE   **AFTER**

Setup and Equipment

A wire electrode is fed through a conducted contact tube before contacting the workpiece base metal. A highly conductive flux is deposited on the workpiece by the wire electrode. A shell or solid slag deposit is left behind on the weld joint.

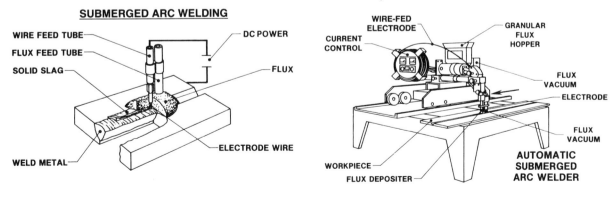

SUBMERGED ARC WELDING

WIRE FEED TUBE
FLUX FEED TUBE
SOLID SLAG
DC POWER
FLUX
ELECTRODE WIRE
WELD METAL

WIRE-FED ELECTRODE
CURRENT CONTROL
GRANULAR FLUX HOPPER
FLUX VACUUM
ELECTRODE
FLUX VACUUM
WORKPIECE
FLUX DEPOSITER
AUTOMATIC SUBMERGED ARC WELDER

Typical Tools and Geometry Produced

A mechanically powered wire roll provides a continuously fed, bare metal, consumable electrode. Granular flux is deposited ahead of the welding arc to protect the weld from atmospheric contamination. Typical geometries produced include butt and lap joints, with many joint variations.

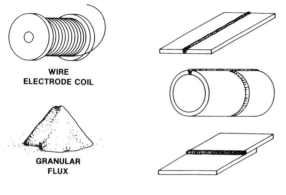

WIRE ELECTRODE COIL

GRANULAR FLUX

Tool Style

There are two main styles used in this process. A single electrode holder is the more common of the two and is used for narrow weld seams and for small deposition rates. Tandem electrodes are used where large deposition rates are required.

DESCRIPTION	STYLE	APPLICATION
SINGLE ELECTRODE		SMALL WELDING JOBS NARROWER WELDING SEAMS SLOWER DEPOSITION RATE
TANDEM ELECTRODES		LARGER WELDING JOBS FAST DEPOSITION RATE LARGER WELDING SEAMS

Geometrical Possibilities

Some different possible joint styles are shown. The joint must be designed to hold flux for maximum surface–flux interaction. Feasible workpiece thickness for flat stock ranges from 0.060 in. to 8 in. with the typical range from 0.250 in. to 1 in.

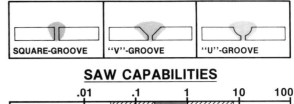

SQUARE-GROOVE   "V"-GROOVE   "U"-GROOVE

SAW CAPABILITIES

	.01	.1	1	10	100

THICKNESS (IN) FLAT STOCK

▦ TYPICAL RANGE   ▧ FEASIBLE RANGE

Toolholding Method

The electrode is automatically fed through the typical electrode holder shown here.

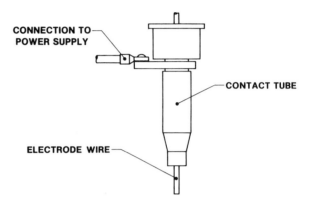

CONNECTION TO POWER SUPPLY

CONTACT TUBE

ELECTRODE WIRE

Workholding Methods

Shown is a device used for large, round workpieces; other workpieces require different arrangements.

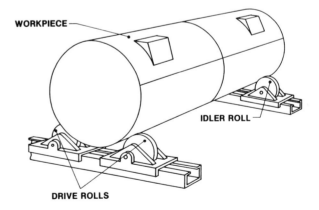

WORKPIECE

IDLER ROLL

DRIVE ROLLS

Surface Finish

Three of the most common weld defects with possible causes are as follows: small cracks may result from mechanical stress, thermal stress, or impurities in alloying elements; porosity may result from poor flux coverage or entrapped gas; incomplete fusion results when bad flux is used, or the weld speed is too fast. These defects can reduce weld strength. In the table below, these three common weld defects are given along with possible causes and effects. Two additional defects are also given.

Defect	Possible causes	Effects
Cracks	* Mechanical stress * Thermal stress * Impurities in alloying elements	* Unacceptable weld requires rework
Subsurface porosity	* Dirty surfaces * Entrapped gas	* Reduced weld strength
Rough and ropey bead	* High weld current	* Bad appearance
Incomplete fusion	* Dirty surfaces * Weld speed too fast	* Reduced strength * Produces stress risers
Inclusions	* Trapped slag or dirt	* Reduced strength

Effects on Work Material Properties

Work material properties	Effects of submerged arc welding
Mechanical	* May harden material * May reduce fatigue strength * Shrinkage may occur * Metal may be annealed * May warp material
Physical	* Poor appearance * May create cracks or porosity
Chemical	* May reduce corrosion resistance in welding zone

Typical Workpiece Materials

Most common metals can be welded by SAW. Mild steel, cast iron, and stainless steel are the most common because of their higher weldability rating.

MATERIAL	SAW WELDING WELDABILITY RATINGS			
	POOR	FAIR	GOOD	EXCEL
MILD STEEL ALLOY STEELS				

■ TYPICAL RANGE   ▨ FEASIBLE RANGE

Tool Materials

Electrode	Weld position	Flux
E6010	Horizontal	High cellulose, sodium
E6013	Overhead	High titania, potassium
E6027	Horizontal fillets	High iron oxide
E7014	Vertical, horizontal, overhead	Iron powder, titania
E7048	Vertical, down, overhead, horizontal	Low hydrogen, potassium, iron powder

Process Conditions

Shown are process conditions for a low-carbon steel workpiece and a low-carbon steel electrode. Workpiece thickness, electrode feed rate, and electrode diameter yield the welding speed in inches per minute (ipm).

Low-carbon steel workpiece and low-carbon steel electrode

Workpiece thickness (in.)	Electrode feed rate (ipm)	Electrode, diameter (in.)	Welding speed (ipm)
0.063	68	3/32	100 to 140
0.109	51	1/8	70 to 90
0.250	47	3/16	25 to 40
0.500	54	3/16	23 to 27
0.750	42	7/32	20 to 22
1.250	50	7/32	13

Factors Affecting Process Results

Variables that influence weld quality include the following:

* Welding current and voltage
* Type of current used (AC or DC)
* Flux coverage of weld zone
* Cleanliness of welding area
* Welding speed or deposition rate
* Electrode composition and end profile
* Metal conductivity
* Joint design
* Workpiece material composition and properties

Power Requirements

Power requirements depend on welding speed, electrode feed rate, and electrode diameter.

Low-carbon steel workpiece and low-carbon steel electrode

Workpiece thickness (in.)	Electrode feed rate (ipm)	Electrode diameter (in.)	Power demand (amp)
0.063	68	3/32	300
0.109	51	1/8	400
0.250	47	3/16	750
0.500	54	3/16	650 to 850
0.750	42	7/32	700 to 1000
1.250	50	7/32	1000 to 1125

Beveled butt joint.

Cost Elements

Cost elements include the following:

* Setup time
* Load/unload time
* Welding time

* Tool costs
* Direct labor rate
* Overhead rate
* Amortization of equipment and tooling

Time Calculation

Shown are workpiece thickness, workpiece length, and direction of welding; all of these determine the welding rate. Time parameters for SAW are heating time, workpiece length, setup time, welding rate, and welding time. To obtain welding time, the workpiece length is divided by the welding rate and then multiplied by the number of passes.

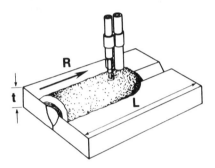

Length (in.) = L
Welding rate (ipm) = R
Thickness (in.) = t
Number of passes = N

$$\text{Welding time} = \frac{L}{R} \times N$$

Wave Soldering

Wave soldering is a process in which a printed circuit board is passed over a wave of molten solder. **The wave is produced by a pump.** As the circuit board is passed over the wave, the solder bonds the printed circuit board paths with the board mounted components.

Process Characteristics

* Produces clean and reliable solder connections
* Is an automated process
* Reuses excess solder and flux
* Requires work inspection, touch-up, and testing
* Increases productivity through speed and efficiency

Process Schematic

In wave soldering, the workpiece is moved on a conveyor, first through a preheating section, then through a mist of flux, and, lastly, through a wave of solder.

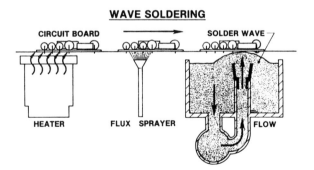

WAVE SOLDERING

Workpiece Geometry

Generally, printed circuit boards are wave soldered.

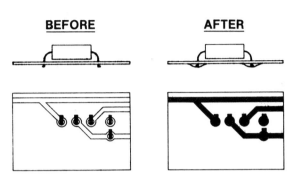

BEFORE   AFTER

Setup and Equipment

Wave soldering equipment consists of a molten pan of solder, a pump to produce the wave, a conveyor to move parts, a flux sprayer, and a preheating pad.

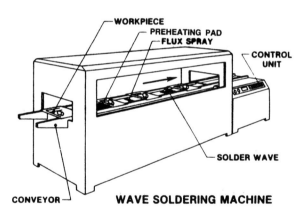

WAVE SOLDERING MACHINE

Typical Tools and Geometry Produced

The source metal in wave soldering is a mixture of tin and lead with small traces of antimony. The percentage of these metals is typically 50% tin, 49.5% lead, and 0.5% antimony.

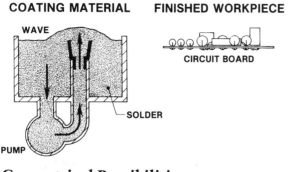

COATING MATERIAL   **FINISHED WORKPIECE**

WAVE

SOLDER

PUMP

CIRCUIT BOARD

Geometrical Possibilities

The workpiece is generally a printed circuit board. The limiting factor of this process is the size of workpiece that a specific machine can handle. Given are some general ranges.

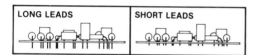

LONG LEADS   SHORT LEADS

WAVE SOLDERING CAPABILITIES

	0	.001	.01	0.1	1	10	100	1000
WIRE DIAMETER								
PC BOARD (IN)								

▥ TYPICAL RANGE ▨ FEASIBLE RANGE

Solder Finish and Quality

Quality of soldering depends on proper heating temperatures and properly treated surfaces. Given is a list of defects and possible causes and effects on the final product.

Defect	Possible causes	Effects
Cracks	* Mechanical stress	* Loss of conductivity
Cavities (porosity)	* Contaminated surface * Lack of flux * Insufficient preheating	* Reduction in strength * Poor conductivity
Wrong solder thickness	* Wrong solder temperature * Wrong conveyor speed	* Susceptible to stress * Too thin for current load * Undesired bridging between paths
Poor conductor	* Contaminated solder	* Product failure

Tool Style

There are several types of tools that form various wave shapes. These tools come in the form of removable inserts. The normal waveform tool is for horizontal soldering at medium speeds. The second type of tool forms a cascade wave for inclined soldering at high speeds. The tool with extenders forms a flat wave and has some of the same capabilities as the two previous tools.

DESCRIPTION	STYLE	APPLICATION
NORMAL WAVE		MEDIUM SPEED LONG LEADS HORIZONTAL SOLDERING
CASCADE WAVE		HIGH SPEED SHORT LEADS INCLINED SOLDERING
FLAT WAVE WITH EXTENDERS		MEDIUM TO HIGH SPEED LONG LEADS INCLINED TO HORIZONTAL SOLDERING

Workholding Methods

Various size brackets and fixtures are used to position workpieces so that they will be properly transferred through the process.

Effects on Work Material Properties

Side effects of wave soldering are mainly caused by heat. However, corrosive fluxes and poor material selection may result in other problem areas. Special precaution needs to be taken against heat affecting the electronic components being soldered to the board.

Work material properties	Effects of wave soldering
Mechanical	Excess heat may cause * Circuit board delamination * Board embrittlement * Warpage of materials
Physical	* Little effect
Chemical	* Harmful corrosive fluxes

Typical Workpiece Materials

Although copper is the most popular base metal, other metals are used. Depending on the metal used, different solder compositions can improve solderability.

MATERIAL	WAVE SOLDERING SOLDERABILITY RATINGS			
	POOR	FAIR	GOOD	EXCEL
COPPER				
ALUMINUM				
NICKEL				
STEEL				
TUNGSTEN				
BRASS				

▦ TYPICAL RANGE   ▨ FEASIBLE RANGE

Conveyor Feed Rates

Feed rates range from 120 in./min to 180 in./min (ipm), depending on the soldering material. At these rates, an average board takes about 2 sec to 3 sec to be soldered.

Material (lead base)	Melting point (°F)	Feed rate (ipm)
1.5% silver 1% tin	589	170
5% tin	594	175
20% tin	531	155
50% tin	421	120
1% antimony	608	180
9% antimony	509	150

Workpiece Cooling

The cooling rate is important. If the workpiece cools too slowly, the soldering can be excessively brittle. If the workpiece cools too rapidly, the board may warp and the solder may have small fractures. A moderate cooling speed will result in an optimum solder finish.

Cooling speed	Cause or method	Effects
Fast	Submersion	* Warpage and deformation of solder and workpiece
Medium	Fine water spray, forced air cooling	* Minimum warpage * Best possible solder joint
Slow	Air cooling	* Excessive alloying resulting in embrittlement * Prolonged heat exposure may harm workpiece components

Solder Composition

There are many different chemical combinations used to form a solder, depending on the desired characteristics. Given are several of the most popular with their composition and general applications. Fluxes are very important. The flux chemically treats or cleans the surfaces to be soldered.

Solder composition	Applications
99.8% tin	* High corrosion resistance * High melting point
60% tin 1% alloys 39% lead	* Most popular * Strong and ductile * Low melting range * Melts sharply and sets quickly
11% tin 37% lead 42% bismuth 10% cadmium	* Low melting point for sensitive components

Flux

Tool materials	Applications
Noncorrosive	* Parts that are precleaned * Where low acidity is required
Corrosive	* Quick action * Requires little precleaning

Factors Affecting Process Results

Variables that influence solderability include the following:

* Workpiece cleanliness
* Temperature of workpiece
* Desired deposit thickness
* Thermal conductivity
* Expansion and contraction
* Solder type
  - Melting point
  - Conductivity
  - Malleability

Power Requirements

Power required to melt different soldering materials is given in the form of Btu's/pound for each degree Fahrenheit. Depending on the material, power requirements range from 29 to 38 Btu/lb-°F.

Material (lead base)	Melting point (°F)	Required power (Btu/lb-°F)
1.5% silver and 1% tin	589	29
	—	—
5% tin	594	30
20% tin	531	32
50% tin	421	38
1% antimony	608	29
9% antimony	509	30

Cost Elements

Cost elements include the following:

* Initial heatup time
* Load/unload time
* Soldering time
* Cooling time
* Conductor and flux costs
* Direct labor rates
* Overhead rates
* Amortization of equipment and tooling costs

Time Calculation

When dealing with automated equipment and conveyor belt processes, time parameters are determined by the selected speed of the conveyor belt. Each segment of the process has a predetermined time. Another factor to take into consideration is whether or not multiple workpieces can be processed at the same time. Main parameters are conveyor velocity and length. Total time can be calculated by dividing conveyor length by the conveyor velocity. This quotient multiplied by the production size, N, plus the initial setup time determines the total run time required.

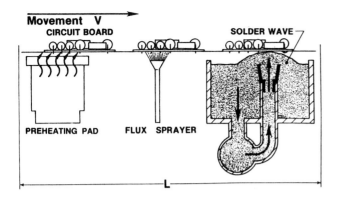

Conveyor velocity (ft/min) = V
Conveyor length (ft) = L
Production size = N

$$\text{Total time} = \left(\frac{L}{V} \times N\right) + \text{setup time}$$

Safety Factors

The following safety risks should be taken into consideration:

* Personal
  - Contact with hot workpiece or equipment
  - Contact with corrosive chemicals
* Environmental
  - Fumes

Chemical Joining

Adhesive Bonding

Adhesive bonding is a joining process where a substance, in a liquid or semiliquid state, is applied to adjoining workpieces to provide a permanent bond. Bondable materials are virtually unlimited. Adhesives can be made from natural and/or artificial compounds and can exist in many forms.

Process Characteristics

* Involves the use of consumable substances in a liquid or semiliquid state
* Allows metals and nonmetals to be joined
* Usually requires controlled time and temperature
* Is ideal for joining thin materials to other materials
* Provides for vibration dampening and an electrical nonconductive joint
* Can greatly reduce assembling costs

Process Schematic

Adhesives are applied to one or both surfaces and the workpieces are aligned. Pressure is applied to aid adhesion and to remove entrapped air pockets. In most cases, the bond will be permanent unless an insufficient amount of time is allowed for proper curing.

ADHESIVE BONDING

APPLICATION OF ADHESIVE — FORCE — WORKPIECES — ADHESIVE BOND

Workpiece Geometry

Adhesive bonding is used extensively in making joints for wood and other relatively lightweight materials. Shown are some of the more successful connections. Generally speaking, joints are strongest where workpiece and adhesive contact area is greatest.

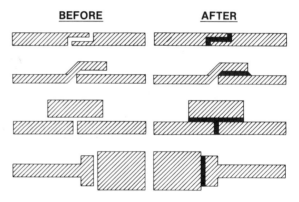

BEFORE   AFTER

Setup and Equipment

Essential equipment includes hot glue guns and caulking guns, both of which are used to apply adhesives essentially in a paste, or semiliquid, form. Spray applicators are used to apply liquid adhesives and can be automated, whereas brushes are manipulated manually.

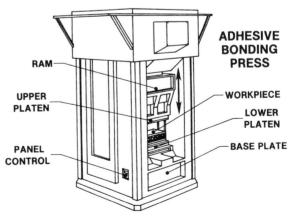

ADHESIVE BONDING PRESS

RAM — UPPER PLATEN — PANEL CONTROL — WORKPIECE — LOWER PLATEN — BASE PLATE

Typical Tools and Geometry Produced

Adhesives come in a wide variety of forms and can be applied in numerous ways. Typically, adhesives exist as liquids, pastes (semiliquids), and even solids, such as glue sticks and adhesive tapes. Applicators are designed according to the adhesive being used and the size of the area to be covered. Thin sheets or foils may be joined to heavy-gage workpieces. Component parts can be mounted on an assembly, and fabrics can be joined to solids.

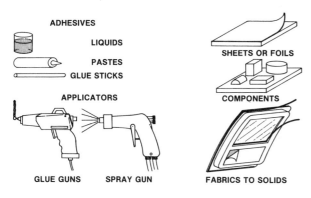

Geometrical Possibilities

The possibilities for adhesive bonding are virtually unlimited, with applications in such areas as automotive, construction, aerospace, and medical industries. With proper adhesives, most any combination of materials may be joined. Bonding pressure may range from 10 psi to 1000 psi, and resultant shear strength, depending on the adhesive used, may range from 900 psi to 13,000 psi.

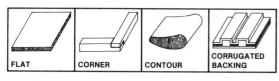

Workholding Methods

Workpieces are held by platens or shaped dies. A workpiece with critical dimensions may be held in a specially designed fixture or vise on platens. Otherwise, guide pins, stops, and alignment blocks may be used for bonding flat workpieces.

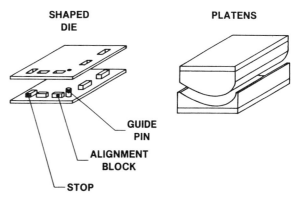

Tool Style

A spray gun may be used to apply adhesive to a workpiece for low to medium production requirements. Roll coaters are an automatic device for high production of both web (non-woven fabrics) and rigid panels. Manual application methods may be more appropriate if output requirements are low or if the workpiece bond is critical. A press may utilize flat platens or a shaped die for the bonding of contour parts. The die then becomes the heat source, if necessary, as well as the medium for pressure application. Steam and water are the preferred methods of heating a die in a large press because the heat is distributed more evenly.

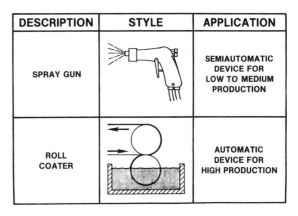

Adhesive Bonding ——————— **401**

DESCRIPTION	STYLE	APPLICATION
FLAT HONEYCOMB PANEL		FLAT SURFACES
FORMED DIE		CONTOUR OR IRREGULAR SHAPES

Medium	Application method
Rollers	* Manual
Brush	* Manual
	* Automated
Extrusion	* Caulking guns
Spray	* Air spray
	* Airless spray
Roll coater	* Bench
	* Pressure roll
	* Dip roll
Knife coater	* Automatic
Film or pellet	* Manual
	* Robotil

Toolholding Methods

Toolholding methods consist of a press ram holding the upper platen or shaped die and a die shoe that holds a lower platen, shaped die, or base plate. Selection of the tool depends on workpiece design, production requirements, and process economies.

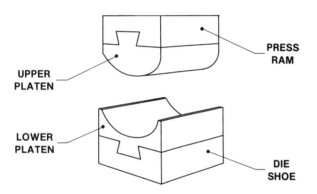

Method of Application

Adhesives may be applied with rollers, brushes, extrusion methods, sprayers, coaters, knife coaters, or film. The method of application depends on the workpiece material, adhesive material, workpiece and production requirements, and process economies.

Effects on Work Material Properties

Adhesive bonding may increase tensile, shear, compression, and impact strengths. Exposure to sunlight or heat may deteriorate the bonding joint, as will some solvents.

Work material properties	Effects of adhesive bonding
Mechanical	* May increase tensile, shear, compression, and impact strengths
Physical	* Exposure to sunlight and heat may deteriorate adherents
Chemical	* Solvents deteriorate adherents

Adhesive Materials

Six main classes of adhesives are shown. Thermosets are quick-set, high strength, solvent-resistant materials. Evaporative melts are both flexible and oil-resistant, and hot melts are flexible and water-resistant. Film adhesives are mainly used in the aerospace industry for honeycomb construction. Pressure sensitive materials are versatile and inexpensive, whereas delayed-tack materials provide a short-term bond.

Adhesive materials	Properties and applications
Thermoset	* Quick-set * High strength * Solvent-resistant
Evaporative	* Flexible * Oil-resistant
Hot melt	* Flexible * Water-resistant
Film	* Honeycomb construction
Pressure sensitive	* Versatile * Inexpensive
Delayed-tack	* Short-term bond

Factors Affecting Process Results

Bond quality depends on the type of work material and adhesive as well as surface preparation, application of adhesive, curing, and equipment used to position heat and clamp/pressurize the workpieces.

* Material and adhesive selection
* Control of the bonding process
  - Surface preparation
  - Adhesive application
  - Curing
* Equipment

Typical Workpiece Materials

Bondability ratings are based on the compatibility of adhesive and adherent material. Aluminum, steel, plastic, and wood products are rated in the good to excellent range. Brass adheres weakly. Copper based materials are difficult to bond due to rapid oxidation. Inadequate surface treatment or the use of adhesives designed for other materials can result in poor quality bonds.

MATERIAL	ADHESIVE BONDABILITY RATINGS			
	POOR	FAIR	GOOD	EXCEL
ALUMINUM				
BRASS				
STEEL				
PLASTIC COMPOSITES				
WOOD				

■ TYPICAL RANGE   ▨ FEASIBLE RANGE

Process Conditions

Process conditions are determined by tensile strength, crystallinity, and Young's modulus of elasticity. Differences in crystallinity and Young's modulus may seriously affect the success of bonding adhesion.

Adhesive	Tensile strength (psi)	Crystallinity	Young's modulus (70°F)
Rubber	3000	Low	Low
Epoxy	6000	Low	Medium to high
Phenolic	5000	Low	High
Nylons	8000	High	Medium to high

Curing

Curing changes the physical properties of an adhesive by producing a chemical reaction through the action of a catalyst, pressure, and/or heat.

Adhesive	Tensile strength (psi)	Typical curing	
		Pressure (psi)	Temperature (°F)
Rubber	3000	50 to 100	300
Epoxy	6000	Clamping	68 – 165
Phenolic	5000	250	350
Nylons	8000	10	350

Power Requirements

Power requirements for various forms of adhesive sprays are given. Heating the adhesive lowers the viscosity and decreases the atomization energy required. This allows a heavier film coating, and aids in correcting humidity-related problems resulting from uneven temperatures.

Spray method	Viscosity (cm/sec)	Temperature (F°)	Energy* (hp/gal/min)
Conventional air	25	70	9.3
Hydraulic spray	22	70	1.1
Hot spray	150 25	70 150	5.7
Hot airless	150 22	70 170	0.32

* Spray atomization energy required.
cm/sec = centimeters per second.

Cost Elements

Cost elements include the following:

* Setup time
* Adhesive application time
* Load/unload time
* Bonding time
* Curing time
* Material costs
* Direct labor rate
* Overhead rate
* Amortization of equipment and tooling

Time Calculation

The adhesive bonding setup of a honeycomb core sandwich shows the upper and lower face sheet with adhesive-impregnated scrim cloth sandwiching the expanded honeycomb core. The velocity of the platen and the time required for adhesive application and bonding are depicted. Operational elements include adhesive application time, load/unload time, platen velocity, bonding time, and curing time. Total time can be calculated by adding adhesive application time, load/unload time, bonding time, curing time, and platen travel distance divided by platen velocity.

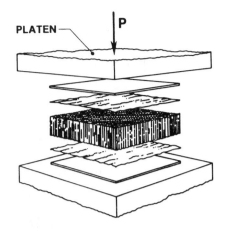

Adhesive application time (sec) = A
Load/unload time (sec) = L
Platen velocity (in./sec) = P
Bonding time (sec) = B
Curing time (hr) = C
Platen travel distance (in.) = D

$$\text{Total time} = A + L + B + C + \frac{D}{P}$$

Safety Factors

The following hazards should be taken into consideration:

* Personal
  - Irritating solutions
  - Vapors and dust inhalation
  - Burns
  - Noise
* Environmental
  - Material disposal
  - Air pollution
  - Fire
  - Groundwater pollution

Annealing

Annealing

Annealing is a heat-treating process for softening metals through a controlled heating cycle with temperatures above the recrystallization temperature. The result is a homogeneous microstructure.

Process Characteristics

* Is used as a softening process, for example, in cold-working process sequences or to create other desired properties
* Presents no problem with scaling or decarburization
* Provides sufficient ductility and machinability for subsequent machining processes or for intended end use

Process Schematic

The schematic shows a cross-section of an oven, detailing the various components. The workpiece is positioned in the work chamber, providing maximum exposure to the circulating heated air. Depending on the purpose, various levels of the annealing process exist, such as full anneal, process anneal, and short-cycle anneal. All apply temperatures above the recrystallization temperature.

ANNEALING

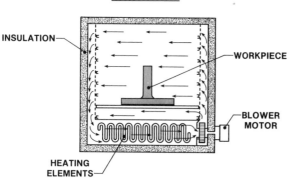

Workpiece Properties

Shown are typical workpieces. The cross-sectional enlargements show examples of work hardness prior to annealing and the softened structure after annealing. Full anneal creates an entirely new homogeneous and uniform structure with good dynamic properties.

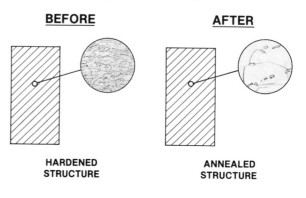

BEFORE — HARDENED STRUCTURE

AFTER — ANNEALED STRUCTURE

Setup and Equipment

Shown is a typical oven used for annealing. The workpieces remain in the oven for controlled cooling. Some alloys are removed and cooled in water, air, or oil.

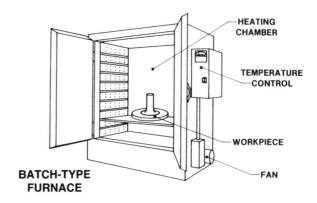

BATCH-TYPE FURNACE

Typical Equipment

Gas fired conveyor furnaces are often used for high volume process annealing. Car-bottom furnaces are also sometimes used if very large parts need to be annealed or high quantities need to be processed.

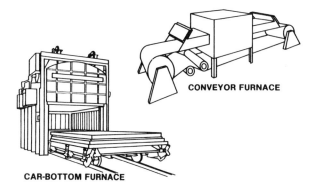

CONVEYOR FURNACE

CAR-BOTTOM FURNACE

Stress Relieving

Stress relieving is a heat-treating process that reduces or eliminates stresses that may be introduced from such operations as cold working, machining, welding, drawing, heading, and extrusion without resorting to complete annealing or normalizing and without greatly affecting other properties.

Process Characteristics

* Relieves stresses without causing significant microstructural changes
* Uses lower temperatures than tempering
* Is less costly than full annealing
* Minimizes warpage
* Helps prevent cracking

Process Schematic

Workpieces are usually heated to a temperature between 300°F and 500°F. A general rule is that workpieces are heated for 1 hour per inch of thickness. The workpiece is held at the desired temperature for a proper time period and then cooled slowly. Heating temperature and time depend on the workpiece material and size and the desired amount of stress relief.

STRESS RELIEVING

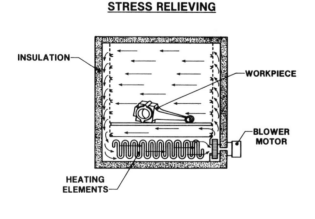

Workpiece Properties

Stress relieving processes are carried out without causing significant microstructural changes that would result in changed mechanical properties or decreased hardness. Removing residual stresses in weldments and castings may prevent distortion, fatigue, loss of strength, or failure.

* The workpiece is held at low temperature (300 to 500°F) for an hour or more, depending on its dimensions; it is then allowed to cool slowly
* No significant change in structure takes place, i.e., no change in properties

Setup and Equipment

These processes are carried out in continuous or batch-type furnaces in which the temperature can be controlled to within 5°F throughout the work zone. Tempering time may vary from less than one hour to several hours.

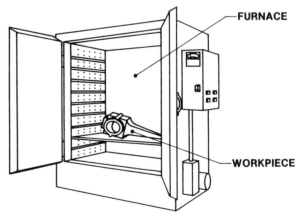

Typical Equipment

Shown are two batch-type furnaces in which the temperature and atmosphere may be closely controlled. In box furnaces, the temperature is controlled in large measure by recirculating the furnace atmosphere with a fan. The temperature in a molten salt bath is controlled by thermocouples in the furnace. Large workpieces may be stress relieved in pit-type, bell-type, or car-bottom furnaces.

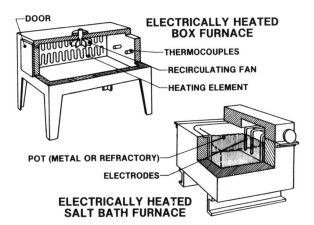

DOOR

ELECTRICALLY HEATED
BOX FURNACE

— THERMOCOUPLES

— RECIRCULATING FAN

— HEATING ELEMENT

POT (METAL OR REFRACTORY)

ELECTRODES

ELECTRICALLY HEATED
SALT BATH FURNACE

Tempering

Tempering is a heat-treating process that improves the toughness and ductility of a previously hardened workpiece. Tempering is accomplished by a controlled reheating of the workpiece to a temperature below its lower critical temperature.

Process Characteristics

* Improves ductility and toughness
* Reduces cracking
* Improves machinability
* Increases impact resistance

Process Schematic

Workpieces that have been subjected to severe machining or heat-treating processes may be tempered by heating them at a desired temperature for about 1 hour per inch of material thickness. Generally, the workpiece is heated to a temperature below the austenite range, removed from the furnace, and cooled in still air.

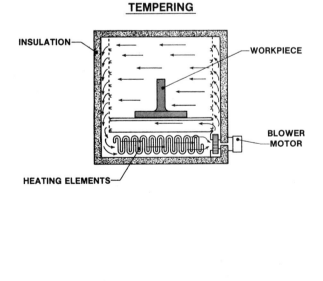

TEMPERING

Workpiece Properties

Shown in this figure are examples of workpiece microstructures before and after tempering. Prior to the tempering process, the steel workpiece primarily has a hard martensitic structure. After tempering, the workpiece is softer and tougher with a more ductile, tempered, martensitic structure.

BEFORE   **AFTER**

HARD STRUCTURE   **TEMPERED STRUCTURE**

Setup and Equipment

Tempering is done in a furnace in which the temperatures can be controlled to within 5°F throughout the heating chamber. Tempering time may range from less than 1 hour to several hours, depending on workpiece size and material.

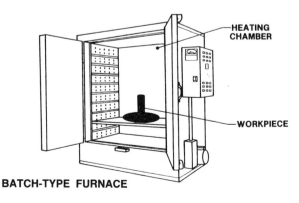

BATCH-TYPE FURNACE

Typical Equipment

Tempering may be done in a batch or continuous furnace. In either case, the workpiece may be heated in a liquid or gas medium. A continuous furnace is equipped with a continuous belt on which the workpiece passes through the furnace at a predetermined rate. The atmosphere is controlled by recirculating the heating medium with a fan. The temperature in the batch-type furnace is monitored and controlled by electrodes in the liquid.

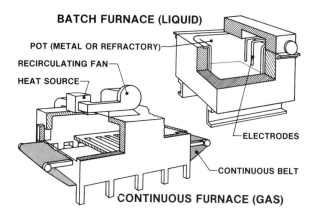

BATCH FURNACE (LIQUID)

POT (METAL OR REFRACTORY)

RECIRCULATING FAN

HEAT SOURCE

ELECTRODES

CONTINUOUS BELT

CONTINUOUS FURNACE (GAS)

Hardening

Age Hardening

Age hardening is a process in which specific nonferrous and ferrous alloys are heated, quenched, and then aged at a relatively low temperature above room temperature, to allow precipitation hardening to occur.

Process Characteristics

* Is primarily for hardening selected nonferrous and ferrous metals and alloys
* Involves workpiece heating, quenching, and aging
* Can increase workpiece strength up to 50%
* Does not affect workpiece ductility
* Causes microstructural changes in the workpiece
* Workpiece refrigeration can discontinue the hardening process

Process Schematic

A workpiece is first heated to a specific temperature, creating a solid solution, held for a required time interval, and rapidly quenched by spray or bath. Finally, the workpiece is reheated to a relatively low temperature and allowed to age. Aging usually requires 12 to 48 hours, depending on the alloy.

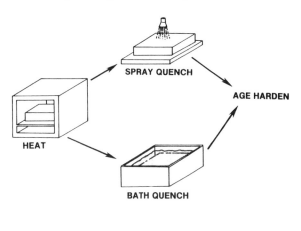

Workpiece Microstructure

Shown is an annealed aluminum alloy with particles of $CuAl_2$ present prior to heat treatment. During heat treatment and quenching, the $CuAl_2$ particles are dissolved in solid solution.

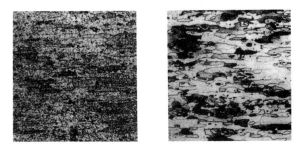

Setup and Equipment

Several of the heating media used include a continuous furnace, a box furnace, and a salt bath furnace. These are commonly used to heat workpieces to required temperatures prior to quenching.

HEATING MEDIUMS

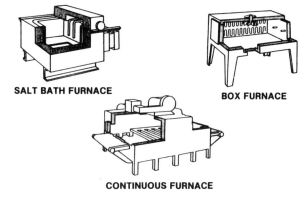

Typical Quench Media and Workpieces

Extruded sections and formed shapes are often age-hardened to increase strength and hardness. Parts are heated and formed while still soft and ductile, then quenched by oil, water, or brine. Selection of quench media depends on the workpiece material. Parts are then reheated to allow for age hardening.

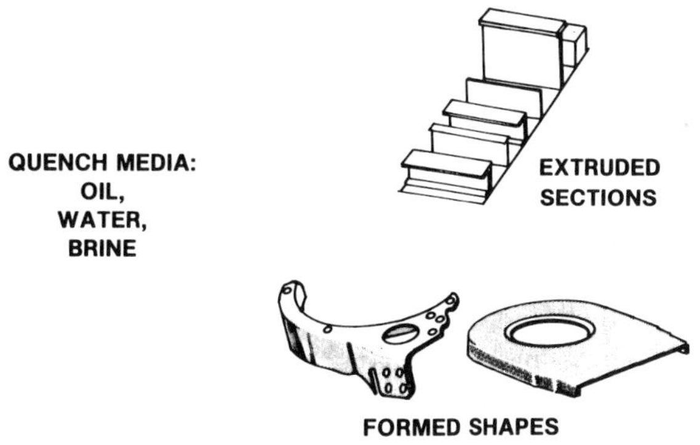

QUENCH MEDIA:
OIL,
WATER,
BRINE

EXTRUDED
SECTIONS

FORMED SHAPES

Carbonitriding

Carbonitriding is a diffusion-hardening process for ferrous metals in which both carbon and nitrogen are introduced into the workpiece surface by heating the workpiece while in contact with a gas or liquid containing carbon and nitrogen. Liquid carbonitriding is commonly called cyaniding.

Process Characteristics

* Surface hardens work materials
* Workpieces are heated in a carbon and nitrogen atmosphere or a cyanide bath
* Produces thin, very hard surface layers of uniform depth
* Little workpiece warpage results
* Case hardness is maintained at elevated operating temperatures
* Has a maximum case depth of 0.030 in.
* Produces better surface wear characteristics than carburizing

Process Schematic

In the carbonitriding process, the workpiece is heated in an appropriate atmosphere where the carbon and nitrogen diffuse into the surface of the workpiece. The workpiece may then be quenched in oil or water. The resulting workpiece has a thin, hard surface with a tough core.

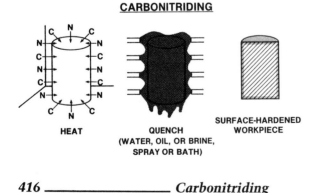

CARBONITRIDING

HEAT

QUENCH
(WATER, OIL, OR BRINE, SPRAY OR BATH)

SURFACE-HARDENED WORKPIECE

Workpiece Properties

The hardened cases produced by carbonitriding are generally shallower and more uniform than cases produced by carburizing. They also maintain hardness at relatively high operating temperatures. The maximum case depth in carbonitriding is about 0.030 in. The ferritic microstructure of the soft structure is converted to a hard martensitic structure after quenching.

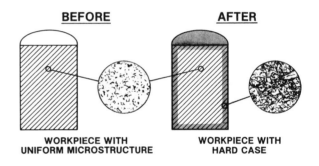

BEFORE

AFTER

WORKPIECE WITH UNIFORM MICROSTRUCTURE

WORKPIECE WITH HARD CASE

Setup and Equipment

Carbonitriding is done in either a gas or liquid atmosphere. The furnace used for gas carbonitriding is usually an enclosed box-type furnace

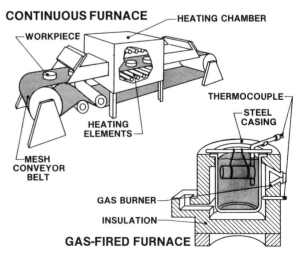

CONTINUOUS FURNACE — HEATING CHAMBER

WORKPIECE

HEATING ELEMENTS

MESH CONVEYOR BELT

THERMOCOUPLE

STEEL CASING

GAS BURNER

INSULATION

GAS-FIRED FURNACE

in which the atmosphere can be controlled. In liquid carbonitriding, the workpiece is usually immersed in a liquid cyanide salt bath.

Typical Hardening Agents

In gas carbonitriding, the carbon and nitrogen are derived from an atmosphere of propane, natural gas, or methane, with a small amount of ammonia. Ammonia supplies the nitrogen, and one of the other gases gives off carbon monoxide as a source of carbon. In liquid carbonitriding, the source of carbon and nitrogen is a sodium cyanide salt bath, which has a high level of nitrogen.

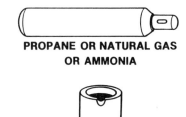

METHOD	HARDENING AGENTS
GAS	PROPANE OR NATURAL GAS OR AMMONIA
LIQUID	SODIUM CYANIDE LIQUID (CYANIDING)

Workholding Methods

Workholding devices vary somewhat with the carbonitriding method. In all cases, the principle objective of the workholding method is to expose the maximum workpiece surface area to the carbonitriding medium.

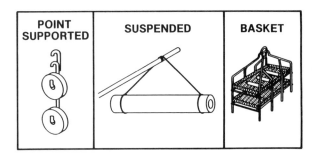

Effects on Work Material Properties

Carbonitriding changes mechanical properties, such as increasing surface hardness and wear resistance. Affected physical properties include possible volume changes. Carbonitriding has little effect on workpiece chemical properties.

Work material properties	Effects of carbonitriding
Mechanical	* Increased surface hardness * Increased wear resistance
Physical	* Some volume change may occur
Chemical	* Little effect

Geometrical Possibilities

In carbonitriding, as in all heat-treating processes, cracking and internal stresses may be avoided by designing parts with uniform shapes, uniform cross-sectional areas, and rounded corners. Depths of hardness achieved are 0.030 in. for gas carbonitrading and 0.015 in. for liquid carbonitriding.

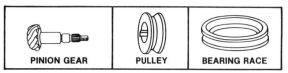

| PINION GEAR | PULLEY | BEARING RACE |

CARBONITRIDING CAPABILITIES
DEPTH OF HARDENING (.001 IN)

	1.0 15 25 30 35
GAS	
LIQUID	

▓▓ TYPICAL RANGE ▨ FEASIBLE RANGE

Dimensional Changes

There are many factors that affect the dimensional stability of a part. The cylindrical 1010 commercial steel part shown underwent a carbonitriding process at temperatures ranging from 1450°F to 1550°F. The process atmosphere consisted of an endothermic carrier gas at 250°F, an enriching gas, and ammonia. Under these conditions, the part experienced a shrinkage of 0.0056 in.

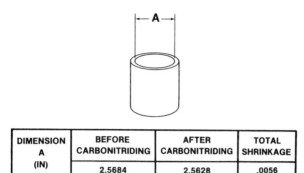

DIMENSION A (IN)	BEFORE CARBONITRIDING	AFTER CARBONITRIDING	TOTAL SHRINKAGE
	2.5684	2.5628	.0056

Gas	Source	Application
Carrier gas	Nitrogen, hydrogen, carbon monoxide, carbon dioxide, methane, oxygen	Prevents air infiltration
Enriching gas	Natural gas, propane, butane	Source of carbon
Ammonia	Anhydrous ammonia 99.9% pure	Source of nitrogen

Typical Workpiece Materials

Steels commonly carbonitrided include those in the 1000, 3000, 4000, and 8600 series with carbon contents up to about 0.25%. Also many steels in these same series with a range of carbon content of 0.35% to 0.50% are carbonitrided to case depths up to about 0.012 in. This provides a combination of reasonably tough core material and a hard, long-wearing surface.

Factors Affecting Process Results

The amount of workpiece distortion depends on the workpiece shape, the method of stacking, type of fixture, quenching temperature, and the chemical composition of the steel.

Temperature Requirements

Shown are typical and possible operating temperature ranges for carbonitriding. The rate of carbon diffusion into the workpiece is proportional to the temperature. At the lower temperatures, carbon diffusion occurs too slowly to be practical in terms of productivity, and at higher temperatures, the carbonitriding furnace may experience accelerated deterioration.

PROCESS TEMPERATURES

Carbonitriding	
Typical temperatures	Feasible temperatures
1375–1600°F	1350–1650°F

MATERIAL AISI NO.	CARBONITRIDING CAPABILITY RATINGS			
	POOR	FAIR	GOOD	EXCEL
1020		▨▨		
1022		▨▨		
1024		▨▨		
3310			▨▨	
4037			▨▨	
4320			▨▨	
8620			▨▨	

▥ TYPICAL RANGE   ▨ FEASIBLE RANGE

Hardening Media

The atmosphere used in carbonitriding generally is composed of a mixture of carrier gas, enriching gas, and ammonia. The carrier gas is supplied to the furnace in sufficient volume to build up a positive pressure in the chamber to prevent the infiltration of air. The enriching gas is the primary source of carbon that is added to the steel surface. The nitrogen absorbed during the process increases the hardenability of the workpiece.

Process Conditions

Gas carbonitriding methods are typically capable of producing case depths of from 0.001 in. to 0.030 in. Case depths produced by carbonitriding are generally more uniform than those produced in carburizing.

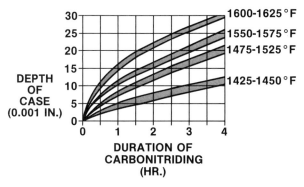

*FOR 1020 STEEL WITH A GAS MEDIUM HARDENING

Quenching Media

Selection of quench media depends primarily on the hardenability of the steel, surface and core hardness requirements, and the amount of allowable distortion.

Quenching media include the following:

* Water
* Brine
* Caustic solutions
* Oil
* Molten salts

Cost Elements

Cost elements include the following:

* Setup time
* Process hardening time
* Direct labor rate
* Overhead rate (maintenance, power, and insurance)
* Material costs
* Batch size
* Amortization of equipment and tooling costs

Productivity Tip

* Advantages of carbonitriding
  - Better hardenability than carburizing
  - Less expensive than carburizing
  - Resistant to softening during tempering
  - Lower temperatures than carburizing
  - Less time than carburizing
* Disadvantages of carbonitriding
  - Limited to case depths of 0.045 in. or less
  - Poor impact strength
  - More distortion than carburizing
  - Difficult to plate

Time Calculation

To calculate the time for the hardening process, first, the hardening time is calculated by dividing the case depth by the hardening rate. The process time, or total time, is then calculated by adding the setup time, the hardening time, and the cooling time.

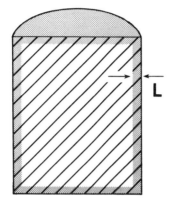

Hardening time (hr) = H
Cooling time (hr) = C
Desired case depth (in.) = L
Hardening rate (in./hr) = R
Setup time = T_s

$$\text{Hardening time, } H = \frac{L}{R}$$

$$\text{Total time} = T_s + H + C$$

Safety Factors

The following risks should be taken into consideration:

* Personal
  - Toxic gases
  - Chemical and thermal burns
  - Explosion
* Environmental
  - Ammonia explosions
  - Chemical leakage
  - Fire

Carburizing

Carburizing is a method of increasing the carbon content of the surface of low-carbon steel through absorption and diffusion. It is done to prepare the steel for quench hardening. Steel is heated above its upper critical temperature in the presence of carbon to allow it to absorb into the metal to the desired depth.

Process Characteristics

* Is applied to low-carbon workpieces
* Workpieces are in contact with a high-carbon gas, liquid, or solid
* Produces a hard workpiece surface
* Workpiece cores largely retain their toughness and ductility
* Produces case hardness depths of up to 0.25 in.

Process Schematic

The workpiece is first heated for several hours during which the carbon is absorbed, producing a high-carbon case. Next, the workpiece is quenched in an appropriate quenching medium (usually oil or water) so that the surface layer of the metal is changed to martensite. A tempering or stress relieving process may follow to reduce residual stresses in the hardened case.

Workpiece Properties

In this process, the surface of the workpiece has been changed to a largely martensitic structure, leaving an essentially unchanged ductile core.

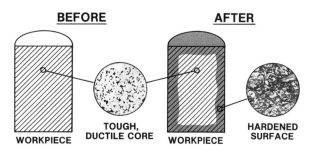

BEFORE   AFTER

WORKPIECE   TOUGH, DUCTILE CORE   WORKPIECE   HARDENED SURFACE

Setup and Equipment

The source of carbon can be in the form of gas, liquid, or solid. (Solid carburizing is more commonly known as pack carburizing.) In each case, a special furnace is used to heat the workpiece and the carbon-containing medium to the transformation temperature. Heat is maintained to a desired depth long enough for the carbon to be diffused into the workpiece. Parts to be pack hardened are held in special airtight containers and surrounded with carbonaceous (carbon-rich) material.

CARBURIZING

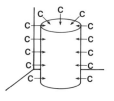

HEAT

QUENCH
(WATER OR OIL,
SPRAY OR BATH)

SURFACE-HARDENED
WORKPIECE

GAS CARBURIZING

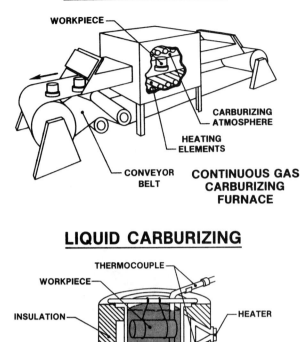

WORKPIECE

CARBURIZING
ATMOSPHERE

HEATING
ELEMENTS

CONVEYOR
BELT

**CONTINUOUS GAS
CARBURIZING
FURNACE**

METHOD	HARDENING AGENTS
GAS	
LIQUID	
SOLID (PACK)	

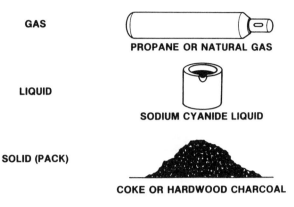

PROPANE OR NATURAL GAS

SODIUM CYANIDE LIQUID

COKE OR HARDWOOD CHARCOAL

LIQUID CARBURIZING

THERMOCOUPLE

WORKPIECE

INSULATION

HEATER

LIQUID BATH

STEEL CASING

**BATCH LIQUID
FURNACE**

Typical Hardening Agents

The principal source of carbon in carburizing is carbon monoxide (CO) gas. In gas carburizing, the CO is given off by propane or natural gas. In liquid carburizing, the CO is derived from a molten salt composed mainly of sodium cyanide (NaCN) and barium chloride (BaCl). In pack carburizing, carbon monoxide is given off by coke or hardwood charcoal.

Workholding Methods

Various methods are used to insure maximum contact between the workpiece surface and the carbon. In gas and liquid carburizing, the workpieces are often supported in mesh baskets or suspended by wire. In pack carburizing, the workpiece and carbon are enclosed in a container to insure that contact is maintained over as much surface area as possible. Pack carburizing containers are usually made of carbon steel coated with aluminum or heat-resisting nickel–chromium alloy and sealed at all openings with fire clay.

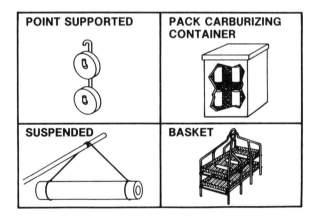

POINT SUPPORTED	PACK CARBURIZING CONTAINER
SUSPENDED	BASKET

Effects on Work Material Properties

The mechanical properties experience an increase in surface hardness, wear resistance, and fatigue and tensile strengths. Physical property changes in a workpiece include grain growth, change in volume, and change in coefficient of heat. Carburizing causes an increase in surface carbon content to 1% carbon or more.

Work material properties	Effects of carburizing
Mechanical	* Increased surface hardness * Increased wear resistance * Increased fatigue/tensile strengths
Physical	* Grain growth may occur * Change in volume may occur
Chemical	* Increased surface carbon content

Geometrical Possibilities

The geometric shapes that may be carburized are limitless. However, as in every process, there are design considerations. Nonuniform sections should be avoided because the cooling difference between the cross-sections may cause excessive stresses in the material and result in breakage. Carburizing methods discussed generally produce compressive surface stresses that improve fatigue life of parts that undergo cyclic loading. Typical depths are 0.007 in. to 0.06 in. for gas carburizing.

TIE ROD	GEAR	MANDREL

CARBURIZING CAPABILITIES

DEPTH OF HARDENING

	0	0.01	0.1	1
PACK (IN) (SOLID)				
GAS (IN)				
LIQUID (IN)				

▓ TYPICAL RANGE   ▨ FEASIBLE RANGE

Dimensional Changes

All parts usually undergo some dimensional changes as a result of carburizing and hardening. These changes depend the workpiece size, shape, and composition, as well as the method of carburizing. For example, the crankshaft shown was normalized for 1 hour at 1750°F, tempered for 1 hour at 1150°F, carburized at 1700°F to a case depth of 0.45 in. to 0.55 in., air cooled, reheated to 1450°F, quenched in salt at 375°F, air cooled, and tempered at 325°F. The result of this process was 0.0011 in. of shrinkage, which is a very small change for heat-treating operations.

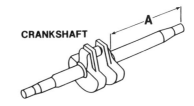

CRANKSHAFT

TEST RESULTS OF 25 IDENTICAL PARTS

DIMENSION A (IN)	BEFORE CARBURIZING	AFTER CARBURIZING	TOTAL SHRINKAGE
	5.4296	5.4285	.0011

Typical Workpiece Materials

Typical materials to be carburized are low-carbon and alloy steels with a carbon content up to 0.30%. In this figure, the most commonly carburized metals are rated according to their carburizing capability. The workpiece surface must be free from contaminates, such as oil oxides and alkaline solutions, which prevent or impede the diffusion of carbon into the workpiece surface.

MATERIAL AISI NO.	CARBURIZING CAPABILITY RATINGS			
	POOR	FAIR	GOOD	EXCEL
1020		▨▨		
1022		▨▨		
1024		▨▨		
3310		▨▨		
4037			▨▨	
4320			▨▨	
8620			▨▨	

Carburizing Temperatures

Shown are typical and feasible operating temperature ranges for carburizing operations. The rate of carbon diffusion into the workpiece is proportional to the temperature. At the lower temperatures shown, carbon diffusion occurs too slowly to be practical in terms of productivity. At higher temperatures, carburizing furnaces deteriorate at an increased rate, and grain growth in the workpiece may result. The costs that result from equipment deterioration often offset productivity gains because of increased carburizing temperatures.

Carburizing method	Typical temperatures	Feasible temperatures
Pack (solid)	1650–1700°F	1550–1750°F
Gas	1650–1700°F	1600–1700°F
Liquid	1550–1650°F	1500–1700°F

Lower temperatures require extended processing time.

Carburizing Media

In gas carburizing, the carbon is obtained from the dissociation of a carburizing gas. In pack carburizing, the carbon is obtained from CO that is released from partially oxidized charcoal. In liquid carburizing (with NaCN or BaCl), the partial decomposition of the cyanide solution releases carbon atoms that rapidly diffuse into the workpiece. Liquid carburizing is often used with small- and medium-sized parts.

Method	Source	Application
Pack	Carbon monoxide (CO)	Large parts, small lots
Gas	Carbon monoxide (CO), natural gas	Small to medium parts, small lots
Liquid	Sodium cyanide, barium chloride	Small and medium parts

Process Conditions

The depth of hardness obtainable with each of the processes varies somewhat because of differences in process conditions and carburizing medium. Shown are case depths for pack carburizing at various temperatures and for various time periods. With each of the carburizing methods, it is possible to achieve a surface carbon content of 0.9% to 1.2%. Before carburizing, a typical steel workpiece has a carbon content of 0.2% to 0.3% and a surface hardness of RC 30–45.

Factors Affecting Process Results

The amount of distortion and warpage of workpieces depends on several factors. Tolerance and surface finish depend upon the following:

* Shape and design of part
* Method of stacking and fixturing
* Growth of surface during carburization
* Quenching temperature
* Chemical composition of steel

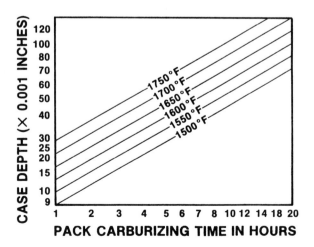

Quenching Media

Most conventional quenching media include water, brine, caustic solution, oil, and molten salts. However, the suitability of each medium must be related to specific parts and will depend primarily on the hardenability of the steel, surface and core hardness requirements, and the amount of allowable distortion.

Medium	Carburizing methods	Application
Water and brine	Liquid or gas	Immersion
Oil	Liquid or gas	Immersion
Salt bath	Liquid or gas	Immersion

Cost Elements

Cost elements include the following:

* Setup time
* Tooling/fixturing costs
* Direct labor rate
* Overhead rate (maintenance, power, and insurance)
* Material costs
* Amortization of equipment and tooling costs

Productivity Tip

Generally, pack carburizing equipment can accommodate larger workpieces than liquid or gas carburizing equipment, but liquid or gas carburizing methods are faster and lend themselves to mechanized material handling. Also, the advantages of carburizing over carbonitriding are greater case depth (case depths greater than 0.30 in. are possible), less distortion, and better impact strength. Disadvantages include added expense, higher working temperatures, and increased time.

Choice of equipment and procedure depends largely on the following:

* Size of workpiece
  – Gas carburizing—large size range
  – Liquid carburizing—small and medium parts
  – Pack carburizing—large parts and individual processing of small lots of parts
* Equipment capacity
* Required case depth

Time Calculation

E.F. Harris has developed a formula for predicting hardening time as a function of desired case depth, hardening rate, and quench time. To calculate the total hardening time, the case depth is divided by the hardening rate and added to the quenching time, which was divided by 3600 to obtain a time in units of hours.

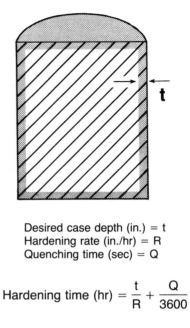

Desired case depth (in.) = t
Hardening rate (in./hr) = R
Quenching time (sec) = Q

$$\text{Hardening time (hr)} = \frac{t}{R} + \frac{Q}{3600}$$

Safety Factors

The following risks should be taken into consideration:

* Personal
  - Toxic gases
  - Fire
  - Explosion
* Environmental
  - Ammonia explosions
  - Chemical leakage
  - Fire

Induction Hardening

Induction hardening is a surface heat-treating process. A ferrous metal workpiece (from medium to high-carbon content) is heated by means of a high frequency electromagnetic field. The heated workpiece is then quickly quenched to produce a hardened structure. This process is generally used to harden the surface only, but through-hardening is possible, depending on workpiece thickness and equipment limitations.

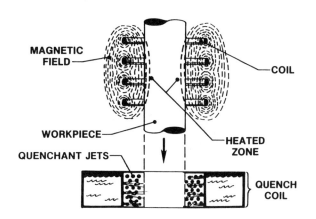

Process Characteristics

* Produces surface hardening on ferrous workpieces. (Carbon content from 0.40% to 1.00%)
* Is rapid and easily controlled
* Heating results from high frequency, induced eddy currents
* Workpieces may be rotated to provide uniform heating/quenching
* Can be used for through-hardening
* Produces little distortion or scaling
* May selectively harden workpiece

Process Schematic

A ferrous metal part is placed inside a coil and subjected to a high frequency alternating current. A rapidly changing magnetic field induces eddy currents in the workpiece surface, thus heating the workpiece by electrical resistance. Heating rates are quite rapid, and efficiency is high. The part is often rotated to insure uniform heating and immediately quenched upon entering a quench coil.

Workpiece Geometry

With induction hardening, selected surfaces of a part may be hardened for wear resistance, and others can remain soft for subsequent machining. The process also can be customized for special shapes, retaining close control and a possibility of complete automation of the hardening and tempering.

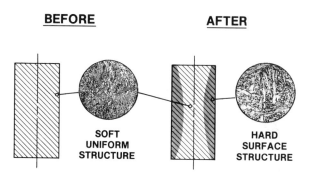

Setup and Equipment

An induction machine normally consists of a generator, induction coil, quench ring, and workpiece holder. The generator provides a high frequency alternating current to the induction coil. Most generators do not exceed 500,000 Hz. For thin cases, high frequencies are used. For intermediate and thick cases, low frequencies give better results.

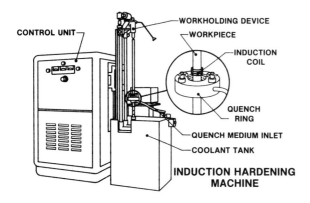

INDUCTION HARDENING MACHINE

Typical Tools and Geometry Produced

An internal coil is used to harden the inner surfaces of workpieces. External and pancake coils harden external surfaces and flat surfaces. Heating coils must be custom designed for a particular type of part.

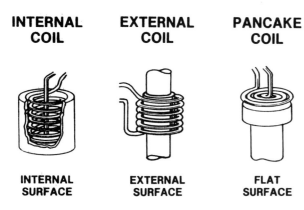

Workholding Methods

Workholding methods include customized horizontal and vertical fixtures and depend on workpiece shape, orientation, and geometry. These fixtures support and sometimes rotate a workpiece while it is being moved from a hardening station to a quenching station.

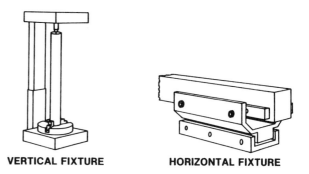

VERTICAL FIXTURE   HORIZONTAL FIXTURE

Effects on Work Material Properties

Mechanical changes include increased surface hardness, wear resistance, and fatigue strength. Compressive stresses are usually present on a hardened workpiece. Microcracking may occur if process parameters are not correct. Physical properties are affected by formation of a martensitic surface layer upon quenching. A chemical effect is the formation of an oxide layer on the workpiece surface.

Work material	Effect of induction hardening
Mechanical	* Increased surface hardness * Wear resistance and fatigue strength * Surface compressive stresses
Physical	* Martensitic surface layer formed upon quenching
Chemical	* Oxide surface layer formed

Geometrical Possibilities

Many special purpose coils are designed for the workpiece shape and required hardening patterns. Water flowing through the coils keeps them from overheating and permits high currents in relatively small coils. This process is used for the selective hardening of gear teeth, splines, crankshafts, and connecting rods. The typical hardening depths range from 0.008 in. to 0.25 in. Feasible depths range from 0.005 in. to 0.500 in.

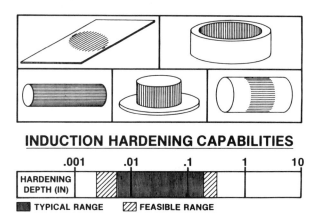

INDUCTION HARDENING CAPABILITIES

	.001	.01	.1	1	10
HARDENING DEPTH (IN)					

■ TYPICAL RANGE   ▨ FEASIBLE RANGE

Dimensional Changes

Dimensional changes are minimal with induction hardening and are directly related to the depth of hardening, workpiece material, and uniformity of heating and quenching.

Typical Workpiece Materials

Induction hardening capability ratings for five ferrous materials typically used in this process are shown.

MATERIAL	INDUCTION HARDENING CAPABILITY RATINGS			
	POOR	FAIR	GOOD	EXCEL
CARBON STEEL		▨▨		
ALLOY STEEL		▨	▨	
TOOL STEEL			▨	▨
DUCTILE IRON		▨	▨	
CAST IRON		▨	▨	

▥ TYPICAL RANGE   ▨ FEASIBLE RANGE

Coil Design

The shape of the coils is determined by the dimensions and configuration of the part to be heated, the desired heating pattern, the number of parts to be heated at one time, and the available power.

STYLE	APPLICATION
	EXTERNAL HEATING
	INTERNAL HEATING
	SPOT HEATING

Factors Affecting Process Results

Depth of workpiece hardening is mainly governed by the operating frequency (lower frequencies are generally more suitable for large parts where through-hardening or a greater hardness depth is required); size, number, and configuration of induction coils; time that the workpiece is exposed to induction current (heating time); power density, measured in kilowatts per square inch of workpiece exposed to induction coils; and degree of inductive coupling between the coil and workpiece.

Depth of hardness for induction hardening depends upon the following:

* Operating frequency range 10 kHz to 2 MHz for 0.010 in. to 0.060 in. depth; 1 kHz to 10 kHz for 0.060 in. to 0.250 in. depth
* Size, number, and configuration of induction coils
* Heating time (5 sec to 254 sec)
* Power density (kW/in.^{2} of workpiece surface exposed to the inductor)
* Inductive coupling between the coil and workpiece

Induction Hardening —————————— *429*

Power Requirements

Shown is how induction frequency and power input relate to depth of hardening. The power density is measured in kilowatts per square inch of workpiece surface. Current density is influenced by generator capacity, optimum metallurgical results, and desired production output.

Frequency (kHz)	Depth of hardening (in.)	Power density kW/in.^{2}		
		Low*	Opt.†	High‡
500	0.015–0.045	7	10	12
	0.045–0.090	3	5	8
10	0.060–0.090	8	10	16
	0.090–0.120	5	10	15
	0.120–0.160	5	10	14
3	0.090–0.120	10	15	17
	0.120–0.160	5	14	16
	0.160–0.200	5	10	15
1	0.200–0.280	5	10	122
	0.280–0.360	5	10	12

* Low kilowatt input may be used when generator capacity is limited.
† Kilowatts for best metallurgical results.
‡ Kilowatts for high production when generator capacity is available.

Process Conditions

The correlation between carbon content, temperature, and workpiece hardness during induction hardening is shown. Time, or frequency duration, is also a variable in determining resultant hardness values and depth. In general, a low-carbon content in ferrous workpieces requires high induction temperatures, producing low hardness values.

Carbon content of metal (%)	Hardening temperature (°F)	Quench medium	Post-quench surface hardness (RC)*
0.30	1650 to 1780	Water	50
0.35	1625 to 1650		52
0.40	1600 to 1650		55
0.45	1600 to 1650		58
0.50	1575 to 1600		60
0.60	1560 to 1600		64
0.60	1560 to 1600	Oil	62

* Rockwell C scale.

Quenching Media

Quenching media include water (most commonly used), soluble oil, air, and brine. Spray quenching is often used for cylindrical parts. Submerged quenching is used for cylindrical or irregular parts. Air quenching with still air is generally used with irregularly shaped parts.

Method	Medium	Applications
Spray quench	Water	Cylindrical parts
Submerged quench	Soluble oil Air Brine	Cylindrical or irregular parts
Air quenching	Still air	Irregular parts

Cost Elements

Cost elements include the following:

* Setup time
* Load/unload time
* Heating/quenching time
* Direct labor rate
* Overhead rate
* Amortization of equipment and tooling

Productivity Tip

Increased heating rates may be achieved by developing a close electromagnetic coupling between the workpiece and generator. This close coupling is done by designing the coil with minimal air gap, appropriate spaces, and turns, and by matching coil inductance and resistance with that of the workpiece.

Other Process Considerations

Generally, the distance between a workpiece and coils is from 1/16 in. to 1/8 in. Coil diameter is usually from 1/8 in. to 1/4 in. Distance between coils is from 1/16 in. to 3/32 in. Heating patterns may vary greatly by changing these distances.

Time Calculation

Time calculations depend on whether the workpiece is held stationary to harden a small section or whether the workpiece is traversed through a heating and quenching coil to harden a long section.

DESIGN RULES FOR INDUCTOR COILS

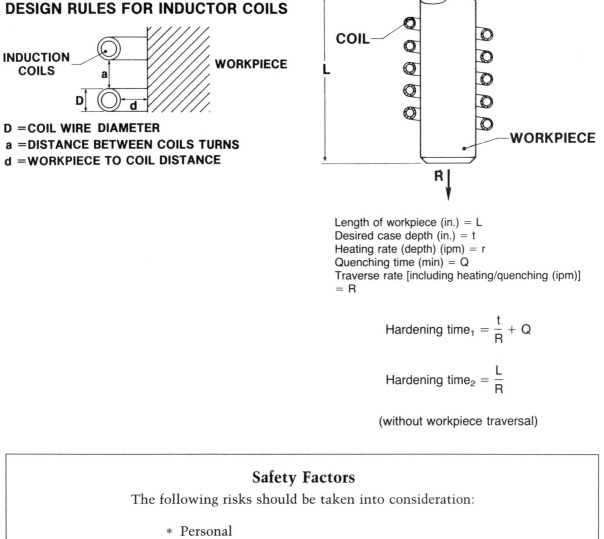

D = COIL WIRE DIAMETER
a = DISTANCE BETWEEN COILS TURNS
d = WORKPIECE TO COIL DISTANCE

Length of workpiece (in.) = L
Desired case depth (in.) = t
Heating rate (depth) (ipm) = r
Quenching time (min) = Q
Traverse rate [including heating/quenching (ipm)] = R

$$\text{Hardening time}_1 = \frac{t}{R} + Q$$

$$\text{Hardening time}_2 = \frac{L}{R}$$

(without workpiece traversal)

Safety Factors

The following risks should be taken into consideration:

* Personal
  - High temperature of workpiece
  - High voltage
  - High current
* Environmental
  - High frequency radiation, which may cause electrical interference

Quench Hardening ⎯⎯⎯⎯⎯⎯⎯⎯⎯⎯⎯⎯

Quench hardening is a process in which steel and cast iron alloys are heated above a certain critical temperature and rapidly cooled to produce a hardened structure. Either surface hardening or through-hardening can result, depending on the cooling rate.

Process Characteristics

* Can be used on ferrous metals and alloys
* Involves the heating and then the quenching of the workpiece
* Is often followed by tempering to reduce brittleness
* Causes microstructural changes in the workpiece
* Requires close control of temperature during heating and quenching

Process Schematic

A workpiece is first heated to a specified temperature, usually between 1500°F and 1650°F, quenched with water, oil sprays, bath, or air for some tool steels, then often tempered to reduce brittleness.

Workpiece Microstructure

Before hardening, a pearlitic grain structure is uniform and laminar. After hardening, martensite forms as a fine, needlelike grain structure, described "like a pile of straw," because of the resemblance.

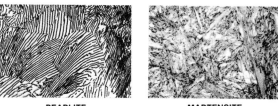

BEFORE HEATING   AFTER QUENCHING

PEARLITE   MARTENSITE

Setup and Equipment

Shown are several types of heat-treatment furnaces including a continuous furnace, a box furnace, and a salt bath furnace.

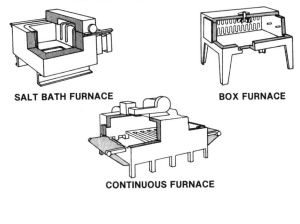

HEAT-TREATING FURNACE

SALT BATH FURNACE   BOX FURNACE

CONTINUOUS FURNACE

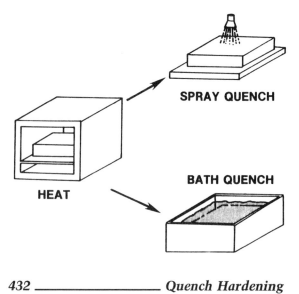

SPRAY QUENCH

HEAT

BATH QUENCH

Typical Quench Media and Workpieces

Typical workpieces include gears, wear blocks, and various shafts. In order of increasing severity of quench, quenching media include oil, water, and brine.

QUENCH MEDIA:
OIL,
WATER,
BRINE

GEAR

WEAR
BLOCK

SHAFT

10.
Surface Preparation

Chemical Degreasing

Chemical degreasing is a cleaning process that uses chemicals to remove foreign substances from the surface of the workpiece. The workpiece is usually submerged in a chemical bath where the chemical attacks the surface of the workpiece. If a workpiece is left in the chemical long enough, its surface will be slightly etched. This process is used for surface preparation.

Process Characteristics

* Removes virtually any foreign materials and contaminants
* Accommodates most workpiece materials and contaminants through the use of different chemicals
* Causes an etching effect on the workpiece surface depending upon the type of chemical and amount of time in contact with the chemical
* Can be used as a cleaning, preparing, or pickling process

Process Schematic

In chemical degreasing, the workpiece is cleaned by being immersed in a chemical solution until surface contaminants are dissolved. Spraying or wiping can also be done for spot cleaning and for degreasing large workpieces. As the chemicals attack the surface of the workpiece, gases are often liberated with some bubbling action, depending on the materials involved. Strong caustic solutions are used for aluminum and magnesium, whereas acidic solutions are used for ferrous, copper, and nickel alloys.

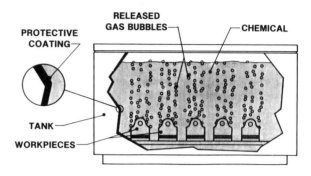

CHEMICAL DEGREASING

Workpiece Geometry

The size of the workpiece to be cleaned is limited only by the size of the equipment. There are essentially no restrictions on the shape of the workpiece due to the ability of the liquid to reach all areas of complex workpiece surfaces. It is sometimes necessary to circulate the chemicals and to rotate workpieces so that the chemical will reach deep recesses and holes. Careful temperature control is also important because high temperatures may cause violent chemical reactions.

BEFORE   **AFTER**

Setup and Equipment

The setup and equipment used in this process are similar to the setup and equipment used in other degreasing processes. Because chemicals used in this process can attack the surfaces of the equipment as well as the workpiece surface, thin rubber or plastic coatings are often used to line the tank. In some cases, materials, such as stainless steels, can be used for the tank, depending on the type of chemicals used and the ability of the materials to withstand the chemicals.

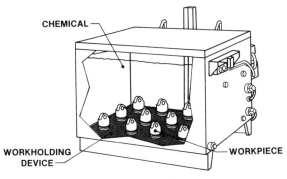

CHEMICAL DEGREASING MACHINE

Typical Tools and Application Methods

A wide variety of chemicals are available for specific applications on different types of workpiece materials. It is important to know what effects the chemicals will have on the workpiece before using them. The application of the chemical can be carried out in a number of ways. Wiping and immersion techniques are often used.

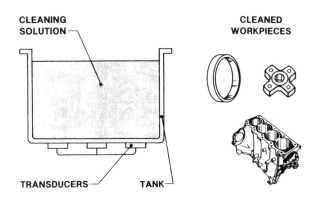

Chemical Degreasing —————— 437

Chromate Conversion _____

Chromate conversion is a process that chemically alters the surface of a workpiece. The process is performed in a solution of chromic acid. The workpiece is immersed in the acidic solution, which decomposes some of the surface of the workpiece and creates a protective film of complex chromium compounds.

Process Characteristics

* Is performed in a tank of chromic acid, which is usually a proprietary chrome salt solution of alodine or iridite
* Changes surface chemical composition
* Typically utilizes aluminum, cadmium, zinc, and zinc base as work materials
* Is a relatively inexpensive way to prepare metal
* Is usually associated with other cleaning and rinsing steps

Process Schematic

In chromate conversion, the workpiece is suspended in the chromic acid. The acid then attacks the workpiece surface, causing a decomposition of some of the surface metal, and a protective film is formed.

CHROMATE CONVERSION

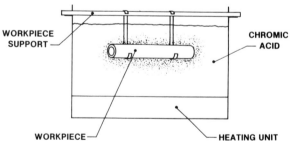

Workpiece Geometry

This process is generally used for surface preparation and does not alter workpiece geometry.

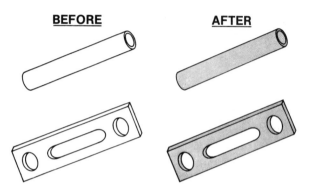

BEFORE   AFTER

Setup and Equipment

Chromate conversion equipment consists of a tank (to contain the acid) and a workpiece holder. A heating source is optional but does accelerate the conversion coating process.

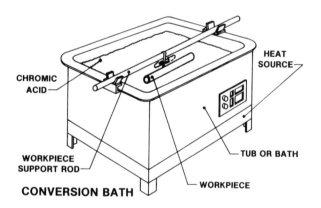

CONVERSION BATH

Typical Tools and Geometry Produced

This process creates a corrosion-resistant surface for nonferrous tools and other components.

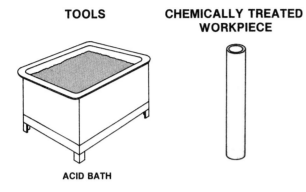

TOOLS

CHEMICALLY TREATED WORKPIECE

ACID BATH

Geometrical Possibilities

This process has many geometrical possibilities such as gears, sheet metal, and extruded parts. The thickness of the converted layer typically ranges from close to 0 in. to around 0.0005 in. Feasible film thicknesses may be as thick as 0.005 in.

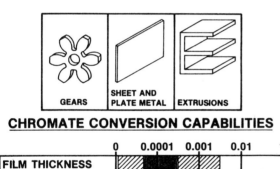

| GEARS | SHEET AND PLATE METAL | EXTRUSIONS |

CHROMATE CONVERSION CAPABILITIES

| FILM THICKNESS | 0 | 0.0001 | 0.001 | 0.01 | 1 |

■ TYPICAL RANGE   ▨ FEASIBLE RANGE

Tolerances and Surface Finish

Chromate conversion tolerances typically run around ±0.0001 in., and the finish is between 9 and 40 microinches.

TOLERANCES		
COATING THICKNESS	TYPICAL	FEASIBLE
	±.0001	±.00005

SURFACE FINISH					
PROCESS	MICROINCHES (A.A.)				
CHROMATE CONVERSION	4	8	16	32	63

■ TYPICAL RANGE   ▨ FEASIBLE RANGE

Workholding Methods

The workholding methods range from small wire baskets for nuts and bolts to large hangers for holding large workpieces. Usually, each workpiece has a special holding device based upon its geometry.

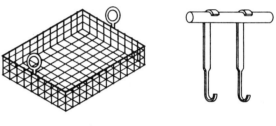

WIRE BASKETS   **HANGERS**

Effects on Work Material Properties

Effects on the work material are improved bonding with organic finishes, such as paint; improved hardness; decreased conductivity; better heat resistance; and superior corrosion resistance.

Work material properties	Effects of chromate conversion
Mechanical	* Improved bonding with organic finishes * Surface hardening
Physical	* Decreases conductivity * Heat resistance improved
Chemical	* Changed surface composition * Increased corrosion resistance

Typical Workpiece Materials

The coatability ratings for aluminum, zinc, and cadmium range from good to excellent. Mild steel has a poor to fair rating, and stainless steel has a poor rating.

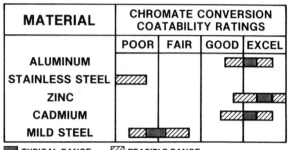

MATERIAL	CHROMATE CONVERSION COATABILITY RATINGS			
	POOR	FAIR	GOOD	EXCEL
ALUMINUM			▨▓▨	
STAINLESS STEEL	▨			
ZINC			▨▓▨	
CADMIUM			▨▓▨	
MILD STEEL		▨▓▨		

▓ TYPICAL RANGE   ▨ FEASIBLE RANGE

Coating Materials

The application of an acid depends on its pH level. An acid with a pH of 1.5 is used for chemical polishing and for creating a surface of medium corrosion resistance. Acids with a pH of 1.0 to 3.5 are used for providing medium to strong corrosion resistance and for color changes. Acids with a pH of 2.5 to 6.0 are used for providing maximum corrosion resistance.

Acids with a pH level	Applications
0.0 to 1.5	* Chemical polishing * Medium corrosion resistance
1.0 to 3.5	* Medium to strong corrosion resistance * Definite color changes
2.5 to 6.0	* Maximum corrosion resistance

Factors Affecting Process Results

Tolerance and surface finish depend upon the following:

* Acid type
* Acid concentration
* Workpiece material
* Workpiece geometry

* Submersion time
* Acid temperature
* Pre- and post-treatments of the workpiece (surface cleanliness)

Bath Composition

This process requires acids in liquid form. There are several kinds of acid baths used. The acid solution used depends on the type of work material.

Solution composition	Oz. per gallon
$Na_2Cr_2O_7 \cdot 2H_2O$	0.99
NaF	0.132
$K_3Fe(CN)_6$	0.66

Solution temperature = 60°F to 130°F.
Treatment time = 5 sec to 8 min.
pH = 1.2 to 2.2.

Process Conditions

Process conditions depend on the pH level of the acid and the type of workpiece material. Different pH levels are used for different types of metals. Immersion times range from 5 sec to 300 sec.

Immersion time (sec)	5 to 20	15 to 45	10 to 180	60 to 300	180 to 300
pH range %	0 to 1.5		1 to 3.5		2.5 to 6
Cadmium	×	×	×	×	—	—
Aluminum	—	—	×	—	—	—
Zinc	×	×	×	×	×	×

Rinsing

Rinsing cools and neutralizes the workpiece surface. If the workpiece becomes too hot, the chromate coating will be damaged and will lose its corrosion resistance.

The rinsing step has three functions.

* Removes acid
* Neutralizes any remaining acid
* Cools the workpiece

Power Requirements

The power requirements to heat the baths are based on operating temperatures. Temperatures range from 60°F to 100°F. The higher the pH number, the lower the power requirements. Power requirements range from 54 to 96 Btu's per pound solution.

pH range	Operating temperature (°F)	Energy requirements (Btu/lb-°F)
0 to 1.5	70 to 100	0.68 to 0.96
1 to 3.5	70 to 90	0.65 to 0.86
	60 to 90	0.54 to 0.86
2.5 to 6	70 to 90	0.62 to 0.83

Cost Elements

Cost elements include the following:

* Setup time
* Load/unload time
* Immersion time
* Cleaning/rinsing time
* Direct labor rate
* Overhead rate
* Amortization of equipment and acid cost

Time Calculation

The process time is obtained by adding immersion, cleaning, and rinse times, then multiplying the total by the number of batches.

CHROMATE CONVERSION SETUP

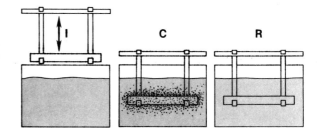

Immersion time (sec) = I
Cleaning (soaking) time (sec) = C
Rinsing time (sec) = R
Number of batches = N

Total time = (I + C + R) × N

Safety Factors

The following hazards should be taken into consideration:

* Personal
    – Contact with acids
    – Contact with toxic vapors
* Environmental
    – Chemical pollution through improper disposal

Phosphate Conversion _____

Phosphate conversion is a surface-altering process that makes a surface corrosion resistant. As the workpiece is immersed in phosphoric acid, the solution decomposes some of the surface metal and forms a protective film.

Process Characteristics

* Changes surface chemical composition to prepare for additional protective coatings
* Is typically used to treat iron and steel
* Forms a thin layer of complex crystalline material on the workpiece surface

Process Schematic

The workpiece is suspended in the phosphoric acid, which decomposes some of the surface metal and forms a protective film.

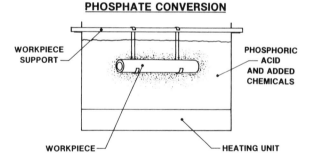

PHOSPHATE CONVERSION

WORKPIECE SUPPORT

PHOSPHORIC ACID AND ADDED CHEMICALS

WORKPIECE

HEATING UNIT

Workpiece Geometery

This process is a surface preparation process and leaves the workpiece geometry unaltered.

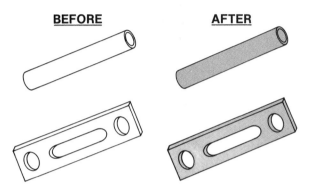

BEFORE   **AFTER**

Setup and Equipment

Phosphate conversion equipment consists of a tank (to contain the acid) and a workpiece holder. A heating source is optional but does accelerate the conversion coating process.

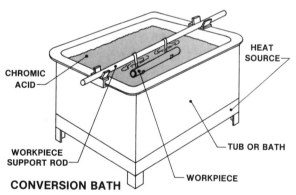

CHROMIC ACID

HEAT SOURCE

WORKPIECE SUPPORT ROD

TUB OR BATH

WORKPIECE

CONVERSION BATH

Typical Tools and Geometry Produced

Typical metals finished with this process are steel and iron. Tools and similar components are often conversion treated and coated with oil. This produces a rust-resistant and lubricated surface. Some piston rings are also treated in this manner.

TOOLS

CHEMICALLY TREATED WORKPIECE

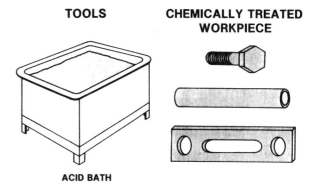

ACID BATH

Solvent Degreasing

Solvent degreasing is a cleaning process that uses the direct application of solvents for the removal of oil, grease, shop dirt, loose chips, and other contaminants. The solvent is sprayed, brushed, or wiped on the surface of the workpiece, or the workpiece is immersed in the solvent.

Process Characteristics

* Removes oil, grease, shop dirt, loose chips, and other contaminants
* Parts are dried at room or elevated temperatures
* Cleans virtually all metals and electrical or electronic assemblies
* Accommodates any size or shape workpiece

Process Schematic

Cleaning is done by soaking the workpiece in a solvent until the soil is broken down. A workpiece may also be sprayed or wiped clean if it is too large to be immersed. Spraying or wiping may also be used when only spot cleaning is required.

PROCESS SCHEMATIC

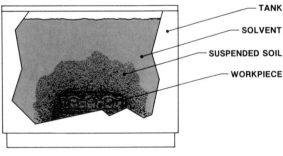

Workpiece Geometry

The size of workpiece to be cleaned is limited only by the size of equipment available. There are almost no restrictions on workpiece shapes due to the ability of the solvent to reach almost all areas of complex shapes.

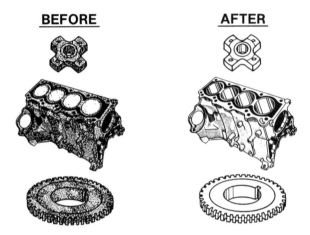

Setup and Equipment

Solvent degreasing equipment is available in a wide range of sizes and capabilities. The main portions of the equipment are the immersion tank, which may have a circulation and filtering system, and an elevating mechanism, which facilitates handing of the workpiece.

Typical Application Methods

Solvents, such as freon and kerosene, are available to allow effective degreasing of all metals. However, many solvents will attack rubber, plastics, and other nonmetals. Most any size or shape workpiece may be cleaned by this process. Solvents do not alter the workpiece geometry.

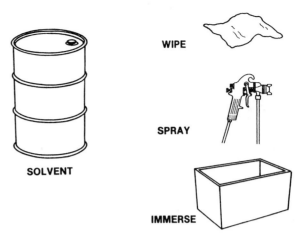

WIPE

SPRAY

SOLVENT

IMMERSE

Safety Factors

It should also be noted that many solvents present a health risk to the operator, so efficient suction, cooling, and so forth, are necessary. Also, many solvents present an environmental risk.

Ultrasonic Cleaning

Ultrasonic cleaning is a surface-cleaning process in which high frequency vibrations create cavitation bubbles in a cleaning solution to scrub away surface contaminants effectively. The cleaning solution can be either water or a mild solvent.

Process Characteristics

* Uses cavitation bubbles to penetrate blind holes, cracks, and recesses
* Thoroughly removes tightly adhering or embedded particles from solid surfaces
* Uses water or a mild solvent for cleaning medium
* Can clean a wide range of workpiece shapes and sizes
* May not require the part to be disassembled prior to cleaning

Process Schematic

In ultrasonic cleaning, the vibrations caused by the ultrasonic transducers create cavitation bubbles that implode with great force (75,000 psi), providing powerful scrubbing action to remove contaminants on the surface of the part.

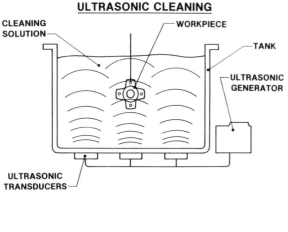

ULTRASONIC CLEANING

CLEANING SOLUTION — WORKPIECE — TANK — ULTRASONIC GENERATOR — ULTRASONIC TRANSDUCERS

Workpiece Geometry

Shown are typical examples of workpieces before and after cleaning. The intricacy of the part generally is not a factor; however, this type of cleaning is limited to materials that are resistant to the collapsing bubbles and the solvent.

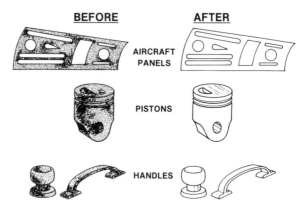

BEFORE — AFTER

AIRCRAFT PANELS

PISTONS

HANDLES

Setup and Equipment

Shown is a self-contained ultrasonic cleaning unit. The unit contains a circulating pump, a cooling system, ultrasonic transducers, and a generator. The workpiece is usually suspended in a tank by means of a wire basket or a hook.

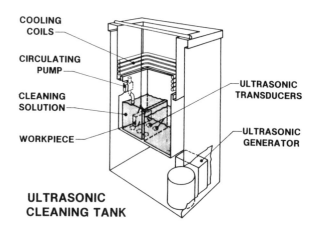

COOLING COILS — CIRCULATING PUMP — CLEANING SOLUTION — WORKPIECE — ULTRASONIC TRANSDUCERS — ULTRASONIC GENERATOR

ULTRASONIC CLEANING TANK

Typical Tools and Applications

There are a variety of tank assemblies. Shown is the basic setup. There are almost no workpiece size limitations.

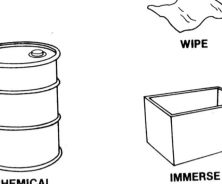

WIPE

CHEMICAL

IMMERSE

Vapor Degreasing

Vapor degreasing is a cleaning process that uses a vaporized or sprayed solvent to remove oil, grease, wax, or other solvent-soluble contaminants.

Process Characteristics

* Thoroughly removes oil, grease, wax, and other solvent-soluble contaminants
* Vapor penetrates blind holes, cracks, recesses, etc.
* Parts emerge from the process dry
* Virtually any size or shape workpiece can be cleaned by this process

Process Schematic

In vapor degreasing, a vaporized solvent atmosphere is created by a boiling solvent. When a cool workpiece is immersed in the hot vapor, the temperature difference causes the vapor to condense on the workpiece, removing contaminants. Spraying the solvent onto the workpiece also helps to remove stubborn soil deposits.

VAPOR DEGREASING

SPRAY — COOLING JACKET

VAPOR ZONE

WORKPIECE

BOILING SOLVENT

HEATING COIL

Workpiece Finish

The size of workpiece to be cleaned is limited only by the size of equipment. There are almost no restrictions on workpiece shapes due to the ability of vapor to reach almost all areas of highly complex shapes. After vapor degreasing, the workpiece is clean and dry.

BEFORE   **AFTER**

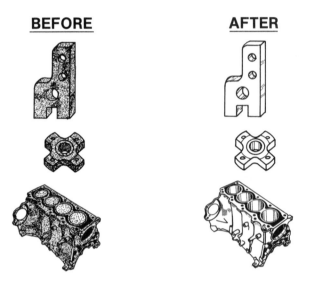

Setup and Equipment

Various types and sizes of vapor degreasing equipment are available to handle most any workpiece. The main unit of the equipment is the tank in which a vapor zone is created. Most vapor degreasers have solvent spraying capabilities. The solvent vapor is heavier than air. A cooling jacket along the top of the tank causes vapors higher in the tank to condense and run back into the bottom of the tank.

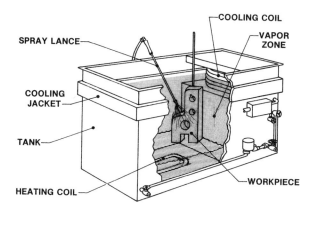

SPRAY LANCE

COOLING COIL

VAPOR ZONE

COOLING JACKET

TANK

HEATING COIL

WORKPIECE

Typical Tools and Geometry Produced

Almost any size or shape workpiece can be cleaned by this process. Workpiece geometry is unchanged by vapor degreasing.

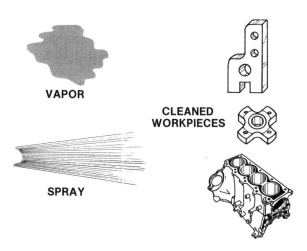

VAPOR

SPRAY

CLEANED WORKPIECES

Safety Factors

Solvents are available to allow efficient vapor degreasing of all metals. However, many solvents will attack rubber, plastics, and other nonmetals. Solvents represent a health and environmental hazard and must be treated accordingly.

Surface Coating

Air Gun Spraying

In air gun spraying, paint is sprayed onto the surface of a workpiece with an air-pressurized spray gun. The paint is directed into an airstream, which reduces it to a fine spray.

Process Characteristics

* Uses compressed air
* Reduces the paint to a fine spray
* Can be a manual or automatic process
* Requires skilled operators for manual operation
* Can achieve a wide variety of spray patterns

Process Schematic

When air gun spraying is used as a manual process, the air gun is manipulated back and forth over the surface to be coated so that the pattern of each stroke overlaps that of the previous one and forms a continuous coating.

AIR GUN SPRAYING

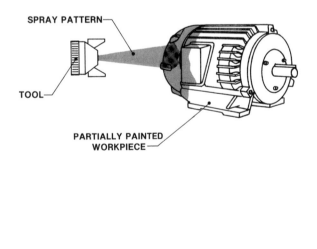

SPRAY PATTERN

TOOL

PARTIALLY PAINTED WORKPIECE

Workpiece Geometry

Workpiece geometry is not limited to size. Paint can be applied in various thicknesses on the workpiece.

BEFORE

AFTER

Setup and Equipment

Equipment consists of an air gun with nozzle, paint, and air compressor. The paint is mixed with the airstream, and a fine spray is produced. The nozzle determines the pattern of spray produced.

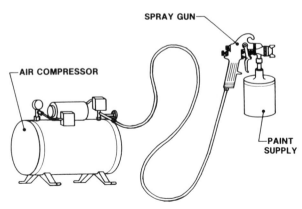

SPRAY GUN

AIR COMPRESSOR

PAINT SUPPLY

Typical Tools and Geometry Produced

There are a wide variety of nozzles that can be used with manual and automatic air gun spraying. The type of nozzle used will depend on the shape of the workpiece to be painted and the consistency of the paint.

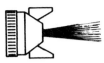

FULL CONE

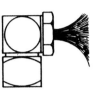

HOLLOW CONE

FLAT

Electroplating

In electroplating, a thin metallic coat is deposited on the workpiece by means of an ionized electrolytic solution. The workpiece (the cathode) and the metallizing source material (the anode) are submerged in solution where a direct electrical current causes the metallic ions to migrate from the source material to the workpiece.

Process Characteristics

* Is performed in a tank of ionized electrolytic solution
* Transfers a metallic deposit via a direct current
* Process variables include time, current, part geometry, and solution contents
* Typically utilizes common metals and alloys
* Must be performed on very clean surfaces
* Is a relatively inexpensive way to apply a desired metallic coating

Process Schematic

A workpiece and source metal are suspended in an ionized electrolytic solution by insulated rods. A direct current is then applied to the anode and cathode, causing metallic ions to migrate from the source metal to the workpiece.

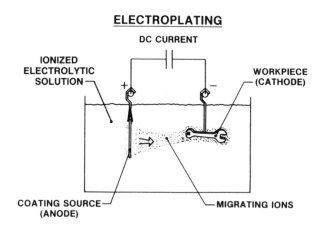

Workpiece Geometry

A wide variety of materials can be electroplated. Plastics and other nonconductive materials must first be treated with some electrically conductive material. Coatings are either decorative or functional.

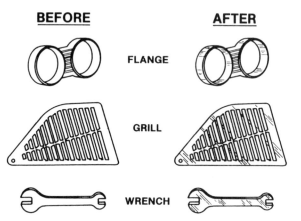

Setup and Equipment

Electroplating equipment consists of an electrolytic solution, a DC power source, and a bath to hold the solution and perform the process. Workpieces hang from conductive rods that are insulated from the tank.

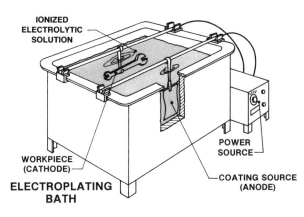

Typical Tools and Geometry Produced

Source metals come in the form of foil, balls, and rods. Source metals are typically common metals and alloys. Gold, silver, copper, zinc, nickel, and tin are among the most common.

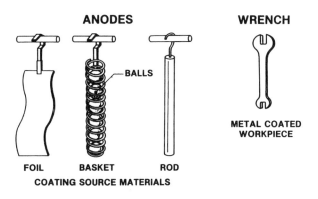

COATING SOURCE MATERIALS

Geometrical Possibilities

The grill received a decorative coating. The flange and wrench have coatings used for corrosion and wear resistance. Workpiece size is limited by tank size. Coating thicknesses range from ±0.0001 in. to around 0.01 in. thick.

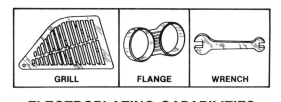

ELECTROPLATING CAPABILITIES

	.0001	.001	.01	.1	1	10	100
COATING THICKNESS (IN)							

▓ TYPICAL RANGE   ▨ FEASIBLE RANGE

Tolerances and Surface Finish

Tolerances range from ±0.0007 in. to as close as ±0.00005 in., depending on the environment of the setup. The surface finish ranges from 1 microinch to around 90 microinches, depending on the material and the workpiece.

TOLERANCES		
ELECTROPLATING	TYPICAL	FEASIBLE
DEPOSITION THICKNESS(IN)	±0.0007	±0.00005

SURFACE FINISH								
PROCESS	MICROINCHES (A.A.)							
ELECTRO-PLATING	2	4	8	16	32	63	125	250

▓ TYPICAL RANGE   ▨ FEASIBLE RANGE

Workholding Methods

Workholding methods largely depend on the type of workpiece being coated. For small workpieces, a porous rotating barrel is used. For larger workpieces, special racks and hangers are used to minimize the support area.

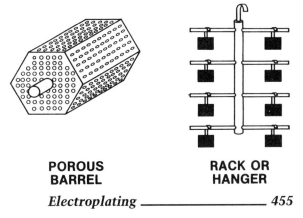

POROUS BARREL   RACK OR HANGER

Electroplating —————— 455

Effects on Work Material Properties

The mechanical effects of electroplating are surface hardening or softening and a change in tensile strength. The physical effects are a change in surface color and composition, dimensional changes, and a change in lubrication properties. The chemical effects are improved corrosion resistance and a change in conductivity.

Work material properties	Effects of electroplating
Mechanical	* Hardened or softened surface * Changed tensile strength
Physical	* Changed surface color and composition * Changed dimension * Increased lubrication surface
Chemical	* Improved corrosion resistance * Changed conductivity

Typical Workpiece Materials

Electroplating coatability ratings range from good to excellent for aluminum, mild steel, copper, brass, and properly treated glass.

MATERIAL	ELECTROPLATING COATABILITY RATINGS			
	POOR	FAIR	GOOD	EXCEL
ALUMINUM			▨▨	
MILD STEEL			▨▨	
COPPER			▨▨	
BRASS			▨▨	
GLASS (TREATED)			▨▨	

▨ TYPICAL RANGE   ▨ FEASIBLE RANGE

Coating Materials

Silver and gold are used to improve conductivity and appearance. Brass, bronze, tin, and aluminum are used to improve corrosion resistance and appearance. Zinc, chromium, and cadmium are used to improve corrosion resistance.

Coating materials	Applications
Silver Gold	* Electrically conductive * Jewelry and decoration
Brass Bronze Tin Aluminum	* Decoration * Corrosion-resistant
Zinc Chromium Cadmium	* Corrosion-resistant

Factors Affecting Process Results

Tolerance and surface finish depend upon the following:

* Workpiece geometry
* Workpiece material
* Surface conditions of the workpiece
* Coating material
* Bath temperature
* Bath concentration and condition
* Operating voltage and currents
* Electrode geometry

Source Geometry

Coating materials come in a variety of shapes and sizes. Most often pellets or sheets are used. The star-shaped anode is used for plating in corners and for other special applications. The last cathode geometry shown is used when large ion transfers are needed—the increased surface area allows rapid disintegration.

STYLE	APPLICATION
	•FOR PLATING FLAT SURFACES •MOST COMMON
	•FOR PLATING IN CORNERS •FOR SPECIAL APPLICATIONS
	•FOR A LARGER SURFACE AREA THUS HIGHER ION TRANSFER

Process Conditions

Process conditions include plating thickness, bath temperature, and the plating time. Plating times range from 5 minutes for a fine coating to as much as 300 minutes for a thick coating. The process conditions given are for plating with chromium.

Plating with chromium

Workpiece and material	Plating thickness (mils)	Bath temp. (°F)	Plating time (min)
Air gage (brass)	0.25	–	15
Cylinder (cast iron)	10	–	300
Bushing (1018 steel)	1	130	45
Cutting tool (tool steel)	0.05	122	5
Gun barrel (steel)	1	130	40
Plug gage (1040 steel)	5	130	150
Valve (steel)	2	–	120

Bath Compositions

Bath compositions and operating conditions for plating copper base with tin–lead alloys are given in the following table.

Material	Ounces per gallon	
	7% Sn to 93% Pb	60% Sn to 40% Pb
Total tin	0.94	8.00
Stannous tin	0.80	7.40
Lead	11.80	3.40
Fluoroboric acid (free)	13.40	13.40
Boric acid (free)	3.40	3.40
Peptone	0.067	0.67
Anode composition	7% Sn to 93% Pb	60% Sn to 40% Pb

Power Requirements

Power requirements are given in the form of amps per square inch of the workpiece. Variables that affect the power requirements are the workpiece material, type of coating, plating rate, and bath temperature. Most current levels range from 2 amps to 4 amps. Power requirements are given for plating with chromium.

Plating with chromium

Workpiece and material	Plating thickness (mils)	Bath temp (°F)	Current density (amps/in.^{2})
Air gage (brass)	0.25	–	2.5
Cylinder (cast iron)	10	–	3.8
Bushing (1018 steel)	1	130	2.1
Cutting tool (tool steel)	0.05	122	2.0
Gun barrel (steel)	1	130	3.0
Plug gage (1040 steel)	5	130	2.5
Valve (steel)	2	–	2.5

Cost Elements

Cost elements include the following:

* Setup time
* Load/unload time
* Solution control time
* Process times
* Direct labor rate
* Overhead rate
* Amortization of equipment and tooling

Time Calculation

The coating time can be calculated by dividing the coat thickness by the deposition rate. The total time can be calculated by adding the set-up time, load time/unload time, and the coating time and multiplying this sum by the number of batches.

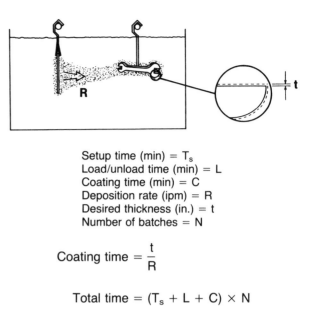

Setup time (min) = T_s
Load/unload time (min) = L
Coating time (min) = C
Deposition rate (ipm) = R
Desired thickness (in.) = t
Number of batches = N

$$\text{Coating time} = \frac{t}{R}$$

$$\text{Total time} = (T_s + L + C) \times N$$

Safety Factors

The following hazards should be taken into consideration:

* Personal
  – Process vapors
  – Harmful chemicals
* Environmental
  – Chemical pollutants
  – Toxic vapors

Electrostatic Coating

Electrostatic coating is the application of charged paint particles to a workpiece with an opposite charge. Charged powdered particles or atomized liquid paint is initially projected toward the conductive workpiece by a mechanical or compressed air spraying method and then accelerated toward the workpiece by the powerful electrostatic charge.

Process Characteristics

* Utilizes a high voltage applied to workpiece and sprayer
* Is 95% paint-efficient due to reduced overspray and better "wrap-around"
* Can be an automatic or manual process
* Materials can be in a powder or liquid form
* Workpieces must be conductive
* Workpieces are usually baked after coating

Process Schematic

The workpiece is conveyed to the paint booth and electrostatically coated with a liquid or powder mixture via a spray nozzle. The powdered paint recovery unit recycles between 95% and 100% of the paint overspray. After coating, the workpiece is transferred to the baking oven for curing.

ELECTROSTATIC COATING

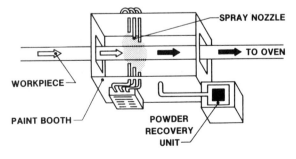

Workpiece Geometry

Workpiece geometry is limited only by the paint booth facility size. Paint may be applied in various thicknesses on the workpiece and is limited only by its tendency to "run." Several thin coats are usually preferred to one thick coat.

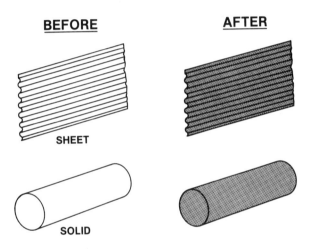

Setup and Equipment

The workpiece may be delivered to the coating booth, and later to the baking oven, by a conveyor or robot. Shown is a highly automated system that determines the coating speed. After the workpiece has been electrostatically coated, it usually proceeds to the baking oven or to the open air for curing.

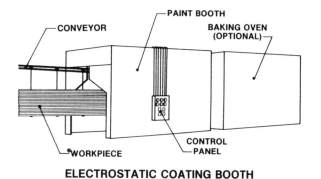

ELECTROSTATIC COATING BOOTH

CONVEYOR

PAINT BOOTH

BAKING OVEN
(OPTIONAL)

CONTROL
PANEL

WORKPIECE

Typical Tools and Geometry Produced

There are a wide variety of spray nozzles available for use in electrostatic coating. The type of nozzle used will largely depend on the shape of workpiece to be painted and the consistency of the paint.

SPRAY
NOZZLE

Hot Dip Coating

In hot dip coating, the workpiece is submerged in a vat of molten metal. When the workpiece is removed from the vat, a layer of the coating metal has bonded to it.

Process Characteristics

* Is performed in a tank of molten metal
* Deposits a layer of the melted source metal on the submerged workpiece
* Typically utilizes source metals with a low melting point
* Is a relatively inexpensive way to coat a workpiece, but coating is thick and uneven
* Requires a clean workpiece surface

Process Schematic

The workpiece is suspended in the molten metal for a short period of time, usually less than one minute. The source metal adheres to the workpiece upon removal from the bath and is air cooled.

HOT DIP COATING

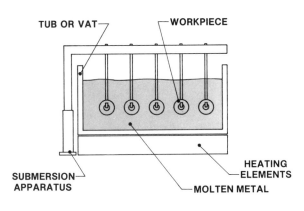

TUB OR VAT — WORKPIECE

SUBMERSION APPARATUS — MOLTEN METAL — HEATING ELEMENTS

Workpiece Geometry

Usually, materials with a higher melting point are hot dipped to prevent workpiece deterioration. The coatings may be either decorative or functional. Functional coatings provide heavy duty corrosion resistance.

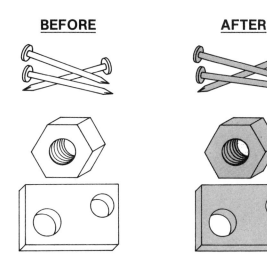

BEFORE   AFTER

Setup and Equipment

Equipment consists of a vat or bath, a heat source to melt the metal, and a workpiece submersion apparatus for suspending articles to be coated.

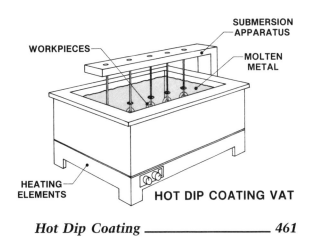

WORKPIECES — SUBMERSION APPARATUS — MOLTEN METAL

HEATING ELEMENTS   HOT DIP COATING VAT

Typical Tools and Geometry Produced

Wire baskets are used for dipping workpieces, such as nails or bolts. The hanger is used for larger workpieces. The workpiece geometry is slightly changed according to the amount of metal applied. Typical coatings are tin, lead, and other metals with low melting points.

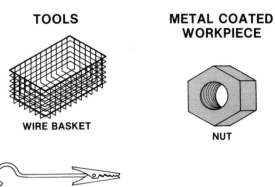

TOOLS

WIRE BASKET

HANGER

METAL COATED WORKPIECE

NUT

Ion Plating

In ion plating, a thin metallic coat is deposited on the workpiece. The workpiece (the cathode) and the source metal (the anode) are submitted to a vacuum chamber where a heat source causes the source material to evaporate and form a vapor. This metallic vapor then condenses on the workpiece.

Process Characteristics

* Is performed in a vacuum chamber using inert gas
* Can produce a uniform metallic deposit on complex tool geometries
* Transfers a metallic deposit via a metal vapor
* Typically utilizes common metals and alloys
* Requires very clean surfaces
* Is a relatively expensive way to apply a desired metallic coating

Process Schematic

In ion plating, the workpiece can either be held in a basket or suspended. The source metal is heated by a filament until it evaporates. This vapor of metal then migrates to the rotating workpiece where it condenses and forms a metallic coating. A shield for the high voltage supply is used to control the electric field.

ION PLATING

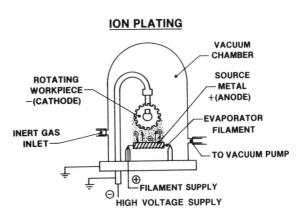

Workpiece Geometry

A wide variety of materials can be ion plated. The coatings are mainly functional (i.e., wear-resistant) rather than decorative due to its high cost.

BEFORE   **AFTER**

Setup and Equipment

Ion plating equipment consists of a vacuum chamber, a power source, a vacuum pump, and a supply of inert gas. The vacuum pump is used to replace the air with inert gas.

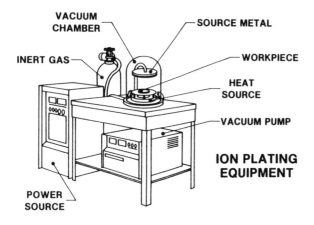

ION PLATING EQUIPMENT

Typical Tools and Geometry Produced

The workpiece is either placed in the holder or suspended. The source metal is commonly placed in a boat and then heated. Most metals and their alloys can be used as source metals.

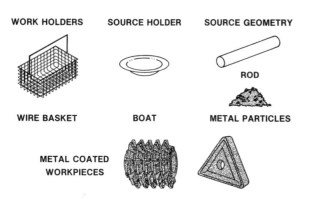

Geometrical Possibilities

Coatings for complex geometries are hard to predict, and the coating may be uneven. Coating thickness ranges from 0.0000001 in. to as much as 0.000008 in. The size of the workpiece is limited only by the size of the vacuum chamber.

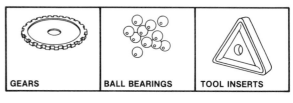

GEARS | BALL BEARINGS | TOOL INSERTS

ION PLATING CAPABILITIES

	0	0.000,000,1	0.000,001	0.000,01
COATING THICKNESS (IN)				

▓ TYPICAL RANGE ▨ FEASIBLE RANGE

Tolerances and Surface Finish

The process can meet precise tolerances as close as ±0.0000001 in., but tolerances typically range around ±0.0000005 in. The surface finish can range from 3 microinches to around 80 microinches.

TOLERANCES

SPUTTERING	TYPICAL	FEASIBLE
COAT THICKNESS	±.000,000,5	±.000,000,1

SURFACE FINISH

PROCESS	MICROINCHES (A.A.)							
	2	4	8	16	32	63	125	250
SPUTTERING								

▓ TYPICAL RANGE ▨ FEASIBLE RANGE

Target Materials

Coating thicknesses range from 0.01 microns to 0.2 microns. Coatings are used to change conductivity, enhance corrosion resistance, and improve wear resistance.

Target materials	Coating thickness (microns)	Applications
Aluminum	0.01 to 0.2	Conductor
Copper	0.01 to 0.2	Corrosion
Gold	0.01 to 0.2	resistance
Lead	0.05 to 0.2	
Nickel	0.05 to 0.2	
Silver	0.01 to 0.2	
Tin	0.05 to 0.2	Superconductor
Tantalum	0.01 to 0.2	Resistor

Workholding Methods

The purpose of the workholder is to position the workpiece above the target material. Wire baskets are the most commonly used workholder. They are used for batches of ball bearings, nuts, and bolts. For larger parts, it may be necessary to use a plate on which the part can be mounted.

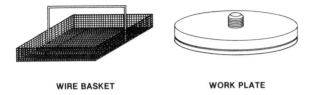

WIRE BASKET   WORK PLATE

Effects on Work Material Properties

The effects of ion plating are increased wear resistance and fatigue strength. Ion plating also improves conductivity, lubrication properties, electromagnetic radiation characteristics, and corrosion resistance.

Work material properties	Effects of ion plating
Mechanical	* Increased wear resistance * Increased fatigue strength
Physical	* Change in conductivity * Improved lubrication at surface
Chemical	* Increased corrosion resistance

Typical Workpiece Materials

The coatability ratings for ion plating are good to excellent for aluminum, chromium, copper, steel, and brass. Glass, plastics, and other materials can be coated but need to be pretreated first.

MATERIAL	ION PLATING COATABILITY RATINGS			
	POOR	FAIR	GOOD	EXCEL
ALUMINUM			▨▓▨	
CHROMIUM			▨▓▨	
COPPER			▨▓▨	
STEEL			▨▓▨	
BRASS			▨▓▨	

▓ TYPICAL RANGE   ▨ FEASIBLE RANGE

Coating Materials

Plating with alumimum, platinum, or silver can change conductivity, increase corrosion resistance, provide a coating for medical instruments, and seal porosity and small cracks. Titanium and tantalum are used to strengthen substrate surfaces and to increase corrosion resistance. Where optical coatings are required, chrome and titanium are used.

Materials	Applications
Aluminum Platinum Silver	* Conductivity * Corrosion resistance * Coating for medical instruments * Seal small cracks and porosity
Titanium Tantalum	* Substrate strengthening * Corrosion resistance
Chrome Titanium	* Optical coatings

Factors Affecting Process Results

Tolerance and surface finish depend on workpiece material, workpiece surface condition, source-to-substrate distance, voltage levels, residual gas type, gas pressure, coating material and emission patterns, type of workpiece holder, vacuum abilities, and equipment efficiency.

Source Geometry

Source materials come in a variety of shapes and sizes, depending on the application. Wire and staples are used when a medium evaporation rate is needed for light to medium coating thicknesses. Pellets are used for slower vaporation for light coats. Powder evaporates very rapidly for heavy coating. Bars or rods are applied for slow evaporation and for close tolerance control.

STYLE	APPLICATION
(arc shapes)	• MEDIUM EVAPORATION • FOR MEDIUM COATINGS
(pellets)	• MEDIUM TO SLOW EVAPORATION • FOR THIN COATINGS
(powder)	• RAPID EVAPORATION • FOR HEAVY COATINGS
(rod)	• SLOW EVAPORATION • FOR THIN AND CLOSE COATING CONTROL

Process Conditions

Process conditions depend on metal type, metal form, evaporation temperature, and exposure time. Exposure time ranges from 10 sec to as much as 300 sec (5 min).

For coating thicknesses from 3 to 5 microns

Metal type	Metal form	Evaporation temperature (F°)	Exposure time (sec)
Aluminum	Wire staples	1830	10 to 20
Chromium	Powder or pellets	2200	60 to 300
Copper	Wire or pellets	2320	60 to 180
Gold	Wire or pellets	2670	60 to 180
Silver	Wire or pellets	1920	60 to 180

Cooling

Cooling is achieved by pumping coolant through veins or pipes that can be mounted on the support plate.

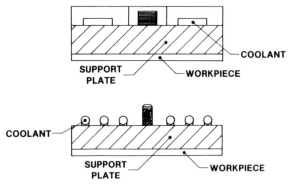

Power Requirements

The power requirements for evaporating metal depend on the type and form of the material and the evaporation temperature. Power requirements are given in calories per mole and range from 67,580 cal/mole for aluminum to 89,440 cal/mole for chromium. System efficiency depends largely on equipment.

Metal type	Metal form	Evaporation temp (F°)	Power requirements (cal/mole)
Aluminum	Wire staples	1830	67,580
Chromium	Power or pellets	2200	89,440
Copper	Wire or pellets	2320	80,070
Gold	Wire or pellets	2670	88,280
Silver	Wire or pellets	1920	74,600

Cost Elements

Cost elements include the following:

* Setup times
* Load/unload times
* Vacuum times
* Plating time
* Direct labor rate
* Overhead costs
* Amortization of equipment and tooling costs

Time Calculation

The plating time can be calculated by dividing the deposit thickness by the deposit rate. The total time can then be calculated by adding the load/unload time, the pumping time, and the plating time; multiplying this sum by the number of batches; and then adding the initial setup time.

Load/unload time (min) = L
Pumping time (min) = P
Plating time (min) = C
Deposition rate (mils/min) = R
Deposition thickness (mils) = t
Number of batches = N

Plating time, $C = \dfrac{t}{R}$

Total time $= [(L + P + C) \times N] + T_s$

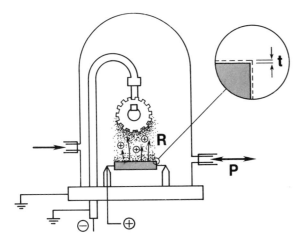

Safety Factors

The following hazards should be taken into consideration:

* Personal
  - High voltages
  - Hot workpieces
  - Toxic vapors

Sputtering

In sputtering, a thin metallic coat is deposited on the workpiece within a vacuum chamber. The source metal (the cathode) is caused to vaporize and accelerate toward the workpiece (the anode), where it is deposited as a thin layer of metal.

Process Characteristics

* Is performed in a vacuum chamber
* Transfers a metallic deposit via ionized gas molecules
* Typically utilizes common metals and alloys
* Must be performed on clean surfaces
* Is relatively slow; it may take hours to acquire a 0.1-micron thick coat

Process Schematic

The workpiece is either held in a tray or by some other device. The workpiece is made positive, relative to the source material, by a power source. When the applied potential reaches the ionization energy of the inert gas (e.g., argon), electrons from the source material form high energy ions with the gas atoms. These ions are then accelerated back to the source material, causing a metal atom to dislodge. This metal atom then accelerates across the electrode gap and is deposited on the workpiece. Because of their high energy (15 to 50 electronvolts), the metal atoms adhere to the workpiece considerably better than in other vaccum coating processes.

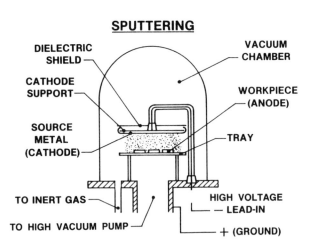

SPUTTERING

Workpiece Geometry

A wide variety of workpiece materials can be successfully coated by sputtering. Due to its high costs, however, sputtering is utilized primarily for functional requirements rather than for mere decorative purposes.

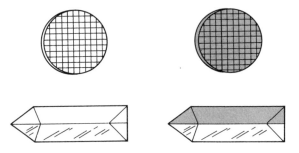

Setup and Equipment

Equipment consists of a vacuum chamber, a vacuum pump, a power source, and a supply of inert gas. The vacuum pump serves to remove the air from the chamber, and it is replaced with inert gas.

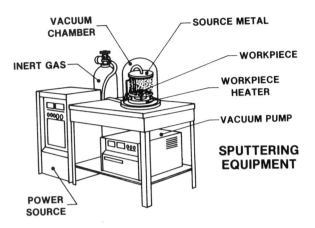

SPUTTERING EQUIPMENT

Typical Tools and Geometry Produced

The source metal typically comes in the form of a disk, as shown. This allows for maximum surface area exposure. Typical coated workpieces include precision parts subject to wear and other parts, such as electronic circuits and special mirrors. The most common coating materials are tungsten, molybdenum, tantalum, oxides, nitrides, and alloys.

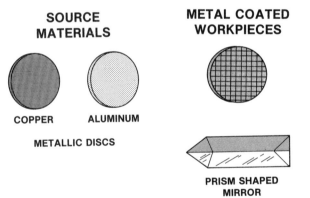

SOURCE MATERIALS

METAL COATED WORKPIECES

COPPER   ALUMINUM

METALLIC DISCS

PRISM SHAPED MIRROR

Geometrical Possibilities

The geometrical possibilities for sputtering are limited to flat and simple geometries. Examples of workpieces are mirrors and electronic grids. Workpiece size is limited by the size of the vacuum chamber. Workpieces range from 0.01 in.^{2} to several hundred square inches. The thickness of the deposit ranges from 0.0000001 in. to 0.000007 in. Typically, the thickness ranges from 0.0000007 in. to 0.000004 in.

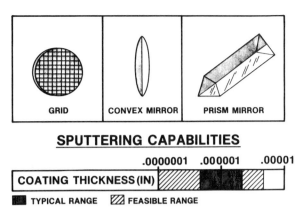

GRID   CONVEX MIRROR   PRISM MIRROR

SPUTTERING CAPABILITIES

	.0000001	.000001	.00001
COATING THICKNESS (IN)			

■ TYPICAL RANGE   ▨ FEASIBLE RANGE

Tolerances and Surface Finish

The thickness tolerances of sputtered coatings typically range around ±0.0000005 in., but tolerances of around ±0.0000001 in. are feasible. The surface finish ranges from 3 microinches to around 80 microinches.

TOLERANCES		
SPUTTERING	TYPICAL	FEASIBLE
COAT THICKNESS	±.000,000,5	±.000,000,1

SURFACE FINISH		
PROCESS	MICROINCHES (A.A.)	
SPUTTERING	2 4 8 16 32 63 125 250	

■ TYPICAL RANGE   ▨ FEASIBLE RANGE

Source or Coating Materials

Shown on the following page are different source or coating materials. Coating thicknesses range from 0.01 microns to 0.2 microns (1 micron = 1×10^{-6} meter). Materials are coated for decoration, increased corrosion resistance for conductors, increased conductivity for superconductors, and reduced conductivity for resistors.

Coating materials	Properties
Aluminum	* Protective and decorative
Tantalum	* Electrically resistive * Withstands high temperatures
Gold Silver Copper	* Electrical conductivity * Decorative
Molybdenum disulfide	* Improves lubrication surface
Chromium	* Corrosion resistance * Electrically resistive
Tungsten	* Withstands high temperatures
Titanium carbide	* Withstands high temperatures
Lead Nickel	* Corrosion resistance
Tin	* Super conductor

Source Metalholding Methods

Three typical source (coating materials) holding methods exist. The threaded peg-type support is used with very brittle materials, such as chromium or cobalt. The threaded-type targets use a threaded plug or peg of the same material as the source to avoid contamination. Ceramics and other materials that cannot be threaded are attached to the support plate using epoxy or by metallizing the back surface of the source material and soldering it to the support plate.

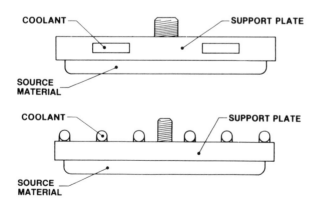

Workholding Methods

A tray and a rotating workpiece holder are common workholding methods. As shown, several workpieces can be coated simultaneously. The rotating motion provides a more uniform coating.

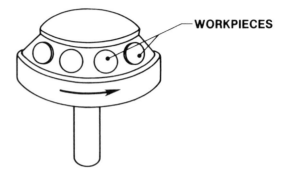

ROTATING WORKPIECE HOLDER

Effects on Work Material Properties

The mechanical effects of sputtering are increased wear resistance; surface hardening or softening, depending on the material applied; and increased lubrication properties. The physical effects are changes in electrical conductivity and color. Composition changes in the surface layer may result in increased corrosion resistance.

Work material properties	Effects of sputtering
Mechanical	* Increased wear resistance * Surface hardening or softening * Increased lubrication
Physical	* Changes electrical conductivity * Changes surface color and composition
Chemical	* Increased corrosion resistance

Typical Workpiece Materials

Commonly coated workpiece materials are steel, aluminum, plastics, glass, and ceramics. Coatability ratings for these materials range from good to excellent.

MATERIAL	SPUTTERING COATABILITY RATINGS			
	POOR	FAIR	GOOD	EXCEL
STAINLESS STEEL				▨▨▩▨
ALUMINUM				▨▨▩▨
MILD STEEL				▨▨▩▨
GLASS				▨▨▩▨
PLASTICS				▨▨▩▨
CERAMICS				▨▨▩▨

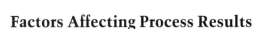

■ TYPICAL RANGE   ▨ FEASIBLE RANGE

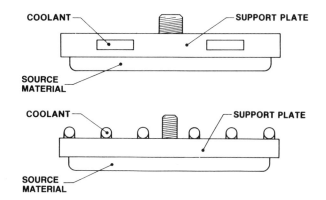

Factors Affecting Process Results

Tolerance and surface finish depend upon the following:

* Workpiece temperature
* Workpiece surface condition
* Workpiece geometry
* Target to workpiece distance
* Type of target material
* Voltage levels
* Residual gas pressure
* Type of residual gas

Process Conditions

Process conditions include metal type, metal form, evaporation temperature, and exposure time. Exposure times range from as little as 10 sec to as much as 300 sec (5 min).

For coating thicknesses from 3 to 5 microns

Metal type	Metal form	Evaporation temperature (F°)	Exposure time (sec)
Aluminum	Wire staples	1830	10 to 20
Chromium	Powder or pellets	2200	60 to 300
Copper	Wire or pellets	2320	60 to 180
Gold	Wire or pellets	2670	60 to 180
Silver	Wire or pellets	1920	60 to 180

Cooling

Cooling of the source material, which is heated by the bombarding gas ions, is achieved by pumping coolant through veins or pipes that are mounted on the support plate.

Power Requirements

The power requirements for sputtering depend on the type of material, the shape of the work-piece, and the evaporation temperature. The relative power requirements needed to vaporize the metals in the table below are given in the form of calories per mole and range from 67,580 cal/mole for aluminum to 89,440 cal/mole for chromium. System design has a large impact on efficient power consumption.

Metal type	Metal form	Evaporation temperature (F°)	Power requirements (cal/mole)
Aluminum	Wire staples	1830	67,580
Chromium	Powder or pellets	2200	89,440
Copper	Wire or pellets	2320	80,070
Gold	Wire or pellets	2670	88,280
Silver	Wire or pellets	1920	74,600

Cost Elements

Cost elements include the following:

* Setup time
* Load/unload time
* Vacuum pumping time
* Deposition time
* Source material costs
* Direct labor rate
* Overhead rate
* Amortization of equipment and tooling

Time Calculation

To calculate the coating time, the desired thickness is divided by the deposition rate. The total time can be obtained by adding setup time, load time/unload time, pumping time/return to ambient pressure time, and coating time.

Load/unload time (min) = L
Coating time (min) = C
Deposition rate (angstroms/min) = R
Desired thickness (angstroms) = t
Pumping/return to ambient pressure time (min) = P
Setup time = T_s

$$\text{Coating time, } C = \frac{t}{R}$$

$$\text{Total time} = T_s + L + P + C$$

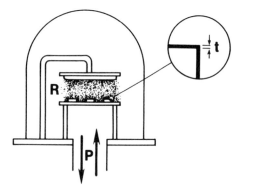

Safety Factors

The principal environmental risk is that of fire, which is rather minimal.
The following hazards should be taken into consideration:

* Personal
  - Hot workpieces
  - Poisonous gases
  - Eye irritation from glow discharge
  - High voltage and current levels
* Environmental
  - Fire risks

Vacuum Metallizing

In vacuum metallizing, a metallic coating material is placed in a vacuum chamber with the workpiece to be coated. Heat is applied to the coating material until it evaporates. The vaporized metal then condenses on the workpiece as a thin metallic film. Rotation of the part aids in uniformity of coating.

Process Characteristics

* Is performed within a vacuum
* Vaporizes the metal
* Can be applied to glass, plastic, metal, ceramic, and paper materials
* Normally uses pure metals
* Workpieces must be very clean
* Produces a very thin coating of metal

Process Schematic

The workpieces are attached to holding devices inside the vacuum chamber. After a vacuum is created in the chamber, heat is applied to the metallizing material until it evaporates. The workpieces are then rotated on the racks so that all sides of them can be coated with the vaporized particles of metal.

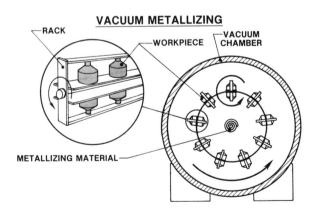

Workpiece Geometry

A wide variety of materials, such as glass, plastics, metal, ceramics, and paper, can be vacuum metallized. The coatings applied are either decorative or functional.

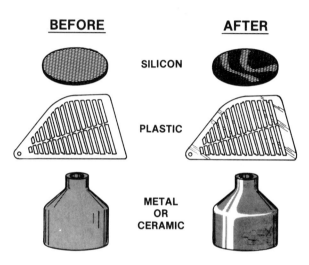

Setup and Equipment

Vacuum metallizing equipment consists of a vacuum chamber, a vacuum pump, and a motor. The workpieces are mounted on racks inside the vacuum chamber. The motor rotates the racks so that the parts can be coated on both sides. The vacuum pump removes the air from the vacuum chamber.

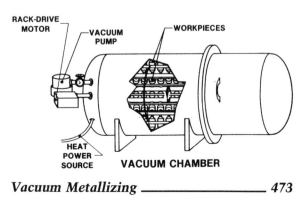

Typical Tools and Geometry Produced

Heater filaments are usually made of tungsten wire and are coated with the desired metal to be vaporized. The helix type is the most commonly used type of filament, and aluminum is the most commonly used material for metallizing.

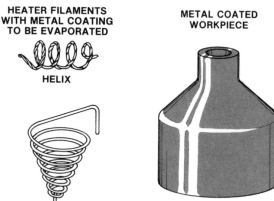

Geometrical Possibilities

Many geometrical shapes can be coated, such as pipes, flat surfaces, complex surfaces, and complex geometries. Feasible coating thickness ranges from nearly zero inches to about 0.00001 in. under extreme conditions. The workpiece size is limited only by the size of the vacuum chamber but generally is limited to a surface area of 1 in.^{2} to 10 in.^{2}.

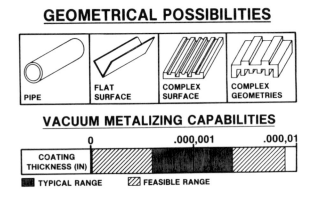

Tolerances and Surface Finish

Typical tolerances are ±0.000001 in., but tolerances as close as ±0.0000005 in. are feasible. Feasible surface finishes range from 2.5 to 40 microinches. The process finish depends largely on the original workpiece finish.

TOLERANCES		
VACUUM METALIZING	TYPICAL	FEASIBLE
COAT THICKNESS (IN)	±0.000,001	±0.000,000,5

SURFACE FINISH		
PROCESS	MICROINCHES (A.A.)	
VACUUM METALIZING	2 4 8 16 32 63 125 250 500	

■ TYPICAL RANGE   ▨ FEASIBLE RANGE

Target Materials

Coating thicknesses range from 0.01 microns to 0.2 microns (1 micron = 1×10^{-6} meter). Coating a workpiece changes its conductivity, enhances its corrosion resistance, and improves its appearance.

Target materials	Coating thickness (microns)	Applications
Aluminum	0.01 to 0.2	Decoration
Copper	0.01 to 0.2	Conductor
Gold	0.01 to 0.2	Corrosion
Lead	0.05 to 0.2	resistance
Nickel	0.05 to 0.2	
Silver	0.01 to 0.2	
Tin	0.05 to 0.2	Superconductor
Tantalum	0.01 to 0.2	Resistor

Workholding Methods

Rotating racks are used to obtain even coatings and to coat complex geometries. A wire basket can be utilized for smaller parts with less critical surface finishes.

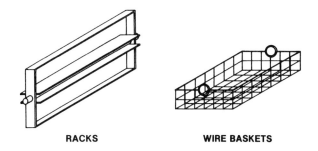

RACKS   **WIRE BASKETS**

Effects on Work Material Properties

The most common mechanical effects on the workpiece are increased wear resistance and fatigue strength. The physical effects are a change in conductivity and improved lubrication properties. The main chemical effect is an increase in corrosion resistance.

Work material properties	Effects of vacuum metalizing
Mechanical	* Increased wear resistance * Increased fatigue strength
Physical	* Change in conductivity * Improved lubrication at surface
Chemical	* Increased corrosion resistance

Typical Workpiece Materials

Aluminum, chromium, copper, steel, and brass are some of the higher rated materials, and all have good to excellent coatability ratings with the proper surface treatment.

MATERIAL	VACUUM METALIZING COATABILITY RATINGS			
	POOR	FAIR	GOOD	EXCEL
ALUMINUM				▨▨
CHROMIUM				▨▨
COPPER			▨▨	
STEEL			▨▨	
BRASS			▨▨	

▓ TYPICAL RANGE   ▨ FEASIBLE RANGE

Coating Materials

Aluminum, platinum, and silver are used to enhance conductivity and corrosion resistance. Titanium and chromium are used for substrate stengthening, for optical purposes, and corrosion resistance.

Materials	Applications
Aluminum Platinum Silver	* Conductivity * Corrosion resistance * Coating for medical instruments * Seal small cracks and porosity
Titanium Tantalum	* Substrate strengthening * Corrosion resistance
Chromium Titanium	* Optical coatings

Factors Affecting Process Results

Tolerance and surface finish depend on workpiece surface condition, workpiece material, coating material, residual gas type, gas pressure, metal evaporation rate and temperature, workpiece holder type, vacuum capabilities, and equipment efficiency.

Source Geometry

Source materials come in a variety of shapes and sizes depending on the application. Wire and staples are used when a medium evaporation rate is desired for medium coating thicknesses. Pellets are used for a slower evaporation rate and for finer coating thicknesses. Metal powder is used when faster evaporation is required for thicker coating purposes. Rods or bars are used for slower deposit rates.

STYLE	APPLICATION
	•MEDIUM EVAPORATION •FOR MEDIUM COATINGS
	•MEDIUM TO SLOW EVAPORATION •FOR THIN COATINGS
	•RAPID EVAPORATION •FOR HEAVY COATINGS
	•SLOW EVAPORATION •FOR THIN AND CLOSE COATING CONTROL

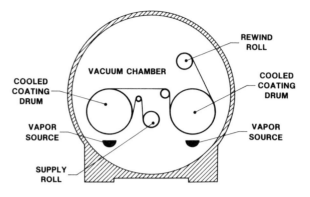

Process Conditions

This example gives the process conditions for coating thicknesses ranging from 3 microns to 5 microns. The metal type, metal form, and the evaporation temperature all determine the exposure time. Exposure times range from 10 sec for aluminum to as much as 300 sec (5 min) for chromium.

For coating thicknesses from 3 to 5 microns

Metal type	Metal form	Evaporation temperature (°F)	Exposure time (sec)
Aluminum	Wire staples	1830	10 to 20
Chromium	Powder or pellets	2200	60 to 300
Copper	Wire or pellets	2320	60 to 180
Gold	Wire or pellets	2670	60 to 180
Silver	Wire or pellets	1920	60 to 180

Cooling

When coating paper or other delicate materials, cooled coating drums are often used. The drums are cooled by pumping coolant through them that increases condensation rates and production rates.

Power Requirements

The power requirements are given in calories per mole. Power requirements range from 67,580 cal/mole for aluminum to 89,440 cal/mole for chromium.

Metal type	Metal form	Evaporation temp (°F)	Power requirements (cal/mole)
Aluminum	Wire staples	1830	67,580
Chromium	Powder or pellets	2200	89,440
Copper	Wire or pellets	2320	80,070
Gold	Wire or pellets	2670	88,280
Silver	Wire or pellets	1920	74,600

Cost Elements

Cost elements include the following:

* Setup time
* Load/unload time
* Vacuum time
* Plating time
* Direct labor rate
* Overhead costs
* Amortization of equipment and tooling costs
* Material evaporation time

Time Calculation

To calculate the plating time, the desired thickness is divided by the deposition rate. The total time can then be calculated by first adding the load/unload time, the pumping/ return to atmosphere time, and the plating time. This is then multiplied by the number of batches necessary. The original setup time is then added to obtain the total time.

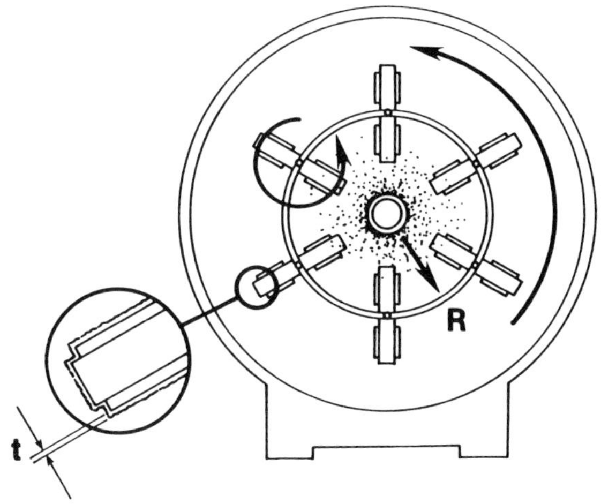

Load/unload time (min) = L
Pumping/return to atmosphere time (min) = P
Plating time (min) = C
Deposition rate (mils/min) = R
Deposition thickness (mils) = t
Number of batches = N
Setup time = T_s

Plating time, $C = \dfrac{t}{R}$

Total time = $[(L + P + C) \times N] + T_s$

Safety Factors

The following hazards should be taken into consideration:

* Personal
  - High voltages
  - Hot workpieces
  - Toxic vapors

Surface Modification

Barrel Tumbling

In barrel tumbling, abrasive media, the workpieces, water, and a cleaning compound are rotated in a drum for cleaning, descaling, deburring, and polishing workpieces.

Process Characteristics

* Utilizes relative movement between the workpieces and abrasive media
* Is used for cleaning, descaling, and deburring
* May use a wide range of synthetic, metallic, or natural finishing media
* Typically requires a cleaning, abrasive, or rust inhibitive compound
* Is suitable for a wide variety of workpiece shapes and materials

Process Schematic

The rotation of the barrel causes the workload to flow downward from the top of the barrel through an abrasive area, known as the effective sliding zone. The workload consists of the workpiece material, water, cleaning compounds, and the abrasive media.

Workpiece Geometry

An abrasive material is selected according to the workpiece material and desired surface finish. These factors also determine rotation speed and time. The versatility of this process allows for a workpiece to be cleaned, descaled, pickled, neutralized, chemically treated, deburred, and burnished in one operation.

BEFORE   **AFTER**

ALUMINUM RING

STEEL INDEX ARM

ALUMINUM FILTER

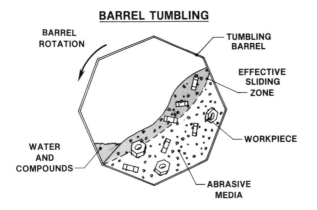

BARREL TUMBLING

Setup and Equipment

An automatic barrel tumbling system consists of a part loading chute, finishing barrel, solution tank, and part removal chute. Automatic units are used to descale, rust inhibit, and polish batches of small- to average-sized parts. Non-automatic units do not use the entrance and removal chute as shown in this illustration. Both horizontal and tilt-type barrels are used in tumbling systems.

Typical Tools and Geometry Produced

Abrasive media are either natural or preshaped. Selection of abrasives depends on the workpiece being processed. The recommended volumetric ratio of abrasive media to workpieces is 2 to 1 or 3 to 1. Many types of workpiece geometries, such as smooth steel valves and aluminum fittings, are processed by barrel tumbling.

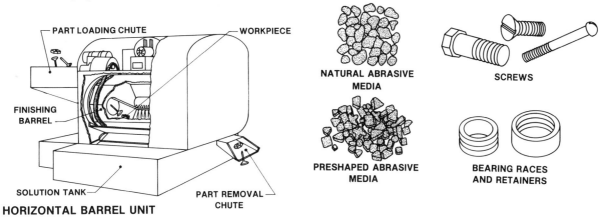

HORIZONTAL BARREL UNIT

References

Allen, Dell K. *Metallurgy Theory and Practice.* Homewood, IL: American Technical Publishers, 1969.

Allen, Dell K. and Paul R. Smith. "Process Classification." Monograph No. 5. Computer Aided Manufacturing Laboratory Brigham. Provo, Utah: Brigham Young University, January 1980.

Alting, Leo. *Manufacturing Engineering Processes.* New York: Marcel Dekker, 1982.

Aluminum Finishes Process Manual. Richmond: Reynolds Metals Company, 1973.

American Society of Tool and Manufacturing Engineers. *High-Velocity Forming of Metals.* Englewood Cliffs: Prentice-Hall, Inc., 1964.

Amstead, B.H., Phillip F. Ostwald, and Myron L. Begeman. *Manufacturing Processes*, 8th ed. New York: John Wiley & Sons, Inc., 1987.

ASM Committee on Gas Carburizing. *Carburizing and Carbonitriding.* Metals Park, OH: American Society for Metals, 1977.

Avner, Sidney H. *Introduction to Physical Metallurgy*, 2nd ed. New York: McGraw-Hill Book Company, 1974.

Beeley, P.R. *Foundry Technology.* Boston: Newnes-Butterworths, 1982.

Bolz, Roger W. *Production Processes: The Producibility Handbook*, 5th ed. New York: Industrial Press Inc., 1981.

Brown, R.L.E. *Design and Manufacture of Plastic Parts.* New York: John Wiley & Sons, Inc., 1980.

Burns, R.M. and W.W. Bradley. *Protective Coatings for Metals*, 3rd ed. Washington: American Chemical Society, 1975.

Cary, Howard B. *Modern Welding Technology*, 2nd ed. Englewood Cliffs: Prentice-Hall, Inc., 1989.

Casting Design Handbook. Metals Park, OH: American Society for Metals, 1962.

Casting Kaiser Aluminum, 2nd ed. Oakland: Kaiser Aluminum & Chemical Sales, Inc., 1965.

Clauser, H.R., Robert Fabian, Donald Peckner, and Malcolm W. Riley, eds. *The Encyclopedia of Engineering Materials and Processes.* New York: Reinhold Publishing Corporation, 1963.

DeGarmo, E. Paul, J. Temple Black, and Ronald A. Kohser. *Materials and Processes in Manufacturing.* 7th ed. New York: Macmillan Publishing Company, 1988.

DoAll Band Tool Manual. Des Plaines: The DoAll Company, 1953.

Drozda, T., C. Wick, J. Benedict and R. Veilleux. *Tool and Manufacturing Engineers Handbook.* 6 vols. Dearborn: Society of Manufacturing Engineers, 1983–1992.

Hausner, Henry H. *Handbook of Powder Metallurgy.* New York: Chemical Publishing Co., Inc., 1973.

Heat Treating Aluminum Alloys. Louisville: Reynolds Metals Company, 1954.

Huebner's Machine Tool Specs, 1st ed. 5 vols. Solon: Huebner Publications, Inc., 1980.

Kaiser Aluminum Rod, Bar and Wire: Product Information, 3rd ed. Oakland: Kaiser Aluminum & Chemical Sales, Inc., 1965.

Kane, George E. *Modern Trends in Cutting Tools.* Dearborn: Society of Manufacturing Engineers, 1982.

Kibbe, Richard R. *Grinding Machine Operations.* New York: John Wiley & Sons, Inc., 1985.

Kibbe, Richard R. *Lathe Operations.* New York: John Wiley & Sons, Inc., 1985.

Kibbe, Richard R. *Milling Machine Operations.* New York: John Wiley & Sons, Inc. 1985.

Koistinen, Donald P. and Neng-Ming Wang. *Mechanics of Sheet Metal Forming: Material Behavior and Deformation Analysis.* New York: Plenum Press, 1978.

Lambert, Brian K. *Milling: Methods and Machines.* Dearborn: Society of Manufacturing Engineers, 1982.

Laue, Kurt and Helmut Stenger. *Extrusion: Processes, Machinery, Tooling.* Metals Park, OH: American Society for Metals, 1981.

Machining Data Handbook, 3rd ed. 2 vols. Cincinnati: Machinability Data Center, 1980.

McKee, Richard L. *Machining with Abrasives.* New York: Van Nostrand Reinhold Company, 1982.

Metals Handbook. 18 vols. Metals Park, OH: American Society for Metals, 1982–1993.

Metzbower, Edward A. *Applications of Lasers in Materials Processing.* Metals Park, OH: American Society for Metals, 1979.

New Trends in Materials Processing. Metals Park, OH: American Society for Metals, 1976.

Oberg, Erik, Franklin D. Jones, Holbrook L. Horton, and Henry H. Ryffel. *Machinery's Handbook*, 24th ed. New York: Industrial Press Inc., 1992.

Olivo, C. Thomas. *Advanced Machine Technology.* Belmont, CA: Wadsworth, Inc., 1982.

Polymeric Materials: Relationships Between Structure and Mechanical Behavior. Metals Park, OH: American Society for Metals, 1975.

Shot-Peening Applications, 7th ed. Vernon, CA: Metal Improvement Co., 1992.

Skeist, Irving. *Handbook of Adhesives*, 3rd ed. New York: Van Nostrand Reinhold Company, 1989.

Springborn, R.K. *Non-Traditional Machining Processes.* Dearborn: American Society of Tool Manufacturing Engineers, 1967.

Value Control Design Guide. New York: Value Analysis, Inc., 1963.

White, Warren T., John E. Neely, Richard R. Kibbe, and Roland O. Meyer. *Machine Tools and Machining Practices.* 2 vols. New York: John Wiley & Sons, Inc., 1977.

Index